# HEREDITY, GENETICS AND GENETICAL DISEASES

# HEREDITY, GENETICS AND GENETICAL DISEASES

[FOR THE STUDENTS OF MEDICAL SCIENCE, VETERINARY SCIENCE, AGRICULTURAL SCIENCE, LIFE SCIENCE, BIOCHEMISTRY AND BIOTECHNOLOGY]

Dr R N Roy

BSc (Hons), MBBS (Cal), MSc (Gold Medallist), PhD (Med) (Cal)

Formerly Research Fellow, Medical College and Hospital, Kolkata
Formerly Research Fellow, Institute of Child Health, Kolkata
Lecturer, Presidency College, Kolkata
Faculty Member, Medical Electronics, Central Calcutta Polytechnic, Kolkata
Examiner, West Bengal Council of Technical Education

New Central Book Agency (P) Ltd

LONDON

HYDERABAD ERNAKULAM BHUBANESWAR
DELHI KOLKATA PUNE

**NCBA**

**REGD OFFICE**

8/1 Chintamoni Das Lane, Kolkata 700 009, India
email: ncbapvtltd@eth.net

**OVERSEAS**

NCBA (UK) Ltd, 149 Park Avenue North, Northampton, NN3 2HY, UK

**EXPORT**

NCBA Exports Pvt. Ltd, 212 Shahpur Jat, New Delhi 110 049
email: ncbaexp@ncbapvtltd.com

**BRANCHES**

208B Peacock Lane, Shahpur Jat
**New Delhi** 110 049
email: ncbadel@ncbapvtltd.com

House No. 3-1-315, 1st Floor
Nimboliadda, Kachiguda, **Hyderabad** 500 027
email: ncbahydb@ncbapvtltd.com

Shop Nos. 3 & 4, Vinayak Towers
681/B Budhwar Peth, Appa Balwant Chowk
**Pune** 411 002
email: ncbapune@ncbapvtltd.com

GSS Shopping Complex, 1st Floor
Opposite College Post Office, Convent Road
**Ernakulam** 682 035
email: ncbaernk@ncbapvtltd.com

Shop No. 15, Bharati Towers, Ground Floor
Forest Park, **Bhubaneswar** 751 001
email: ncbabub@ncbapvtltd.com

HEREDITY, GENETICS AND GENETICAL DISEASES



First Published: July 2011

**PUBLISHER AND TYPESETTER**
New Central Book Agency (P) Ltd
8/1 Chintamoni Das Lane, Kolkata 700 009

**PRINTER**
New Central Book Agency (P) Ltd
Web-Offset Division, Dhulagarh, Sankrail, Howrah

**COVER DESIGNER**
Sournen Paul

**COVER PRINTER**
Biswanath Studio, Kolkata

**TECHNICAL EDITOR**
Pradip Kr Chowdhury

**PROJECT TEAM**
Swapan Sarkar, Kakoli Banerjee, Sarab Chatterjee, and Prabhat Jas

ISBN: 978-81-7381-658-1

Price: ₹ 350.00

To

MY WIFE

For her impetus, zeal, interest and devotion

and

my beloved granddaughter

RANJABATI RAY

for her cooperation and making everything worthwhile

# CONTENTS

## PART IV
## BIOCHEMICAL GENETICS—INBORN ERROR IN METABOLISM



## PART V
## DIFFERENT FACTORS AND GENETICS

## PART VI
## IMMUNOGENETICS

# PREFACE

I take great pleasure to introduce this volume for the benefit of the students of Genetics. Main purpose is to keep the primary emphasis on the elementary principles and, at the same time, to make the students aware of the latest developments in the subject.

Genetics is the branch of science which is an essential item under the medical curriculum as well as for the undergraduate and postgraduate students of Veterinary Science, Agricultural Science, and General course of Physiology, Zoology, Biochemistry, and Botany. In this outline, I have undoubtedly incorporated more experiments than the average to cater to the needs of the students of various disciplines of different universities.

Information regarding heredity has also been elaborated considerably. In general, this has been dealt with general principles of transmission mechanism of development.

Inheritance of human traits has been incorporated wherever possible since the methods of study of human traits are difficult and animal and plant experiments differ. I have also included a separate chapter on twins as material for studying the relative effects of heredity and environment.

Genetics has been divided according to the matter related to it, e.g. Developmental Genetics, Haematological Genetics, Environmental Genetics, Biochemical Genetics, Molecular Genetics, etc. The chapters have been written in simple language and in lucid form, and with illustrations. The chapter on Metabolic Genetics has been discussed properly to make the students understand the Biochemistry of metabolism avoiding structural formulas. The chapter on Molecular Genetics has been discussed with Biochemistry and structural formula of DNA and RNA, their biological role, mechanism of transmission of heredity, replication, gene expression, protein biosynthesis, etc.

A separate chapter on Genetic Engineering with recent clinical developments in progress has been included.

Another chapter on Genetics in forensic medicine, e.g. DNA fingerprint technology, gene therapy, Human Genome Project, human cell cloning in therapeutics, individual genetic diversity and determination (ROMA) has also been incorporated for the students to cope up with the recent developments in Genetics.

Preparation of the volume may lead to inevitable errors of omission and commission; the author will appreciate having these brought to attention. Any suggestion is earnestly solicited.

I am greatly indebted to late D. N. Roy and late T. N. Banerjee for inspiring me to prepare this volume.

I must owe a special debt of gratitude to Mrs S. Roy for consistency of inspiration, encouragement and interest while preparing the volume and also for correction of proof, preparation of index and illustrations. I wish to extend special thanks to Miss Ranjabati Ray for timely assistance in preparation of the volume, and preparation of manuscript with excellent illustrations.

Thanks are due to Mr Amitabha Sen, Director, New Central Book Agency (P) Ltd, Mr Prabhat Jas and other production-staff of New Central Book Agency (P) Ltd for their cooperation in preparing this volume.

**R N ROY**

"GEETANJALI" Apartment
19/1A/91A Seal Lane
Kolkata 700 015
India

**PART-I**

# HEREDITY

# 1

# HEREDITY

1. Introduction, 2. Variation, 3. Law of inheritance, 4. Inheritance, 5. Single-factor inheritance, 6. Multifactorial inheritance, 7. Genotype and phenotype, 8. The back-cross, 9. Incomplete dominance.

## Introduction

Heredity is a genetic relation between the parent and offspring. Offspring resemble their parents, and this fact is popularly expressed by the saying "like begets like." The resemblance involves physical or morphological characters such as skin colour, stature of the body and physiological characters or mental traits, expressed or latent, based on descent.

## Variation

Not withstanding the resemblances, there are differences in the offspring among themselves, and from their parents in varying degree. The difference or departure from the parent form is known as "variation". The variation may involve one or more characters. There may be differences in coat colour or in sizes as are commonly noticed in a litter of kittens or puppies or like the sudden appearance of a red-eyed-Drosophila from a white-eyed stock. The variation may be:

1. Slow and gradual or
2. Sharp and sudden or
3. At random or
4. Fluctuating
5. In a definite direction.

The causes of variation may be:

1. Environmental factor, e.g. food, temperature are said to be *somatic variation* or "modification."
2. Zygote factor, i.e. during zygote formation or rather prior to it known as "germinal variation or congenital or blastogenic variation."

The germinal variation are born and deep-seated. They originate by recombination of the germinal materials or genes or by mutation of genes, and it is held that the hereditary genes (germ plasm) maintain continuity from one generation to the next and are not the product of the body (soma) which merely carries or houses these ancestral genes.

- **Erroneous idea**

There are many erroneous ideas about heredity. One of these is maternal impressions or experiences of the mother during pregnancy, which were supposed to influence the character of the child. The mother's health or nutrition may affect the growth of the embryo but this is a physiological change and not a change in the germinal constitution of the child.

Another is telegony which means the supposed influence of the first sire on subsequent progeny that was obtained by mating of the female animal with males other than the first. Animal breeders also often believed in prepotency of one parent implying thereby its ability to transmit its characters to the offspring in exclusion of the characters of the other parent.

## Law of Inheritance

The inheritance of physical characters and mental traits obeys biological laws called *law of inheritance*. These laws are based on records of observations and experimental results of crossing and their statistical analysis.

1. **Galton's Laws of Ancestral Inheritance:** The offspring receives inherited faculties from parents, grandparents and previous ancestral generations in relative proportions to their remoteness. Thus, the two parents contributed each on an average and half inherited faculty, the four grand-parents each contributing $\frac{1}{4}$ and so on.
2. **Law of Filial Regression:** The second law called *law of filial regression* states that, on an average, the offspring tend to swing like a pendulum nearer the mean of their race than their parents; the offspring of talented parents swing back to average mediocrity. Nature, therefore, does not favour enjoyment of special traits in a single family.
3. **Mendel's Law:** The modern study of heredity dates from Mendel's crossing or hybridizing experiments on the pea plants.
   (a) **Mendel's first law:** In the germ cells of hybrids there is a free separation of factors contributed by two parents without regard to which parent supplied them. This is known as "law of segregation".
   (b) **Mendel's second law:** Two or more allelomorphic pairs of factors segregate independently and may recombine in all possible combinations governed by change alone.

## Inheritance

There are regions on the chromosomes called *genes* which consists of DNA having large group of organic bases linked in a particular *sequence* in the chromosome. This DNA is a molecule of heredity. The *purine bases* (Adenine, Guanine) and *pyrimidine bases* (Cytosine, Thymine) *carry genetic information,* whereas the sugar and phosphate groups perform a structural role. The *sequence* of these *bases* is very important factor in heredity. The gene may consist of a particular sequence of up to 1000 *base pairs* in a DNA molecule.

In some cases, the presence of a single gene may determine the appearance of one characteristic, e.g. eye colour of *Drosophila* but most human characteristics are controlled by more than one gene. This *multifactorial* inheritance and difficult to do human experiment, make it difficult to supply genetical information about human beings and most of the informations are based on animal experiments and some clinical observations.

Inheritance may be:

I. *Single-factor inheritance*

II. *Multifactorial inheritance.*

Most human characteristics are controlled by more than one gene. The multifactorial inheritance and the impossibility with humans of breeding experiments, make it difficult to collect and present simple clear-cut genetical information about man.

## Single-factor Inheritance

If homozygous black mouse is mated with a pure-breeding brown mouse, the offspring will be black without any intermediate colour. The gene for black fur is *dominant* to that for brown fur because baby mice must carry genes for both blackness and brownness but only the blackness is expressed. The gene for brown fur is *recessive.* These black mice are called the *first filial* or *$F_1$ generation.* After maturity if these $F_1$ black mice are mated amongst themselves, their offspring, the *$F_2$ generation* will include both black and brown mice. If the total number of $F_2$ generation are added up, the ratio of black to brown mice will be approximately 3 : 1. This ratio 3 : 1 occurs only when large numbers of individuals are considered. The appearance of brown fur in the second generation is evident by the fact that the $F_1$ black fur mice carried the recessive gene for brown fur even though it did not find expression in their observable features (Figs. 1.1, 1.2).

- **Explanation**

It can be explained that a pure-breeding black mouse carries a pair of genes controlling the production of black pigment on homologous chromosomes and represented by the letters BB, the capital letters signify dominance in the Fig. 1.1. Corresponding chromosomes in brown mice, there are the genes bb for brownness. The genes B and b are called *allelomorphic genes* or *alleles.* The process of meiosis during the formation of gametes, separates the homologous chromosomes, so that the gametes will contain only one gene from each pair. All the sperms from the pure-breeding black parent will carry the factor B and all the eggs from the brown parent will carry the factor b. When the gametes fuse, the zygote will contain both B and b factors but since B is dominant to b, only the former gene is expressed, and the offspring will be black. When these black $F_1$ mice will produce gametes, the process of meiosis will separate the chromosomes carrying the B and b factors (Fig. 1.2) so that half the sperms of the male parent will carry B and half will carry b. Similarly, half the ova from the female will contain B and half b. At fertilization there are equal chances that a B-carrying sperm will fuse with either an egg carrying the B gene or an egg with the b gene so producing either a BB or a Bb zygote. Similarly, there are equal chances of a b-carrying sperm fusing with either a B- or a b-carrying ovum to give bB or bb zygote. The result might

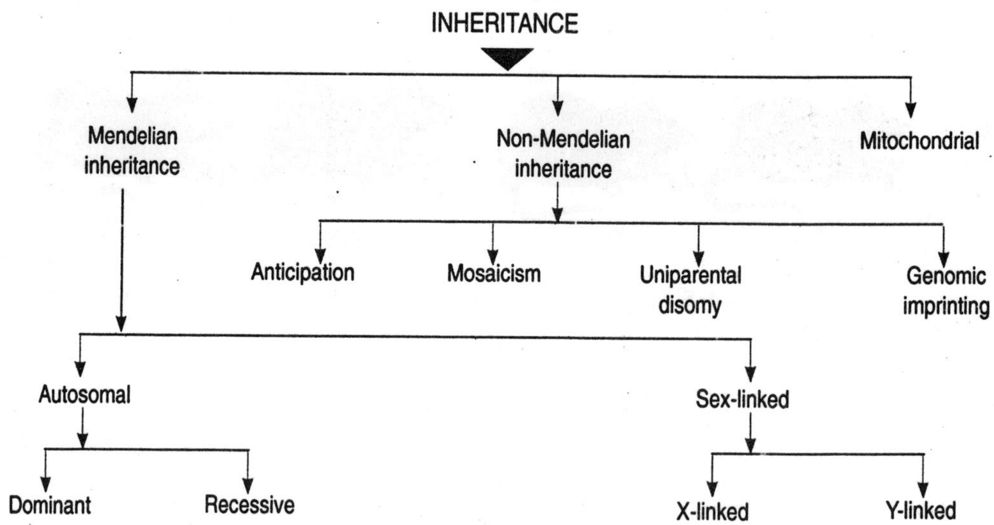


**Fig. 1:** A Flow chart showing types of inheritance.

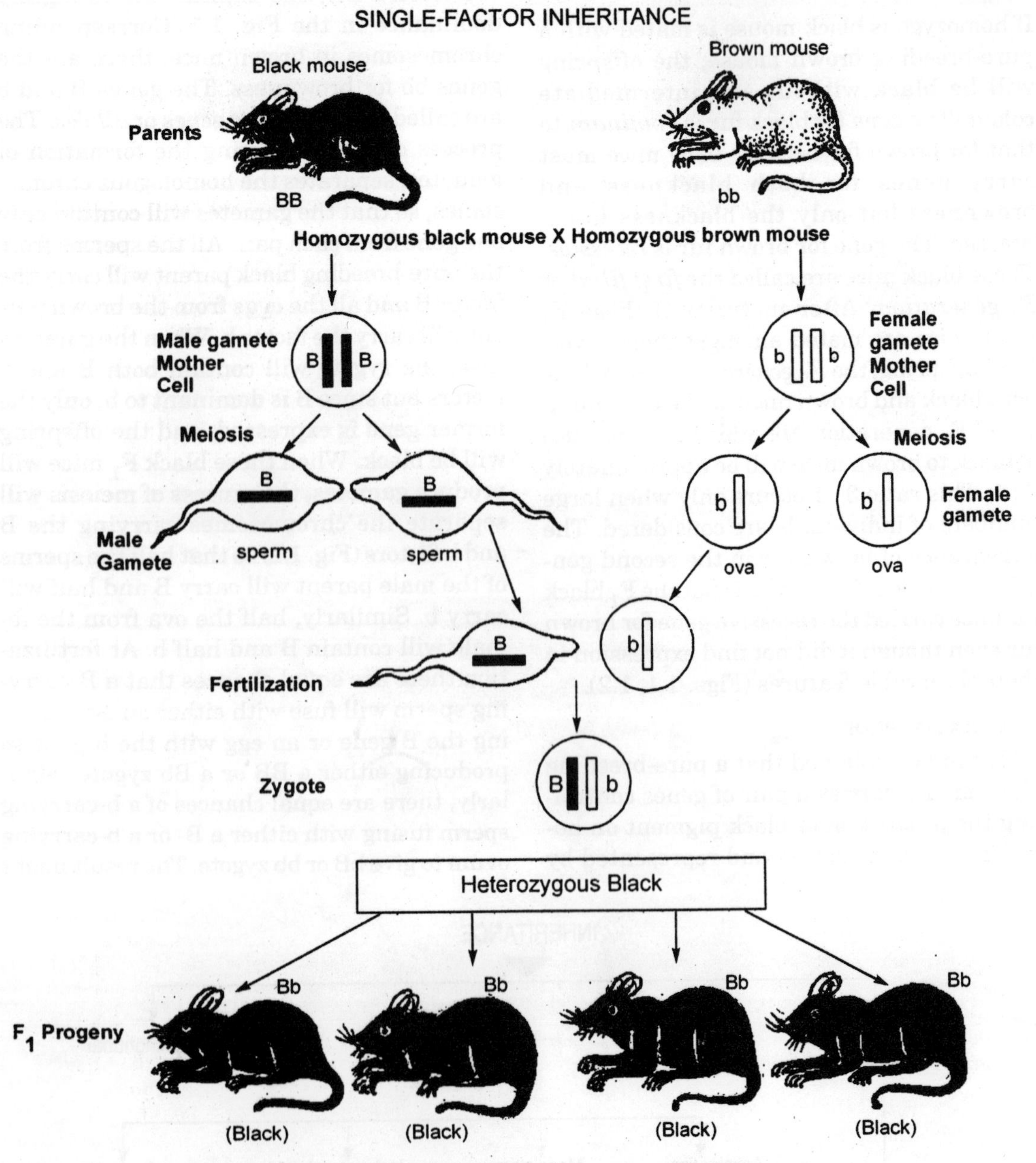


**Fig. 1.1:** Single-factor inheritance for coat colour in mice.

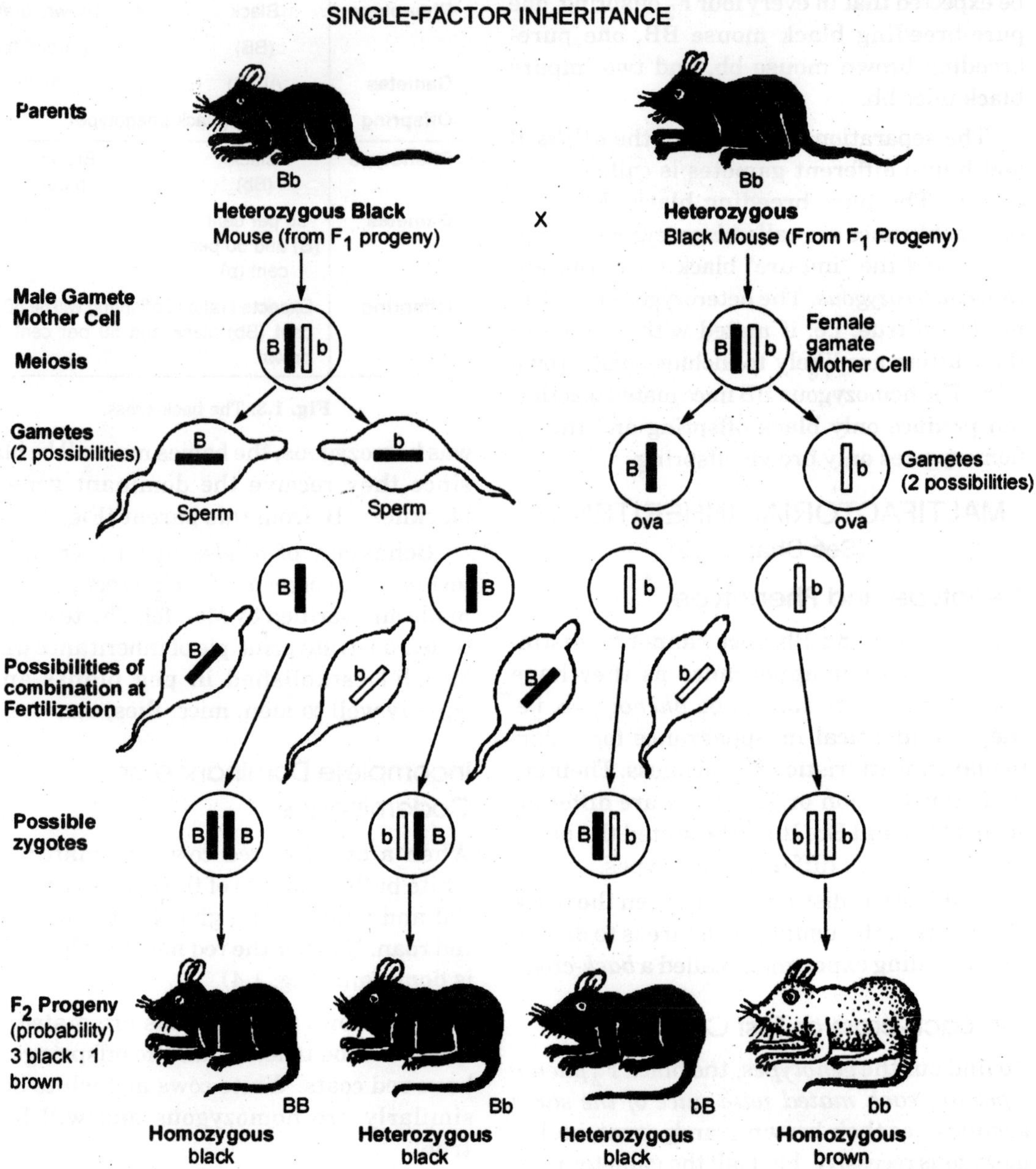


**Fig. 1.2:** Single-factor inheritance for coat colour in mice of parents of heterozygous black from $F_1$ progeny and probability in case of $F_2$ progeny.

be expected that in every four $F_2$ offspring, one pure-breeding black mouse BB, one pure-breeding brown mouse bb, and two impure black mice bb.

The separation at meiosis of the *alleles* B and b into different gametes is called *segregation.* The pure-breeding black (BB) and brown (bb) mice are called *homozygous* for coat colour and the "impure" black mice (Bb) are called *heterozygous.* The heterozygous mice will not breed true, i.e. if mated with each other their litters are likely to include some brown mice. The homozygous BB mice mated together can produce only black offspring and the bb homozygotes only brown offspring.

## MALTIFACTORIAL INHERITENCE

(See Chapter 33)

### Genotype and Phenotype

The BB mice and Bb mice cannot be distinguished in their appearance as they have black fur and are said to be *phenotypes*, i.e. they are identical in appearance for a particular characteristic i.e. blackness. Their genetic constitution or *genotypes* are different such as BB and Bb. It can be stated the black phenotypes has different genotypes.

To make the distinction between the black phenotypes, the usual procedure is to do further breeding experiment called a *back-cross.*

### The Back-cross or Test Cross

To find out the *genotypes,* the black $F_2$ *phenotypes are each mated with mice of the same genotypes* as their brown grandparent, i.e. homozygous recessive, bb. Half the gametes from the heterozygous black mice Bb will carry the B gene and half will carry the b gene.

The gametes from the homozygous black mouse, BB, will not carry the B gene. Similarly, the gametes of homozygous recessive brown mouse will all carry the b gene. Thus, when the black parent is heterozygous, one can expect the back-cross to yield approximately equal numbers of black and brown babies in the litters but if the black parent

| | | | |
|---|---|---|---|
| **Phenotype** | Black (BB) | × | Brown "grand-parent" (bb) |
| **Gametes** | All (B) | | All (b) |
| **Offspring** | All (Bb) (black phenotype) | | |
| **Phenotype** | Black (Bb) | × | Brown "grand parent" (bb) |
| **Gametes** | 50 per cent (B) and 50 per cent (b) | | All (b) |
| **Offspring** | Expected ratio of offspring will be 50 per cent (Bb) black and 50 per cent (bb) brown | | |

**Fig. 1.3:** The back-cross.

was homozygous, the babies must all be black since they receive the dominant gene for blackness, B, from this parent (Fig. 1.3).

Behaviour of *alleles* in back-crosses or test-crosses corresponds to correct prediction made and verified by Mendel. The test serves to demonstrate principle of inheritance which Mendel established in pea plants apply equally well to men, mice, flies, etc.

### Incomplete Dominance or Codominance

When a breed of red cows are mated with white bulls the coats of the calves carry both red and white hairs, giving a Colour called red roan. Neither the red nor the white factor is dominant (Fig. 1.4).

If red cows and red bulls are mated, the breed will be true, i.e. all the offsprings will have red coats. White cows and white bulls, similarly are homozygous and will breed true.

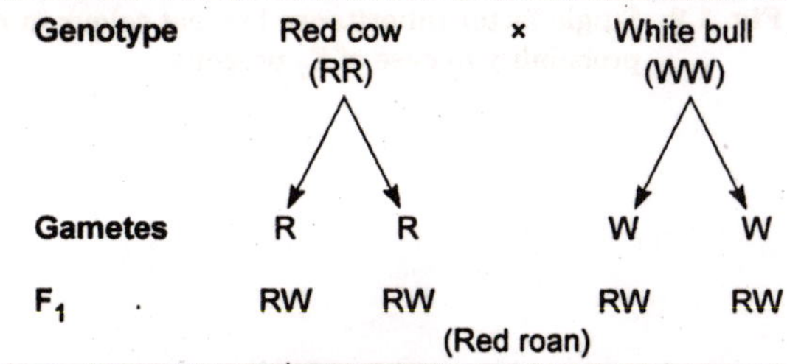


**Fig. 1.4:** Incomplete dominance or codominance.

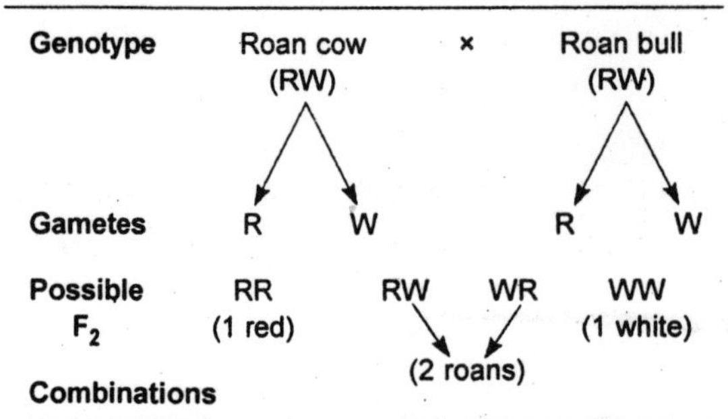


**Fig. 1.5:** Breeding between $F_1$ roan cattle.

The $F_1$ roan cattle, however, are heterozygous and will not breed true. Their progeny will include red calves, white calves and roans (Fig. 1.5).

- **Example of Incomplete Dominance**

In man, the instance of inheritance of A, B, O blood groups is incomplete dominance.

People are classified into four major blood groups, A, B, AB and O which is important for blood transfusion. The blood group is controlled by three genes, A, B, O acting at the corresponding site on homologous chromosomes. One person inherits two of these genes, one from each parent. Gene O is recessive to both A and B, *but A and B* are *codominant,* i.e. if a person inherits gene A from one parent and gene B from the other, he will be group AB, neither gene being dominant to the other. It follows that group O people must have the genotype OO while group A persons could be AA or AO and group B individuals BB or BO. The following table shows the blood group of newborn baby of father of group A and mother of group B and both of them are heterozygous (Table 1.1).

**Table 1.1:** Inheritance of blood groups in man—incomplete dominance

| | Heterozygous | |
|---|---|---|
| | Man | Women |
| **Phenotype** | Group A | Group B |
| **Genotype** | AO | BO |
| **Gametes** | A and O | B and O |
| **$F_1$ Genotype** | AB AO | OB OO |
| **Phenotype** | Group AB<br>Group A | Group B<br>Group O |

## QUESTION

1. Write briefly on heredity.
2. What are the laws of inheritance?
3. What do you mean by inheritance? What are the types of inheritance? Describe briefly the single-factor inheritance.
4. What do you mean by genotype and phenotype?
5. What is back-cross? Describe briefly.
6. What do you mean by codominance or incomplete dominance?

# 2

# LAWS OF HEREDITY

1. Monohybrid inheritance, 2. Punnett square, 3. Probability, 4. Mendel's first law, 5. Breeding true and pure line, 6. Monohybrid inheritance in man, 7. Dihybrid inheritance, 8. Mendel's second law, 9. The back-cross, 10. Patterns of inheritance, 11. Incomplete dominance.

Genetic information is transmitted through a structure called gene and located in the chromosomes. Further research has unravelled the chemical structure of the genes, mode of transmission of hereditary characters, gene transplantation, etc. But the study of heredity and genetics which explain the similarities and differences between parents and offspring was brought to light by Gregor Mendel in 1856 before the genes and chromosomes has been discovered.

He had selected garden pea for his research because it has a number of sharply contrasting and easily recognizable characteristics, e.g. long and short stem, red and white flowers, smooth and wrinkled seeds, etc. With such differences it was possible to cross or self-pollinate certain plants and observe the characteristics of their offspring.

## Monohybrid Inheritance

Mendel studied the inheritance of just one pair of contrasting characteristics which is known as *monohybrid inheritance.* He had selected a pure-breeding tall pea plant and crossed it with a dwarf plant. This was done by taking pollen from one plant and dusting on the stigma of the other, precaution being taken to ensure that the latter was not contaminated with its own pollen or that of any other plants. The seeds produced from the cross were collected and sown. It was observed that all the seeds gave rise to tall plants, no dwarf plants being produced at all. These plants are called *first filial generation*, $F_1$ (Fig. 2.1).

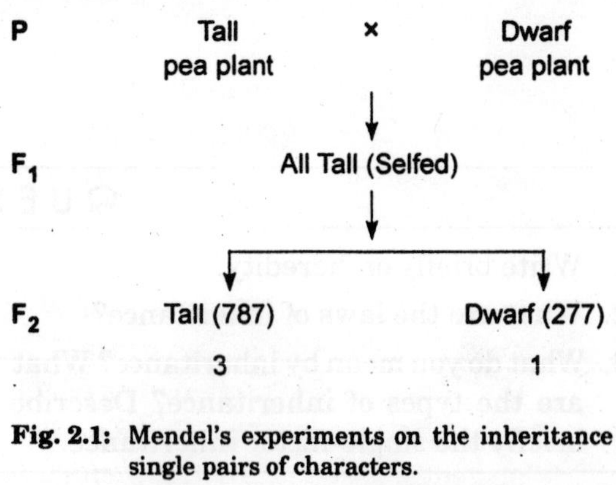


**Fig. 2.1:** Mendel's experiments on the inheritance of single pairs of characters.

One of these tall $F_1$ plants was then self-pollinated, precautions being taken to prevent its being pollinated by any other kind of pollen. The seeds produced were sown and the offspring, i.e. second filial generation, $F_2$ were examined. It was found that the $F_2$ plants were a mixture of tall and dwarf plants. In one particular cross he counted 787 tall plants and 277 dwarf plants, i.e. three-quarters were tall and one-quarter dwarf, i.e. a ratio of 3 : 1 approximately. Other monohybrid crosses produced the same results which are given below (Table 2.1).

3. There are no dwarf plants in the $F_1$ generation but present in the $F_2$ generation though one of the parent plants was dwarf. For this two explanations can be given:
   (a) Though the $F_1$ plants are tall, they must have received a factor from dwarf parent which remained suppressed or hidden in the $F_1$ generation without any external appearance until the $F_2$ generation.

**Table 2.1:** Results of monohybrid crosses

| Character selected | Cross | $F_2$ generation | Ratio |
|---|---|---|---|
| 1. Form of seed | Smooth, Wrinkled | 5474 Smooth, 1850 Wrinkled | 2.96 : 1 |
| 2. Colour of cotyledons | Yellow, Green | 6022 Yellow, 2001 Green | 3.01 : 1 |
| 3. Colour of seed coat | Grey-brown, White | 705 Grey-brown, 224 White | 3.15 : 1 |
| 4. Form of pods | Inflated, Constricted | 882 Inflated, 299 Constricted | 2.95 : 1 |

- **Conclusion from the Monohybrid Experiment:**

1. In $F_1$ and $F_2$ generations, there are no intermediate-sized, i.e. medium-sized plants between the tall and dwarf. It can be concluded that the characters of two parents are not blended to produce an intermediate result. But it is a process in which definite structures or particles are transmitted from parents to offspring which may or may not show themselves in the external appearance of the organism.
2. Such particles exist may be combined in one generation but separated in the next generation ($F_2$ generation) and the inheritance is described as *particulate.* What we now call these particles as genes, Mendel called it *factors.*

   (b) The factor for dwarfness fails to display itself in the $F_1$ generation but in $F_2$ generation in some mechanism it swamped by the factor for tallness. At present, we call the gene for tallness as *dominant* and the gene for dwarfness *as recessive.*

Although Mendel did not know about chromosomes and genes but he postulated that *factors* are transmitted from parents to offspring through the gametes. *Discussion on Mendel's Mono-hybrid Inheritance in the light of Modern Concept—Genes and their Trans-mission.*

The genes for tallness is designated as T and genes for dwarfness as t. Tall plant each has a pair of identical genes TT and tt in case of dwarf plant. Organism having identical genes of these types called *homozygous*. These TT and tt are the genetic constitution and genetic constitution of an organism is called *genotype*, i.e. genotype TT for tall plant and tt for dwarf plant. The outward appearance, i.e. tall or dwarf is known as *phenotype*. Each of the gametes produced by the tall will contain *one* T *gene*, and produced by the dwarf will contain t gene. After fertilization T and t genes are brought together. Now all the offspring of $F_1$ have the genotype Tt. Phenotypically they are tall because all contain T gene which is dominant to t gene. And at this stage it is called *heterozygous* because the organism contains two dissimilar genes. Here T gene is dominant and expresses as tall in the phenotype and t gene is recessive and suppressed by T gene. A *dominant* gene can express itself *even in the heterozygous condition* and a *recessive gene* can express itself *in homozygous* condition. An individual homozygous for a recessive gene called the *double recessive*.

The gene-controlling size in the pea plant are of two types. One functions normally and is responsible for production of tall plant. The other is responsible for development but if two are present together, a dwarf plant is produced and such genes are called *alleles*.

Each of $F_1$ plants, being heterozygous, produces two types of gamete: half will contain T gene, the other half the t

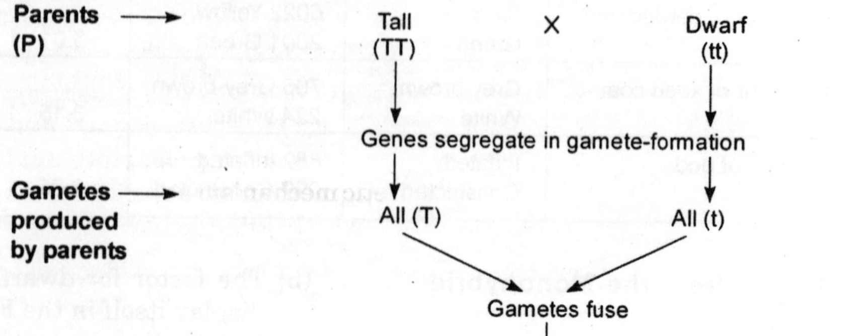


**Fig. 2.2:** Genetic diagram of monohybrid inheritance. Result is shown when pure-breeding tall pea plant is crossed with a pure-breeding dwarf plant. Mendel's first law, the law of segregation is based on this experiment.

gene. On self-pollination of $F_1$ plants, the two types of gametes will fuse randomly to produce offspring of three genotypes TT, Tt and tt. As a result the $F_2$ generation will produced $\frac{3}{4}$ tall and $\frac{1}{4}$ dwarf (Fig. 2.2).

## Punnett Square—Graphic Method of Expression of Result of Fusion of Gametes

A convenient graphic method to employ the Punnett square after the name of Punnett, a geneticist to show the result of fusion of gamete. The gametes of one parent are written along top of series of boxes and the gametes of the other parent are written along the down series of boxes. The products of the various fusions are written in the proper boxes and their relative numbers are estimated.

| ♀ Gametes | | ♂ Gametes | |
|---|---|---|---|
| | | $\frac{1}{2}$ T | $\frac{1}{2}$ t |
| | $\frac{1}{2}$ T | $\frac{1}{4}$ TT | $\frac{1}{4}$ Tt |
| | $\frac{1}{2}$ t | $\frac{1}{4}$ tT | $\frac{1}{4}$ tt |

Fusions

**Fig. 2.3:** Punnett square, a method to show the fusion of $F_1$ gametes.

In the above discussion (Fig. 2.3), the gene T gene is present in the half of the male gametes and in the half female gametes. The proportion of zygotes receiving TT will therefore be $\frac{1}{2}\times\frac{1}{2}=\frac{1}{4}$. Similarly, in case of t gene, half the male gametes and half the female gametes contain t gene, so the proportion of zygotes receiving tt will be $\frac{1}{2}\times\frac{1}{2}=\frac{1}{4}$. In case of the heterozygous, Tt offspring there are two possibilities:

1. Half the male gametes are T and half the female gametes are t, so the proportion Tt zygotes will be $\frac{1}{2}\times\frac{1}{2}=\frac{1}{4}$.
2. Equally, half the male gametes are t and half the female gametes are T, which give a proportion of Tt zygotes of $\frac{1}{4}$.

Thus, the total proportion of Tt zygotes will be $\frac{1}{4}\times\frac{1}{4}=\frac{1}{2}$. The $F_2$ offspring will be $\frac{1}{4}$TT, $\frac{1}{2}$Tt + $\frac{1}{4}$tt, e.g. $\frac{3}{4}$ will be tall and $\frac{1}{4}$ will be dwarf.

## Probability

Mendel put forward the basic law and principles of heredity from his experiment on pea-plant by applying the laws of probability. Probability is the ratio of a specified event to total events. Limits of probability are from 0 if an event never occurs, to 1 if it always occurs.

The laws of probability apply to the genetic mechanism as well as to other processes in which uncertainty exists. The figures which is expressed that $\frac{3}{4}$ of the offspring are tall and $\frac{1}{4}$ is dwarf, it is applied when there is large number of offspring. If $F_2$ plant is selected at random, there is a 3 in 4 chances of its being tall, and a 1 in 4 chances of its being dwarf, i.e. *probability* of its being tall is $\frac{3}{4}$, and dwarf $\frac{1}{4}$.

• **The Role of chance and Sampling Error:** In genetic experiment, if $F_2$ plants are expressed that $\frac{3}{4}$ are tall which is an approximate and are expected. It is not certain that $\frac{3}{4}$ will be tall because chance plays an important role in reproduction. It is fact that allelic genes are separated from each other during gametogenesis, but there is no certainty that they are distributed in equal numbers between the viable gametes. Many gametes may die or become inactive. And the union between the gametes occur randomly. Hence, it is a matter of sheer chance that a particular male gamete will unite a particular female gamete. For this reason, the observed ratios obtained in the experiment do not approximate the expected ratios.

The observed ratios may be deviated from the expected ratio when the sample is of small number. The difference between the observed ratio and the expected ratio is known as *sampling error*. The error is minimum when the sample is greater. In Mendel's experiment, the number of offspring in the $F_2$ generations was larger hence the sampling error comparatively minimum.

## Mendel's First Law: the Law of Segregation

Mendel's conclusion in general is known as Mendel First Law or the Law of Segregation. He stated that the internal factors determine the characteristics of organism and these internal factors are in pairs. Only one pair of such internal factors can be represented in a single gamete.

It can be stated in the light of modern concept, *the characteristics of an organism are controlled by genes occurring in pairs. Of a pair of such genes, only one can be carried in a single gamete.*

- **Explanation of Mendel's First Law**

The Mendel's First Law can be explained by the process of *meiosis*. In meiosis homologous chromosomes segregate from each other, as a result of which a gamete receives only one of each type of chromosome instead of the normal two. Genes also present in pairs, each is located on one of the two homologous chromosomes. When these homologous chromosomes segregate in the process of meiosis they take their genes with them, and thus the gametes receive only one of a pair of alleles because they receive only one of a pair of chromosomes (Fig. 2.4). It is

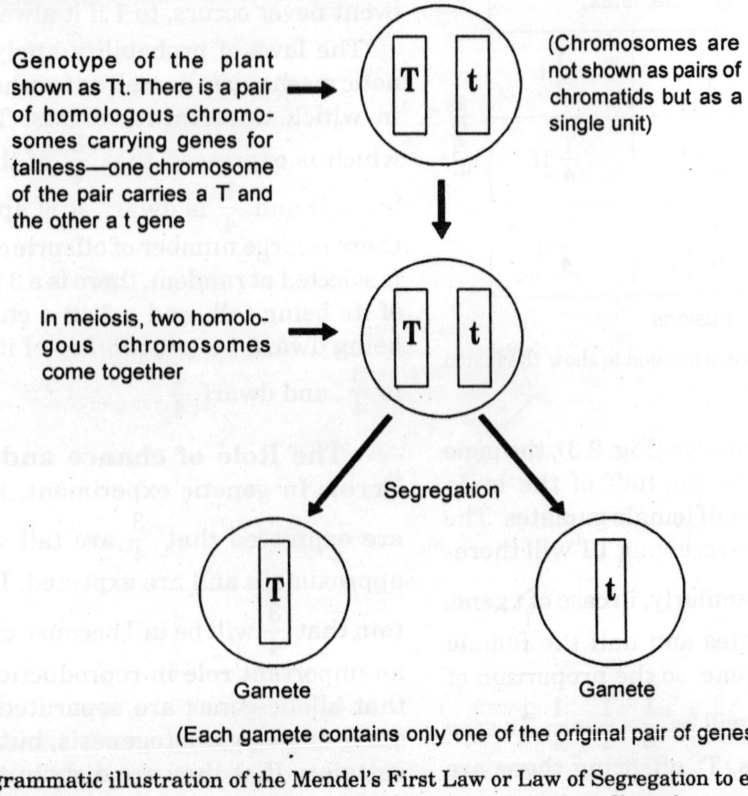


**Fig. 2.4:** Diagrammatic illustration of the Mendel's First Law or Law of Segregation to explain by the process of meiosis. The segregation of genes corresponds to the segregation of homologous chromosomes in meiosis.

observed that there is similarity between the segregation of Mendel's internal factors in inheritance and the segregation of homologous chromosomes in meiosis. It is further evident that the genes are located on the chromosomes.

### • Test-cross or Back-cross

*Drosophila melanogaster* has a whole series of contrasting features, one of which is size of the wings. Large numbers have long wings and they are called *wild type.* Some have vestigial wings which are short, stumpy

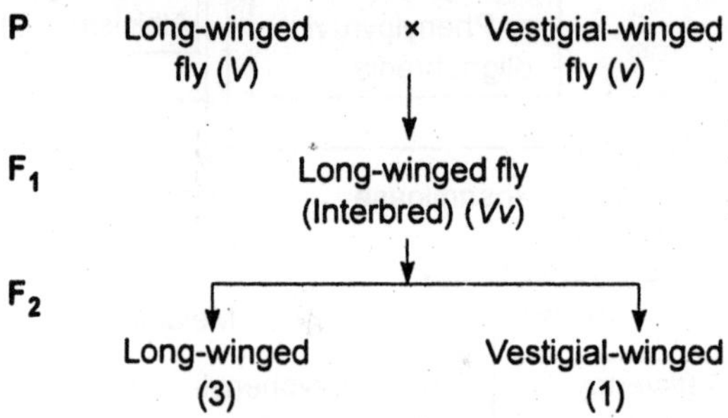


**Fig. 2.5:** Shows the result of test-cross.

and functionless. The gene for long wing is *V* which is dominant to the gene for vestigial wing is *v*. If a homozygous long-winged fly is mated with a vestigial-winged fly, the $F_1$ offspring are all long-winged with a genotype *Vv* (Fig. 2.5).

If two of these $F_1$ flies mate with each other, a mixture of long-winged and vestigial-winged flies are produced in the proportions of 3 : 1 (Fig. 2.5).

This result corroborates to the experiment done by Mendel with tall and short peas.

To determine the $F_2$ long-winged fly is homozygous (VV). or heterozygous (Vv), it should be crossed with vestigial-winged (vv) fly the double recessive. If the unknown fly is VV then after crossing with vestigial-winged fly (vv) will produce long-winged flies. If the unknown fly is Vv, then the cross will give a mixture of long and vestigial-winged flies in equal proportions (Fig. 2.6).

This test-crosses with double recessive are the routine method to determine the

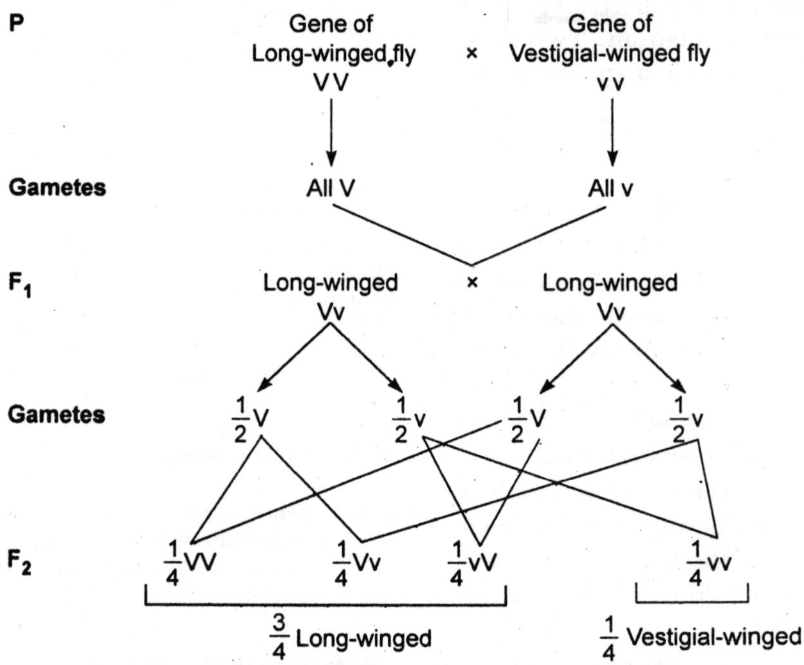


**Fig. 2.6:** Test-cross of monohybrid inheritance in Drosophila showing the genotype.

genotype of the organism, whether it is capable of self-fertilization or not.

The law of segregation can be applied to many organisms besides peas. The *Drosophila* demonstrated the principles of monohybrid inheritance which Mendel showed in peas and the principles equal applicable in flies, mice, men, molds, etc.

## Breeding True and Pure Line

When a homozygous organism is self-fertilized, offspring are produced, all of which are identical with the parent. It happens if a homozygous organism is crossed with another homozygote. This is called *breeding true.* When the offspring are self-fertilized or crossed amongst themselves, another generation of identical organisms is produced and if it is continued a *pure line* is established.

## Monohybrid Inheritance in Man

Certain conditions occur to be associated with a single pair of genes which are transmitted in men in Mendelian fashion. Two rare hereditary diseases are albinism and xeroderma pigmentosum.

1. *Albinism* is due to presence in double dose of a recessive gene and the second to a dominant gene. Albinism is an example of monohybrid.
2. Albinism, complete or nearly complete in which there is complete absence of black pigment melanin in the skin, eyes and hair.
3. It is due to absence of *tyrosinase* enzyme in the melanocyte (Fig. 2.7).
4. There are various types of albinism in which pigment is only lacking from certain parts of the body, such as the eye, areas of the skin, and areas of the hair.
5. When exposed to intense sunshine they may suffer dangerous sunburns. If albinos protect themselves from sunburn they suffer little inconvenience from their inherited peculiarities.
6. It is found occasionally in mammals.
7. In man albino individuals occur, with some rarity, in many different races. In English population, about one out of 20,000 babies born is an albino.

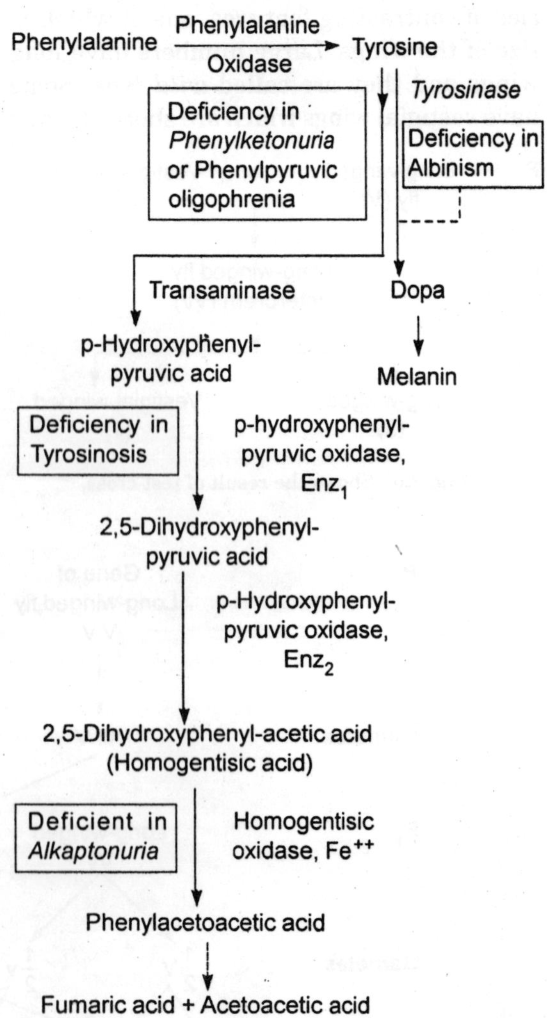


**Fig. 2.7:** Diagram showing hereditary defects in phenylalanine and tyrosine metabolism are albinism, alkaptonuria, tyrosinosis and phenylketonuria (phenylpyruvic oligophrenia).

Pedigree show that albinism is due to homozygosis for a recessive allele, *a.* Its dominant allele, *A,* is necessary for the development of melanin pigment. Nonalbinos are either homozygous, *AA*, or heterozygous, *Aa*.

Marriages of albinos with normally pigmented persons nearly always produce only nonalbino children, since the albino allele is rare and most people do not carry it. Suppose a phenotypically normal couple have an albino child. Then child must be *aa* and both parents must be heterozygous *Aa,* i.e. the parents, though not themselves albinos, carry the albino gene; hence they are known as *carriers.* Naturally, the question arises whether the next child would be albino or not in this couple.

Both parents are heterozygous *Aa* and each produces *A* and *a* gametes in roughly equal numbers. These fuse randomly to produce three types of genotypes *AA, Aa* and *aa.* A 3 : 1 ratio is obtained because two heterozygous organisms are crossed, i.e. it means that the probability of their next child being an albino is $\frac{1}{4}$, one chance in four (Fig. 2.8). The diagram shows the genotype and phenotype of the offspring as a result of union between two carriers of the albino genes.

## Dihybrid Inheritance

Mendel studied the inheritance of two pairs of characteristics which is known as a *dihybrid cross.* Mendel crossed a *pure-breeding tall pea plant having coloured flowers* with a *dwarf plant possessing white flowers.* In the $F_1$ generation, he obtained all the tall plants with coloured flowers. In the $F_2$ generation he obtained four different kinds of plants:

(a) Tall with coloured flowers;

(b) Tall with white flowers;

(c) Dwarf with coloured flowers;

(d) Dwarf with white flowers.

It means that the offspring showed two pairs of characteristics, tall and dwarf, coloured and white (Fig. 2.9).

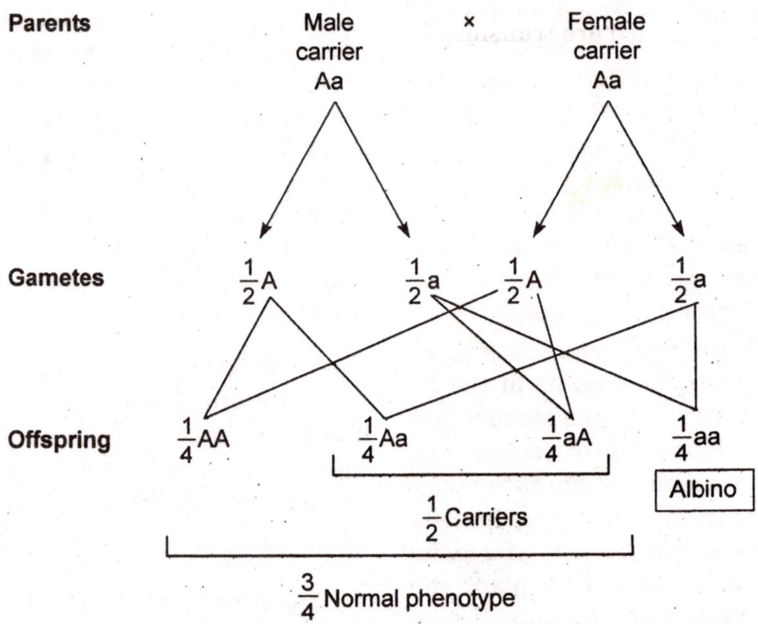


**Fig. 2.8:** Albinism in man, an example of monohybrid inheritance. Genotype and phenotype of the offspring is the result of union between two carriers of the albino genes. It shows a one-in-four chance of any one child being an albino.

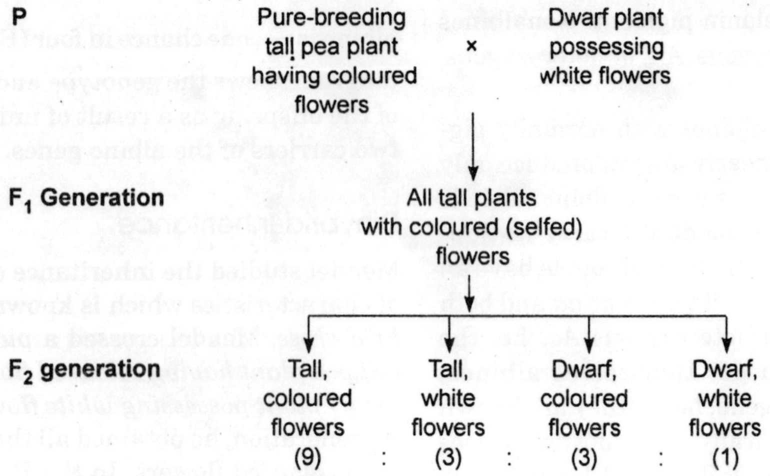


**Fig. 2.9:** Summary of the experimental result of dihybrid inheritance.

## Gene Transmission in Dihybrid Inheritance

All plants of $F_1$ generation are tall with coloured flowers which reveals that tall is dominant to dwarf and coloured flowers to white flowers. Gene for tallness is *T*, *t* for dwarfness, *C* for coloured flowers, and *c* for white flowers. Being pure-breeding, the parent plants must be homozygous for both pairs of alleles. The genotype of the tall plant with coloured flowers must be *TTCC,* and the dwarf plant with white flowers *ttcc*. According to Mendel's First Law, i.e. the Law of Segregation, the gametes produced by the parent plants must be *TC* and *tc* respectively. Then, all the offspring will therefore be *TtCc,* heterozygous for both pairs of alleles.

Now, if all these four combinations of characteristics are to be present in the $F_2$ generation, we can conclude like Mendel that $F_1$ plants produce four kinds of gamete: *TC, Tc, tC* and *tc.* These gametes can fuse in different ways which is shown in the Punnett square which also shows the genotype of the offspring of $F_2$ generation. For tall plant, the genotype must contain at least one *T* gene and for coloured at least one *C* gene. From the Punnett square it observed, that 16 possible fusions can give rise to 9 tall coloured plants, 3 tall white, 3 dwarf coloured, and 1 dwarf white. The observed ratio can be expected, if all the possible fusions occur with equal likelihood (Fig. 2.10).

## Mendel's Second Law, the Law of Independent Assortment

It can be concluded that the two pairs of genes are transmitted independently from parents to offspring and assort freely. This phenomenon was pointed out in *Mendel's. Second Law,* the *Law* of *Independent Assortment* which he stated that each of a pair of contrasted characters may be combined with either of another pair. It can be stated in the light of modern concept that *each member of a pair of alleles may combine randomly with either of another pair.*

- **Probability**

Expressed in terms of probability, transmission of the genes determining the size and colour of the flower are *independent events.* Considering the genes for size, the probability of one $F_2$ tall plant is $\frac{3}{4}$, and for dwarf is $\frac{1}{4}$. Similarly, considering the flower colour genes alone, then the probability of $F_2$ coloured plant is $\frac{3}{4}$ and for white is $\frac{1}{4}$. Then the probability of $F_2$ tall and coloured plant

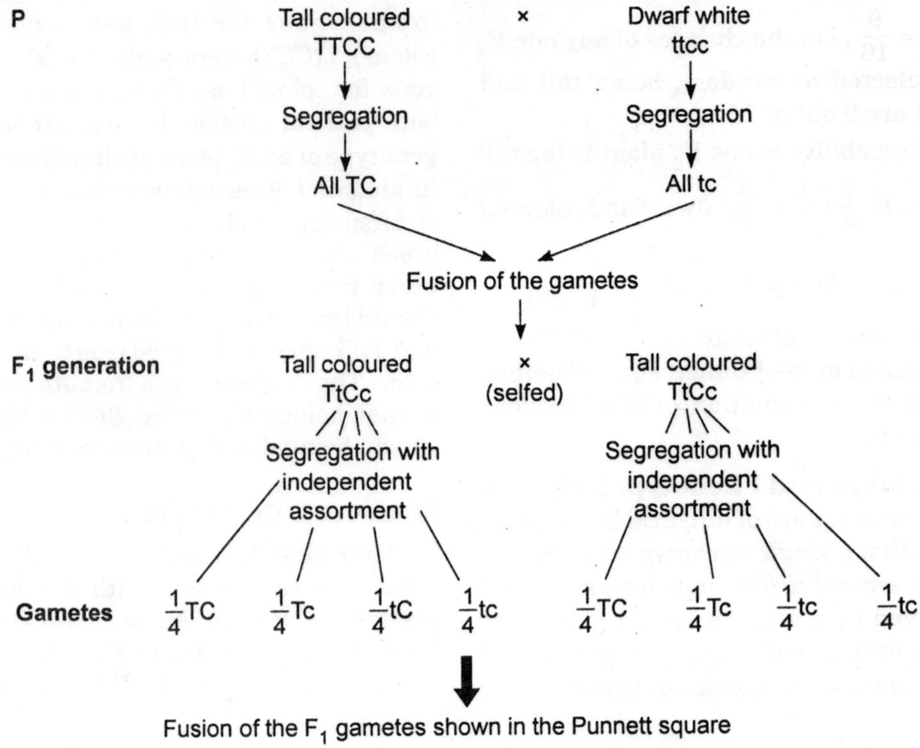


| | | Male gametes (pollen nuclei) | | | |
|---|---|---|---|---|---|
| | | 1/4 TC | 1/4 Tc | 1/4 tC | 1/4 tc |
| Female gametes (egg cells) | 1/4 TC | 1/16 TTCC Tall Coloured | 1/16 TTCc Tall Coloured | 1/16 TtCC Tall Coloured | 1/16 TtCc Tall Coloured |
| | 1/4 Tc | 1/16 TTcC Tall Coloured | 1/16 TTcc Tall Coloured | 1/16 TtcC Tall Coloured | 1/16 Ttcc Tall White |
| | 1/4 tC | 1/16 tTCC Tall Coloured | 1/16 tTCc Tall Coloured | 1/16 ttCC Dwarf Coloured | 1/16 ttCc Dwarf Coloured |
| | 1/4 tc | 1/16 tTcC Tall Coloured | 1/16 tTcc Tall White | 1/16 ttcC Dwarf Coloured | 1/16 ttcc Dwarf White |
| $F_2$ generation | | 9/16 Tall Coloured | 3/16 Tall Coloured | 3/16 Dwarf Coloured | 1/16 Dwarf Coloured |

**Fig. 2.10:** Genotype and phenotype of dihybrid inheritance.

is $\frac{3}{4} \times \frac{3}{4} = \frac{9}{16}$, i.e. the changes of any one $F_2$ plant, selected at random, being tall and coloured are 9 out of 16.

The probability of one $F_2$ plant being tall and white is $\frac{3}{4} \times \frac{1}{4} = \frac{3}{16}$; dwarf and coloured $\frac{1}{4} \times \frac{3}{4} = \frac{3}{16}$; dwarf and white $\frac{1}{4} \times \frac{1}{4} = \frac{1}{16}$.

These findings of ratios agree with the results obtained in the Punnett square method and with the observation of the Mendel's experiments.

• **Genotype and Phenotype Relationship:** Punnet square of dihybrid inheritance revealed that *a single phenotype* may be produced by *several different genotypes.* It will be observed that a tall coloured plant may have one of four possible genotypes, *TTCC* (homozygous for both pairs of genes), *TTCc* (homozygous for tall, heterozygous for colour), *TtCC* (heterozygous for tall, homozygous for colour), or *TtCc* (heterozygous for both pairs of alleles). If we are to know the genotype of an $F_2$ plant, it should be self-pollinated and observation is done for the type of offspring produced. Such as, *TTCC* can breed true to both characters, *TTCc* should breed true to tall but not to coloured, *TtCC* should breed true to coloured but not to tall, and *TtCc* should not breed true to either character. The only genotype from its phenotype is that belonging to the short white plant which, being double recessive, must be *ttcc.*

## Back-cross or Test-cross

To determine the genotype of a tall coloured plant is to test-cross it with the double recessive *ttcc.* If the unknown plant is homozygous for both characters, the offspring will all be tall and coloured. If it is heterozygous

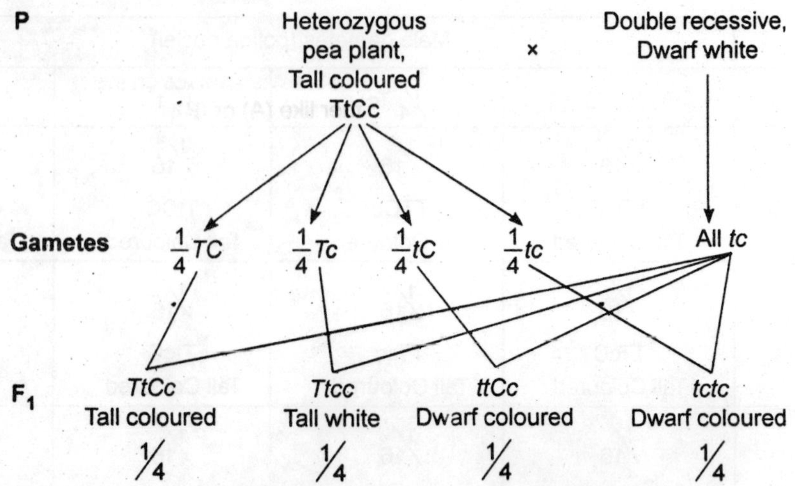


Fusion of $F_1$ gametes shown in the Punnett square

| | | Gametes | | | |
|---|---|---|---|---|---|
| | | TC | Tc | tC | tc |
| Gametes | ↑tc | TtCc<br>Tall coloured<br>1/4 | Ttcc<br>Tall white<br>1/4 | ttcC<br>Dwarf coloured<br>1/4 | ttcc<br>Dwarf white<br>1/4 |

**Fig. 2.11:** Results of test-cross.

for both pairs of alleles, it will produce four types of gametes: *TC, Tc, tC* and *tc*. The double recessive should produce one type of gamete: *tc*. The fusion of the gametes indicates that four types of offspring should be produced in approximately equal proportions, tall coloured, tall white, dwarf coloured and dwarf white.

Mendel performed this experiment and observed in one experiment the following results: 47 tall coloured, 40 tall white, 38 dwarf coloured, and 41 dwarf white. Two other possible genotypes for a tall coloured plant are, *TTCc* and *TtCC* (Fig. 2.11).

- **Discussion on Mendel's Second Law**

With the help of meiosis, independent assortment of genes, like segregation can be explained. The genes determining size are located on one pair of chromosomes and the genes determining flower colour are located on another pair of chromosomes. In metaphase of first meiotic division, homologous chromosomes line up side by side on the spindle prior to separation. Different pairs of homologous chromosomes behave independently of each other. The way one pair of homologous chromosomes arrange themselves on the spindle and subsequently separate does not have any impact on the behaviour of any other pair of chromosomes. The Fig. 2.12 shows how the four different types of gametes, viz. *TC, Tc, tC* and *tc* can be formed from a heterozygous tall plant with coloured flowers (*TtCc*). The behaviour of genes as observed in genetical crosses and

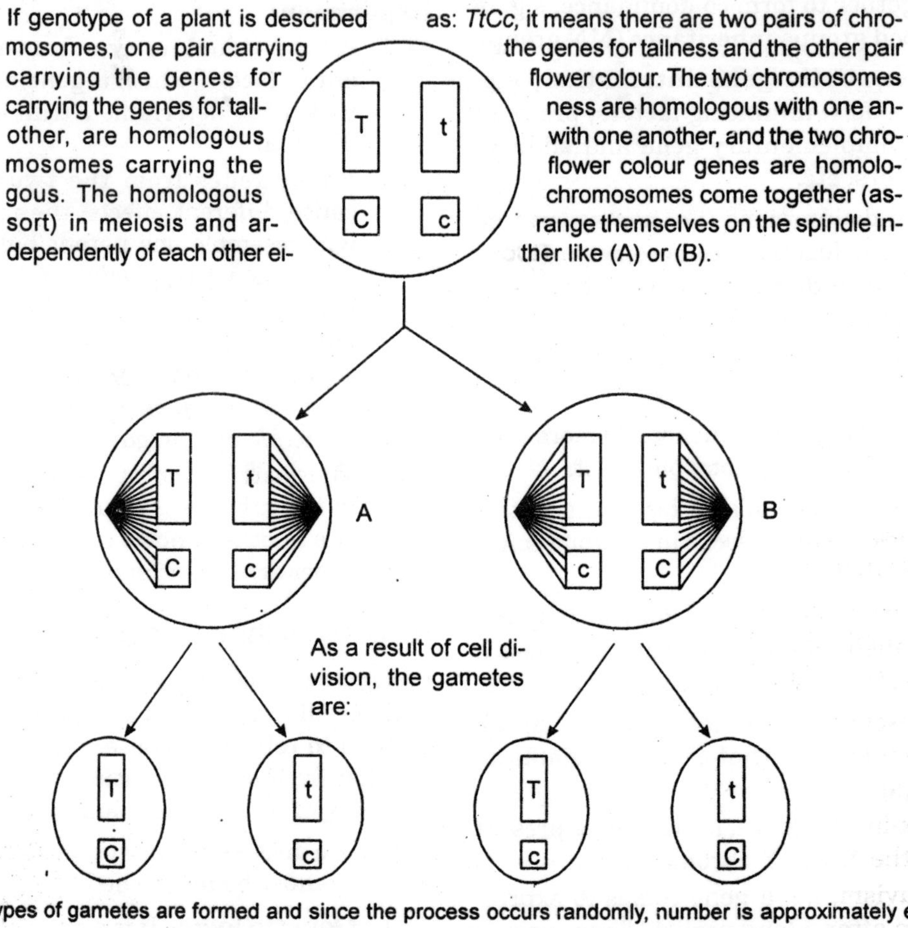


**Fig. 2.12:** Mendel's Second Law, the Law of Independent Assortment whose explanation is provided by the Meiosis. The free assortment of genes in heritance corresponds to the free assortment of chromosomes in meiosis.

the behaviour of chromosomes in meiosis revealed that the genes are located on the chromosomes.

## Major Limitations of Mendel's Principles

Principles of heredity advocated by Mendel have many exceptions. For this reason, it cannot always be applied for determining the inheritance pattern of all the traits or genes in human. The major limitations are due to the following facts:

1. Dominance did not occur in every case of contrasting characters.
2. Blending type of inheritance may happen in many crosses.
3. Occasionally two dominant come together to form co-dominance, e.g. in blood groups inheritance (MN group).
4. Non-allelic gene interaction, e.g. epistasis, inhibiting factors, presence of complementary gene and additive factors, etc.
5. Non-disjunctions (i.e. non-separation of homologous chromosomes) of chromosomes during the time of formation of gametes during meiosis.
6. During linkages between different genes.
7. A gene having influences on more than one trait, i.e. pleotropic effect.
8. The presence of a series of alternative genes of a given locus in a chromosome, i.e. allelism.
9. Multiple genes, that may or may not be allelic, for a single trait (skin colour, height, weight) and their inheritance.
10. Presence of a lethal genes, i.e. an allelic of a gene that renders inviable an organism.
11. Sex-linked genes, i.e. the genes present in the X and Y chromosomes.
12. Atavism, i.e. a gene comes to expression after a long period of non-expression because of the presence of different effects.
13. Presence of phenocopy, i.e. the similarity between the phenotype of a particular genotype trait with the phenotype of another genotype. This phenomenon occurs due to the environment influences or hormonal influences.
14. Position effects of the gene, i.e. a single gene may have differential activity or expression in different position in a single chromosome.
15. Differential environmental influences on the genetic effect of a particular phenotype.
16. Differential penetrance power (complete or incomplete) and expressivity of a given gene in different time.

- **Conclusion of Mendel from the Experiment**

1. An individual receives from each parent a "factor" which is gene nowadays which can determine a trait of the individual.
2. When genes from the two parents cause different effects, the individual will resemble one parent but not the other known as *principle of dominance.*
3. During the formation of gametes, the parental gene for each characteristic segregates from the maternal gene for that same characteristic known as *principle of segregation.*
4. The distribution of segregated genes in the gametes is random in respect to one another known as *principle of independent assortment.*

## Patterns of Inheritance

For investigation of genetics of a particular trait or disorders in humans, it is usual method to rely on: (a) observation on transmission from one generation to another or; (b) study of its frequency among relatives

Before discussing the actual subject some glossary must be mentioned.

1. **Genetic locus:** It is a specific position or location on a chromosome. Usually, locus is used to refer to a specific gene.

2. **Alleles:** These are alternative forms of gene, or of a DNA sequence, at a given locus. If both alleles at a locus are identical, he or she is *homozygous* at that locus; if they are different, he or she is *heterozygous*. Such individuals are called *homozygotes* or *heterozygotes* respectively.
3. **Compound heterozygote:** An individual with two different mutant alleles at a given locus is *compound heterozygote.*
4. **Double heterozygote:** An individual with one mutant allele at each of two different loci is a *double heterozygote.*
5. **Genotype:** It is the genetic constitution or composition of an individual. Genotype also can refer to the alleles at a specific genetic locus.
6. **Phenotype:** The phenotype is the observed result of the interaction of the genotype with environmental factors more specifically, the observable expression of a particular gene or genes.
7. **Single mutant gene:** Mendelian diseases as diseases that are the result of *single mutant gene* that has a large effect on phenotype and that are inherited in simple patterns similar to or identical with those described by Mendel for certain discrete characteristics in garden peas.
8. **Autosomal:** Mendelian diseases are *autosomal* if they are encoded by genes on one of the 22 pairs of autosomes, or non-sex chromosomes.
9. **X-linked:** If encoded by a mutant gene on the X chromosome.
10. **Dominant:** These conditions that are expressed in heterozygotes, i.e. individuals who have one copy of a mutant allele and one copy of a normal, or wild-type, allele.
11. **Recessive:** Those conditions that are clinically manifest only in individuals homozygous for the mutant allele or compound heterozygotes for two different mutant alleles, i.e. carrying a double dose of an abnormal gene. Dominance and recessivity refer to traits, or phenotypes, and not to genes.

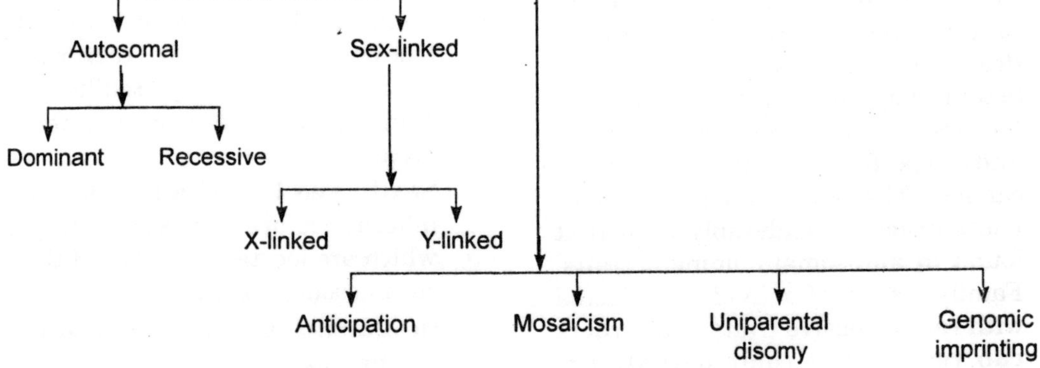


**Fig. 2.13:** Flow chart of patterns of inheritance.

## I. Mendelian Inheritance

A. **Autosomal Dominant Inheritance:** An autosomal dominant trait is one which manifests in the heterozygous state, i.e. in a person possessing both the abnormal or mutant allele and the normal allele. It is often possible to trace a dominantly inherited trait or disorder through many generations of a family. Each gamete from an individual with a dominant trait or disorders will contain either the normal allele or the mutant allele. Autosomal dominant traits can involve only one organ or part of the body.

The clinical features show striking variation from person to person is referred to an *variable expressive.* In some cases heterozygous for certain autosomal disorders, the presence of the mutation can be undetected clinically is referred to reduced penetrance which is thought to be the result of the modifying effects of other genes. In autosomal dominant disorders an affected person will usually have an affected parent. The rarity of most autosomal dominant disorders and diseases means that they usually only occur in the heterozygous state.

B. **Autosomal Recessive Inheritance:** Autosomal recessive traits and disorders are only manifest when the mutant allele is present in a double dose, i.e. homozygosity. Individuals heterozygous for a recessive mutant allele show no features of the disorder and are perfectly healthy, i.e. they are carriers. The family tree for recessive traits differs considerably from that found in autosomal dominant traits. Family history of individuals affected with rare recessive traits or disorders can reveal that their parents are related, i.e. *consanguineous.*

If the normal allele is 'A' and the recessive mutant allele is 'a', then each parental gamete carries either the mutant or the normal allele.

If an individual who is homozygous for an autosomal recessive disorder marries a carrier of the same disorder, their children have a 1 in 20, i.e. 50 per cent chance of being affected. Such a pedigree is said to exhibit *pseudodominance.*

Many disorders can be inherited in various ways due to a number of different genes or what is known as *genetic heterogeneity*. When the disorder is inherited in the same manner may be due to mutations in more than one gene, or what is known as *locus heterogeneity.* Families have been described in which all the children born to deaf parents due to autosomal recessive genes have had perfectly normal hearing and known as *double heterozygotes.* Disorders with same phenotype due to different genetic loci are known as *genocopies,* while the same phenotype being due to environmental causes is known as a *phenocopy.*

Individuals have been identified who have two different mutations at the same locus and known as compound heterozygotes. This constitutes what is known as allelic heterogeneity. Most individuals affected with an autosomal recessive disorder are probably, in fact, compound heterozygotes rather than true homozygotes, unless their parents are related when they are likely to be homozygous for the same mutation by dissent.

C. **Sex-linked Inheritance:** Sex-linked inheritance are carried out by genes which are located on either of the sex chromosomes. These are:

(a) *X-linked*: Genes carried on the X chromosome.

(b) *Y-linked*: Genes carried on the Y chromosome, also known as *holandric inheritance.*

i. *X-linked Recessive Inheritance*: An X-linked recessive trait is determined by a gene carried on the X chromosome and usually only manifests in males. A male with a mutant allele on his single X chromosome known as *hemizygous* for that allele.

A male transmits his X chromosome to each of his daughters and his Y chromosomes to each of his sons. If a male with haemophilia has children with a normal female, then all his daughters will be obligate carriers but none of his sons will be affected. A male cannot transmit an X-linked trait to his son with the very rare exception of uniparental heterodisomy. In humans, several X-linked disorders are known in which heterozygous females have a mosaic phenotype with a mixture of features of the normal and mutant alleles, e.g. ocular albinism.

ii. *X-linked Dominant Inheritance*: This is uncommon, but there are *X-linked dominant traits* which are seen in the heterozygous female as well as in the male who has the mutant allele on his single X chromosome, e.g. vitamin D-resistant rickets.

(c) *Y-linked Inheritance*: In this case only males are affected. The affected male transmits Y-linked traits to all his sons but to none of his daughters.

## II. Non-Mendelian Inheritance

Some disorders are not in the pattern of Mendelian inheritance but different mechanisms have been recognised for this type of inheritance. They are:

A. Anticipation
B. Mosaicism
 (a) Somatic mosaicism
 (b) Gonadal mosaicism
C. Uniparental disomy
D. Genomic imprinting

A. **Anticipation:** In few autosomal dominant traits or disorders, e.g. myotonic dystrophy, the onset of the disease occurs at an earlier age in the offspring than in the parents or the disease occurs with increasing severity in subsequent generations and this phenomenon is known as *anticipation*.

B. **Mosaicism:** An individual or a particular tissue of the body, can consist of more than one cell type or line, through an error occurring during mitosis of any stage after conception known as *mosaicism*.

(a) *Somatic mosaicism*: This mechanism is suggested by the features of a single gene disorder being less severe in an individual than usual or by being confined to a particular part of the body in a segmental distribution. Depending on when a mutation arises in development, it can be transmitted to the next generation with full expression if the mutation is also present in all or some of the germ line cells.

(b) *Gonadal mosaicism*: There are many families with various autosomal dominant disorders, e.g. achondroplasia and osteogenesis imperfecta and X-linked recessive disorders, e.g. Duchenne muscular dystrophy and haemophilia, in which the parents are phenotypically and genetically normal by all known tests, but in which more than one of their children have been affected. The most favoured explanation for these observations is gonadal or germ line mosaicism in one of the parents, i.e. the mutation is present in a proportion of the gonadal cells.

C. **Uniparental Disomy:** One person normally inherits one of a pair of homologous chromosomes from each parent. But recently with the help of DNA technology individuals have been recognized who have been shown to have inherited both homologous of a chromosome pair from only of their parents, so-called uniparental disomy. If an individual inherits two copies of the same homologue from one parent, through an error in meiosis II, which is called *uniparental isodisomy*, when the individual inherits the two different homologues from one parent through an error in meiosis I, is referred as uniparental heterodisomy.

D. **Genomic Imprinting:** It has been recognized from the different clinical features can result, depending on whether a gene is inherited from the father or from the mother. This effect of origin is termed as genomic imprinting. It is evident by two dysmorphic syndromes such as:

1. *The Prader-Willi syndrome*: The syndrome is characterized by short stature, obesity, hypogonadism and learning difficulty.
2. *The Angelman syndrome*: It is characterized by epilepsy, severe learning difficulties, an unsteady or ataxic gait and a happy affect.

## III. Mitochondrial Inheritance

Mitochondrial DNA is a circular form of DNA which without noncoding sequences of introus codes for several proteins or enzymes involved in energy producing oxidative phosphorylation. 13 structural genes have been located in the mitochondria. Cell has many thousands of copies of mitochondrial DNA which are abundant in the cell having high energy consumption such as muscle and brain.

1. Mitochondria and their DNA are almost exclusively inherited from the mother through the oocyte.
2. Mitochondrial DNA (mtDNA) has a higher rate of spontaneous mutation than nuclear DNA.
3. Accumulation of mutations in mitochondrial DNA has been suggested as being responsible for some of the somatic effects seen with ageing.
4. Cytoplasmic or mitochondrial inheritance in humans may be a possible explanation for the pattern of inheritance in some rare disorders which affect both males and females but are transmitted only through females known as *maternal or matrilineal inheritance.*
5. Some of rare disorders with unusual combinations of muscle and neurological characteristics, sometimes occurring in association with other disorders such as diabetes or deafness, have been manifested as being due to mutations in mitochondrial genes.
6. The organs most susceptible to mitochondrial mutations are the CNS, skeletal muscle and heart.
7. In major number of persons the mitochondrial DNA from different mitochondria is identical called *homoplasmy.*
8. When mutation occurs in the mitochondrial DNA of an individual, initially there will be two populations of mitochondrial DNA called *heteroplasmy.*
9. The example of such cytoplasmic inheritance is a particular type of hereditary blindness called *Leber's optic atrophy.*

## QUESTION

1. Describe briefly monohybrid inheritance.
2. What is your conclusion regarding monohybrid experiment? How do you explain the concept of Mendel's factors in the light of modern research?
3. What is Punnett Square? Why it is used?
4. Explain the Mendel's First Law.
5. What is test-cross or back-cross?
6. Define breeding true and pure line.
7. What are the diseases transmitted by single pair of genes in Mendelian fashion? Describe the mechanism of occurrence of albinism.
8. What is dihybrid inheritance? State briefly the gene transmission in dihybrid inheritance.
9. Discuss the Mendel's Second Law.
10. State briefly the law of independent assortment.

# 3 CHROMOSOMES

1. Chromosome, number, existence, behaviour at different stages of cell division, 2. Replication of chromosomes, 3. Biochemical composition, 4. Behaviour of chromosomes in meiosis in gamete-forming cell, 5. Linkages and crossing over of genes, 6. Karyotype and Karyogram, 7. Study of human chromosome.

The nucleus of the eukaryotic cells is bounded by a membrane called *nuclear membrane*. Within the nuclear membrane a fluid-like material is called *nuclear sap*. Within the nuclear sap there are darkly staining *chromatin material* or granules (DNA) having diameter of 20 nm in case of somatic cell scattered throughout the sap (Fig. 3.1). During *mitosis* these chromatin materials which are precursors of the

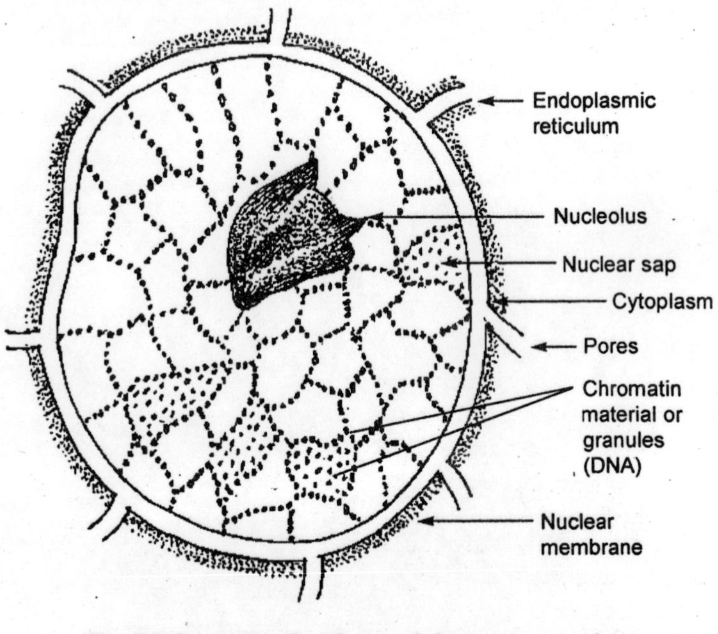


**Fig. 3.1:** Structure of nucleus and chromatin granules.

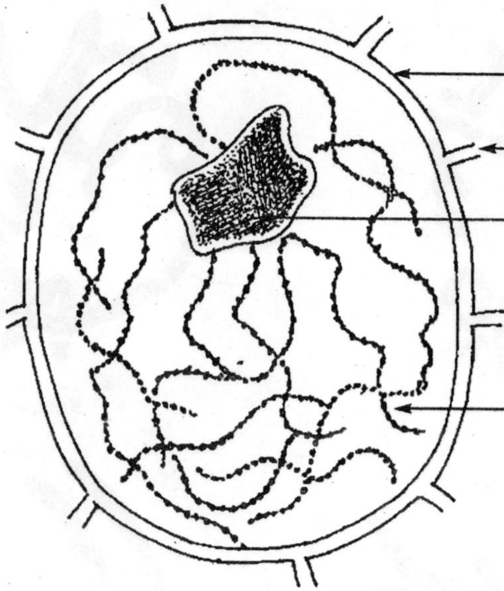


**Fig. 3.2:** Nucleus showing chromosomes at early prophase.

*chromosomes* become readily identifiable as part of the highly structured chromosomes which are thread-like and can be seen under the light microscope (Fig. 3.2). Chromosomes are so called because they take up certain basic stain very readily but they can be *observed* in the *unstained nuclei* of *dividing living* cells. When the cell is not dividing the chromosomes cannot be seen in the nucleus ever after staining. Since they reproduce themselves when they form *chromatids* and the chromatids are shared equally between the cells by the mitosis. It has been calculated that human chromosome contains a DNA molecule about 32,000 μm (3.5 cm) long, looked to form *chromatid* about 6 μm long and 0.8 μm in diameter.

## Number of chromosomes

Counts of chromosomes show that there is a definite number for each species animal or plant, e.g. Human 46, Cray fish 200, Drosophila (fruit fly) 8, Mouse 40, Domestic fowl 36, Kangaroo 12, Honey-bee 16, Corn 20, Frog 26, Hydra 30, Cat 38, Rat 42, Rabbit 44, Chicken 78, etc. (Table 3.1). This indicates that number of chromosomes determine the difference between one species and another (Fig. 3.3).

**Table 3.1:** Diploid chromosome number in human and some animals.

| SPECIES | Number of diploid chromosome |
|---|---|
| 1. *Drosophila melanogaster* (Fruit fly) | 8 |
| 2. *Triturns viridiscens* (U.S.) | 22 |
| 3. *Mus musculus* (Shrew) | 40 |
| 4. *Macaca mulata* (Monkey) | 42 |
| 5. *Mesocricetus auratus* | 44 |
| 6. *Homo sapiens* (Human) | 46 |
| 7. *Chimpanzee* | 48 |
| 8. *Columba livia* (Pigeon) | 80 |
| 9. Cray fish | 200 |
| 10. Mouse | 40 |
| 11. Domestic fowl | 36 |
| 12. Kangaroo | 12 |
| 13. Honey-bee | 16 |
| 14. Corn | 20 |
| 15. Frog | 26 |
| 16. Hydra | 30 |
| 17. Cat | 38 |
| 18. Rat | 42 |
| 19. Rabbit | 44 |
| 20. Chicken | 78 |

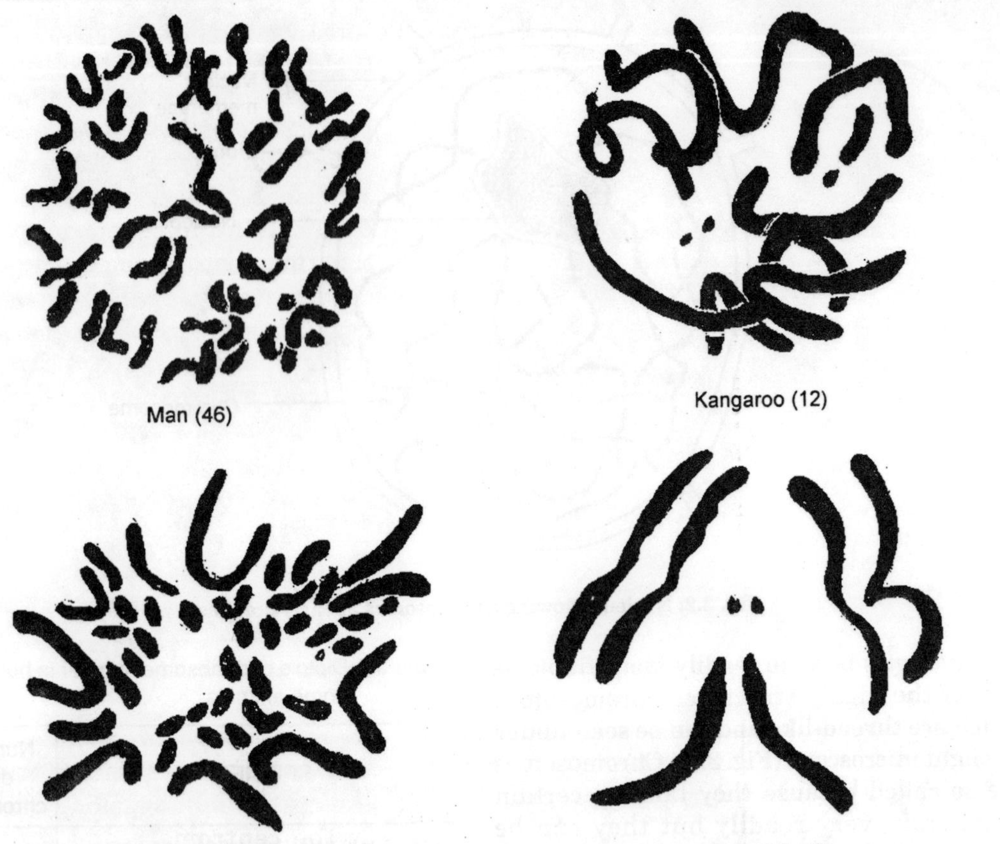

Man (46)

Kangaroo (12)

Domestic fowl (36)

Drosophila (8)

**Fig. 3.3:** Number of chromosomes of different species.

## Existence

Chromosomes exist in pairs, not actually joined together, each pair having a characteristic length and, during anaphase, a characteristic shape governed by the position of the centromere at which the chromatids are pulled apart. Sometimes a "V" shape if the centromere is central or a " $\surd$ " shape if it is close to one end. In other words, human cell nuclei contains 23 pairs of chromosomes. One member of each pair having been derived from the male and one from the female parent. The members of each pair are called *homologous chromosomes,* and the number of chromosomes in the cell is the *diploid number.* The genes in the two chromosomes of each pair are identical or almost identical with each other.

## • Behaviour of Chromosomes at Different Stages of Cell Division (Mitosis)

Prior to cell division, the nucleus of the cell enlarges and in the nucleus there appear a definite number of fine, coiled, threadlike structures called *chromosomes.* The behaviour of these chromosomes during cell division is usually described as a series of stages, e.g. prophase, anaphase, metaphase and telophase. This has been studied by *microphotography* and has shown that the stages of mitosis are not a static event but is a continuous one and different stages do not mean that at a particular stage comes to a halt. Beside that, the different stages do not appear at regular interval because it has been observed in case of embryonic cell of grasshopper, the timing at 38°C is prophase 100 min., metaphase 15 min., anaphase 10 min., telophase 60 min.

Fig. 3.4: Diagram illustrating the way chromosomes appear to become thicker and shorter during prophase.

1. **Prophase:** The chromosomes become more pronounced, i.e. they become short and thick, and react more readily to stains and chemical fixatives (Fig. 3.4), probably by coiling like a helical spring, but with the coils so close to each other that they are not visible at low magnifications (Fig. 3.5).
2. **Metaphase:** Centrioles outside the nucleus, move away from each other and migrate to the opposite ends of the cell. Each centriole radiates protoplasmic fibres which meet and join near the centre of the cell which make a web-like structure called *Spindle*. The chromosomes attach themselves to "*equitorial*" region of the spindle, the larger ones at its edge, and the smaller ones within the equatorial plane. By this time, each chromosome consists of two strands, called *chromatids,* joined in one particular region, the *centromere.* In forming two chromatids, the chromosomes *does not split* but reproduces itself exactly called *replication.* This replication occurs *before prophase,* but is more evident *during metaphase.*
3. **Anaphase:** The two chromatids now separate *at the centromere* and begin to migrate in opposite directions towards either end of the spindle. Experiments show that the spindle fibres play some role in separating the chromatids. The appearance is that of the chromatids first repelling each other at the centromere and then being pulled entirely apart by the shortening spindle fibres (Fig. 3.6, 3.7).

What is the mechanism of moving or migration of chromatids in anaphase? Polariz-

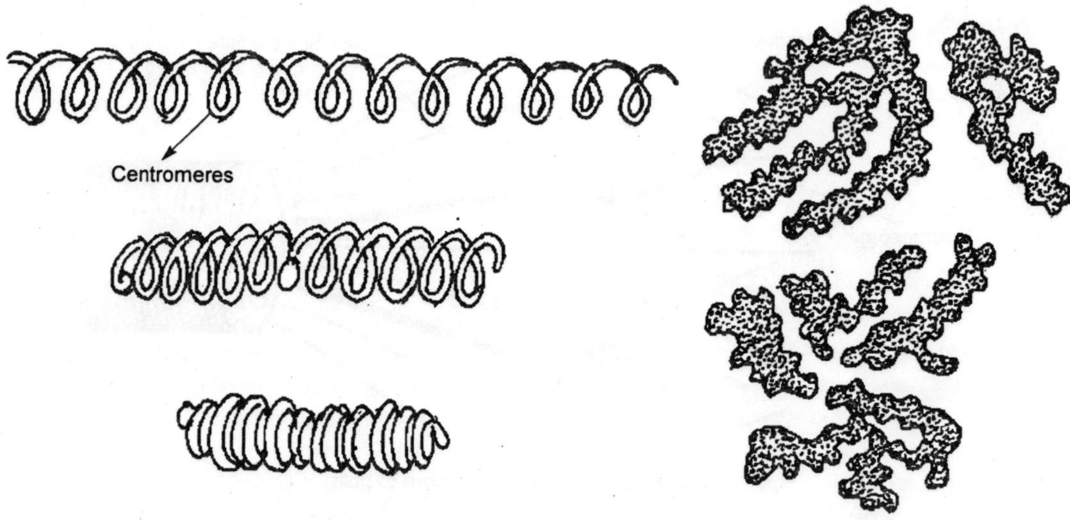


Fig. 3.5: Diagram illustrating the coiling of chromosomes.

**Fig. 3.6:** Each chromosome consist of two chromatids which are attached to the equator of the spindle and are beginning to separate.

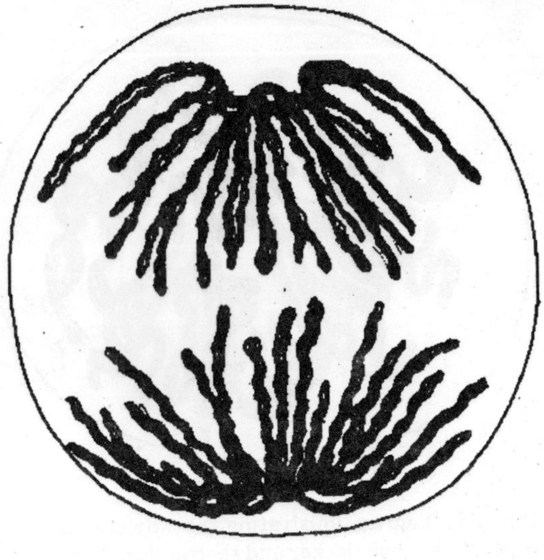


**Fig. 3.7:** The chromatids have completely separated. The spindle is not visible.

ing Microscopy, Electron Microscopy and phase contrast Microscopy during the study of the spindle apparatus have shown that:

1. Certain fibres run from the *centromere* of *each chromosome* to the adjacent pole;
2. Other fibres run from pole to pole (Fig. 3.8).

It is opened that centromere-to-pole fibres *pull* the chromosomes towards the ends of the spindle during anaphase.

The spindle fibres are microtubules and can be disassembled and reassembled. Probably, the that shortening of centromere-to-pole fibres may

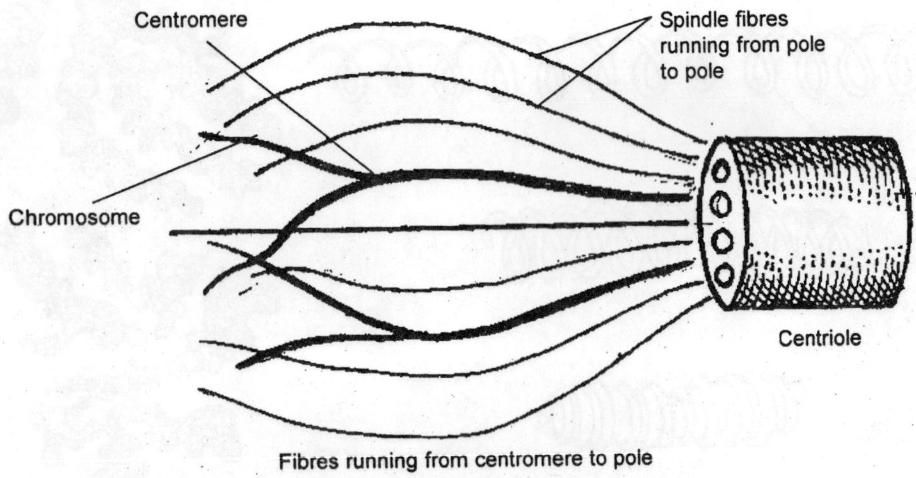


**Fig. 3.8:** Diagrammatic illustration of migration of chromatids.

**Fig. 3.9:** The chromosomes are less distinct.

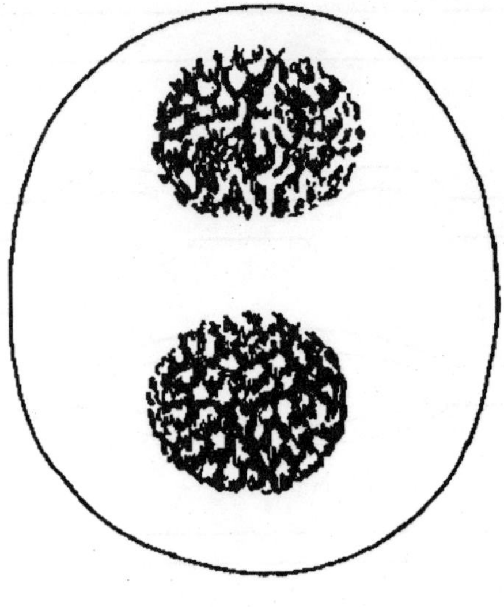


**Fig. 3.9A:** The daughter nuclei are separated by a new cell wall.

occur by progressive dissembly of the protein microtubules. During anaphase, as the chromatids move to opposite poles, the central part of the spindle, the part of the spindle between the sister chromatids, elongates. This could be due to the assembly of extra lengths of microtubules in this region.

It is evident from the recent discovery that the spindle fibres are *interconnected by crossbridges* which may serve as a *rachet* similar to muscle fibre shortens and pulls the chromatid, centromere first, towards the pole (Fig. 3.10). With the centromere in the lead, the rest of the chromatid will tend to trail behind.

4. **Telophase:** The chromatids, now chromosomes, collect together at the opposite ends of the spindle and be-

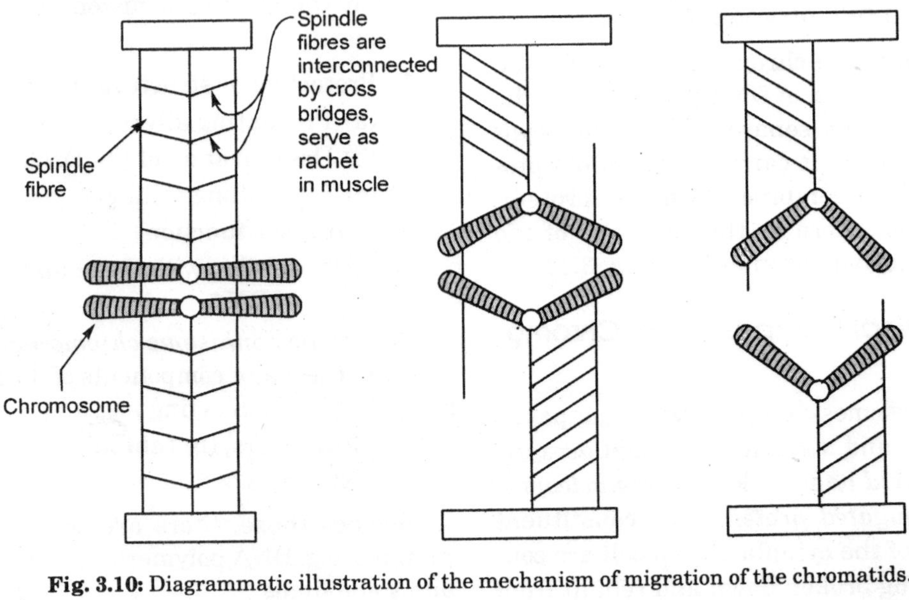


**Fig. 3.10:** Diagrammatic illustration of the mechanism of migration of the chromatids.

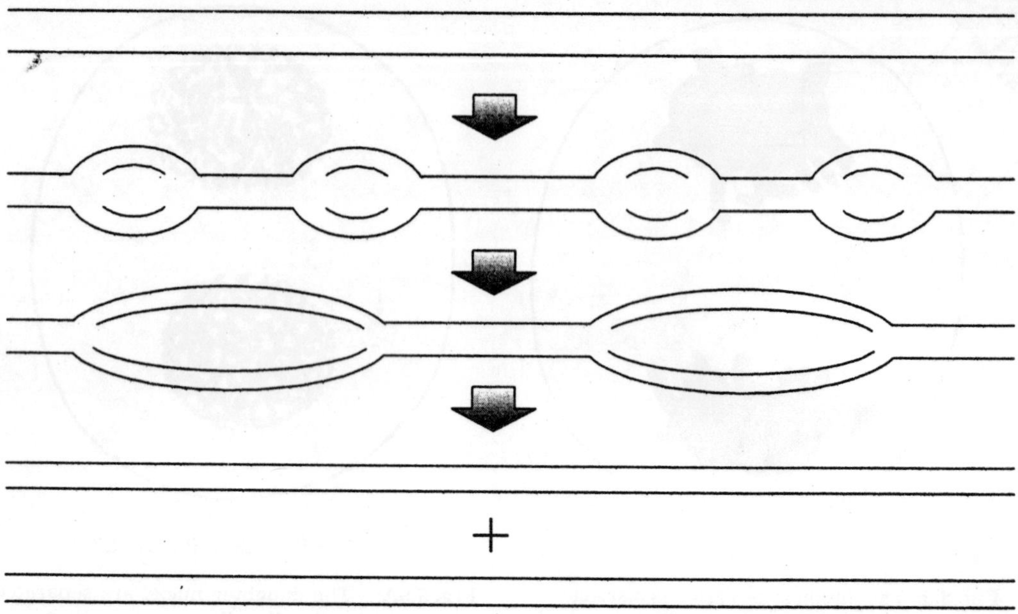

**Fig. 3.11:** Schematic diagram of the replication of the chromosome. Outer line—Parental DNA, Inner—Replicated DNA.

come less distinct, probably by becoming uncoiled and therefore thinner. The one or more nucleoli reappears and a nuclear membrane forms round each group of daughter chromosomes so that there are now two nuclei present in the cell.

## Replication of Chromosomes

Replication of the chromosomes follows the natural replication of DNA strand. When the new double helix separates from the original double helix, it carries some of old protein with it, or combines with new protein, the DNA acts here as the backbone of the newly replicated chromosome (Fig. 3.11).

## Bio-chemical Composition of Chromosomes

The chromosomes consist of two major parts: the *protein* and a substance called *deoxyribonucleic acid* hence it known as *nucleoprotein* a *conjugated protein.* The constituent chemicals of the cytoplasm of a cell are constantly being broken down and rebuilt from fresh material, the chemicals of the chromosomes remain remarkably stable.

The protein is composed of small molecules of which a large part of this is basic *histones* having M.W. 10,000 to 20,000. The function of histone is to serve to *fold* and *compact* the DNA strands into limited sizes. Beside that complexing of histone with DNA may have a:

1. Protective or structural function
   (a) Preventing alteration, or
   (b) Denaturation of the DNA
   (c) Controlling coiling.
2. Repressor function
   (a) Interfering with the template activity of DNA.

The acidic *nonhistone chromosomal proteins* are the major components of the genetic regulatory system acting as:

(a) Activators, (b) Inhibitors,
(c) Enzymes.

Besides these, there are some enzyme proteins, e.g. DNA polymerase, nuclear RNA and some lipids.

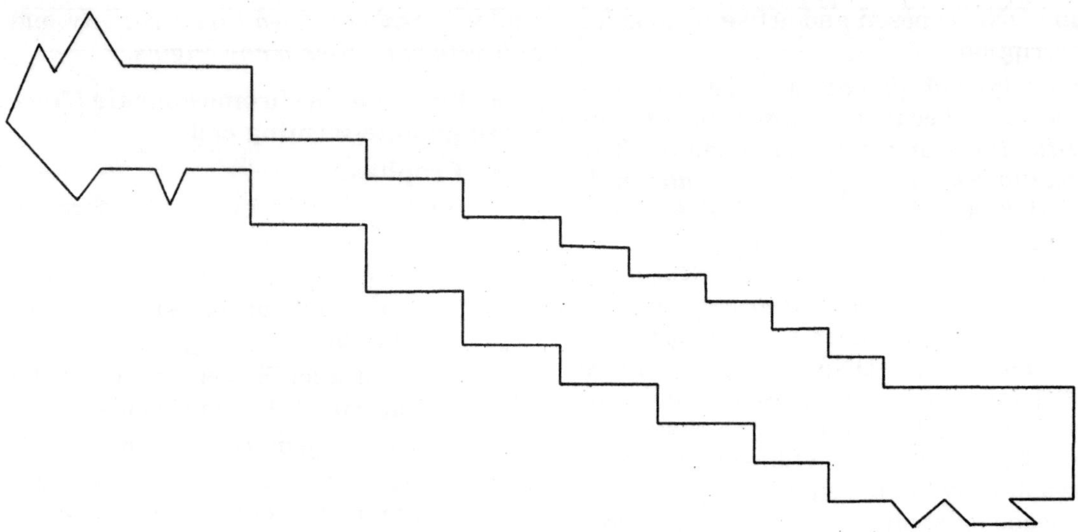

**Fig. 3.12:** Diagram illustrating DNA molecule of Drosophila by autoradiograph.

Recent experiments by *Viscoelastic technique* and *Autoradiography* taking largest chromosome known as giant chromosome of *Drosophila melanogaster* (fruit fly) have shown that DNA of a particular chromosome is arranged in *one long double helix* and that *genes are attached end-to-end with each other.* Such a molecule in the human being has a molecular weight of about 60 billion and if spread out linearly would approximately 7.5 cm. long or several thousand times as long as the diameter of the nucleus. The experiments further showed that this long double helix is folded or coiled like a spring and is held in this position by its linkages to the histone molecules. The measured mass of this largest molecule is $43 \times 10^9$ daltons was revealed by the experiments and it was further observed that the Drosophila chromosome contains *a single uninterrupted molecule of DNA* and is *linear* and *unbranched.*

This DNA is tightly bound to a group of small, basic proteins histones. Histones comprise half of the mass of chromosomes, the other half being DNA. This nucleoprotein chromosomal material is called *chromatin.* Ion-exchange chromatography has shown *5 types* of *histones* differentiated on the basis of their relative of lysine and arginine, viz. H1, H2A, H2B, H3, H4 and they have *high content of positively charged side chain.*

The recurring positive charges of the histones form electrostatic association with the negatively charged phosphate groups of DNA, which render the DNA more stable and flexible. The double helix of DNA, together with the associated histones, is supercoiled and folded back and forth many times in the chromatin fibres. These fibres in turn form a number of loops held together at the centromere. The histones appear to be involved in the supercoiling of the DNA in the eukaryotic chromosomes, either to serve as structural or to cover or repress specific segments of DNA so that they are not transcribed. Such segments can be unwound and thus exposed for transcription in response to certain molecular signals. An Electron Micrograph of one of the 46 chromosomes of a human cell has shown that chromatin fibres contain extended segments of DNA-protein fibres separated by *bands* in which the fibres are transiently enlarged and *highly folded* called *"puffs"* because the DNA in that region is converted from a condensed form into a dispersed one. It has been suggested the "puffs" in the folded regions are loops of bare, unwound DNA exposed and active regions for transcription.

Each type of histone can exist in a variety of forms because of *post-translational modifications* of certain side chains. The *chromatin is made up of repeating units,* each consisting of 200 base pairs of DNA and of two each of H2A, H2B, H3 and H4. This repeating units are known as *nucleosomes.* Most of the *DNA is wound around the outside of a core of histones i.e. histone octamer.* The remainder of DNA, called the *linker DNA,* joins adjacent nucleosomes and contributes to the flexibility of the chromatin fibres. Thus, a chromatin fibre is considered to be a flexibly jointed chain of nucleosomes like beads on string (Fig. 3.13). This model structure of chromatin is supported by the following experimental findings:

1. Electron Microscopy
2. X-ray and Neutron Diffraction
3. Nuclease Digestion
4. Reconstitution.

The winding of DNA around a nucleosome contributes to the *packing of DNA* by decreasing its linear extent and nucleosome is the first stage in the *condensation of DNA.* The nucleosomes themselves form a helical array. Eukaryotic DNA, like other DNA molecules is *replicated semiconservatively* and *bidirectionally from many origins.*

- **Behaviour of chromosomes in Meiosis in gamete-forming cell**

1. **Prophase**
   (a) The diploid number of chromosomes appear.
   (b) Homologous chromosomes pair with each other, shorten and thicken.
   (c) Replication has occurred and the chromatids become visible.
   (d) Homologous chromosomes move apart except at the chiasmata where the chromatids have exchanged portions.
2. **Anaphase**
   (a) A spindle forms and homologous chromosomes move to opposite ends taking the exchanged portions with them.
   (b) Homologous chromosomes separated but not enclosed in nuclear membranes.
3. **Second Meiotic Division:** Spindles form at right angles to the first one and the chromatids separate.

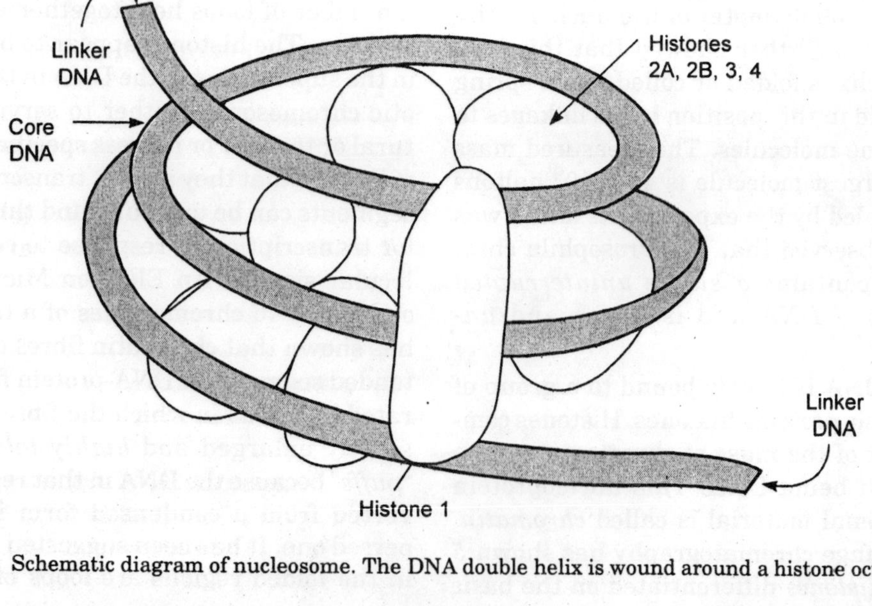


**Fig. 3.13:** Schematic diagram of nucleosome. The DNA double helix is wound around a histone octamer.

4. **Telophase**
   (a) Four nuclei appear, each enclosing the haploid number of chromosomes.
   (b) Cytoplasm divides to form four gametes.

Chiasmata have two functions:

1. **Mechanical:** Two homologous chromosomes together while they manoeuvre themselves onto the spindle prior to segregation.
2. **Genetic:** The chiasmata represent places where the chromatids may break and rejoin, a portion of one chromatid changing places with the equivalent portion of another. This enabls exchange of genes to occur between homologous chromosomes, a process known as "*crossing over*". Since the two homologous chromosomes contain different genes, crossing over promotes genetic variety, which in turn plays an important role in the process by which evolution has taken place.

## Linkages and Crossing over of Genes

During early stage of meiosis, homologous chromosomes pair up, the paternal and maternal chromosomes exchange portions (Fig. 3.14). This causes variability in the gene combinations in the gametes. In the absence of crossing over, the maternal genes A, B, C on the same chromosome would always appear together whatever is the way of assortment of chromosomes in the meiosis. Similarly, the paternal genes a, b, c would always remain together. Suppose, if Drosophila contains genes for black body, purple eye and vestigial wings. After crossing over between the chromosomes, there is every possibility of breaking up these *linkage groups,* so the new combinations, viz. ABc, Abc, aBc, abC, AbC, could be formed in the gametes, e.g. black body with purple eyes and normal wings.

### • Karyotype and Karyogram

Techniques have been developed for examining human chromosomes. Cells in metaphase are spread, and the chromosomes are stained or banded with a trypsin-Giemsa technique for counting or photography and mounting. A systematized array of chromosomes from a single cell is called a *Karyotype.* When the chromosomes silhouettes from the photomicrograph are cut and arranged in order called *Karyogram.* Homologous chromosome, can be identified and paired according to size and binding pattern. Chromosomes are positioned so that their *shorter p arms* point upward and their *longer q arms* point downward. The meaning of karyotype term is usually extended to imply that the chromosomal pattern of the cells examined is typical of all the cells of the individual. To determine accurately the karyotype, however, one must characterize the chromosomes from several tissues, and at least 100 metaphase cells from each tissue should be counted. The karyotype of a normal woman is designated 46, XX; that of a normal male 46, XY. The karyotype of an individual with an extra X chromosome is 47, XXY.

## Localisation of Gene on Individual Human Chromosomes

1. **Chromosome 1:** Rh factor; Histone proteins; Antithrombin III; Coagulation protein V; Elliptocytosis, Congenital cataract; Retinitis pigmentosa.
2. **Chromosome 2:** Lambda light chain of immunoglobulin; Red Cell acid phosphatase; Apolipoprotein B, Protein C (coagulopathy disorder); Collagen 3 (alpha-1 chain).
3. **Chromosome 3:** Acute leukemia antigen; a gene for renal cancer; a gene for small cell lung cancer; protein S (coagulatory disorder).
4. **Chromosome 4:** Huntington's chorea; Mucopolysaccharidosis type I (Hurler syndrome); immunoglobulin kappa light chain; Coagulation factor XI.
5. **Chromosome 5:** Complement factor 6, 7, 9; Spinal muscular atrophy; Familial adenomatous polyposis (polyposis coli); coagulation factor XII.

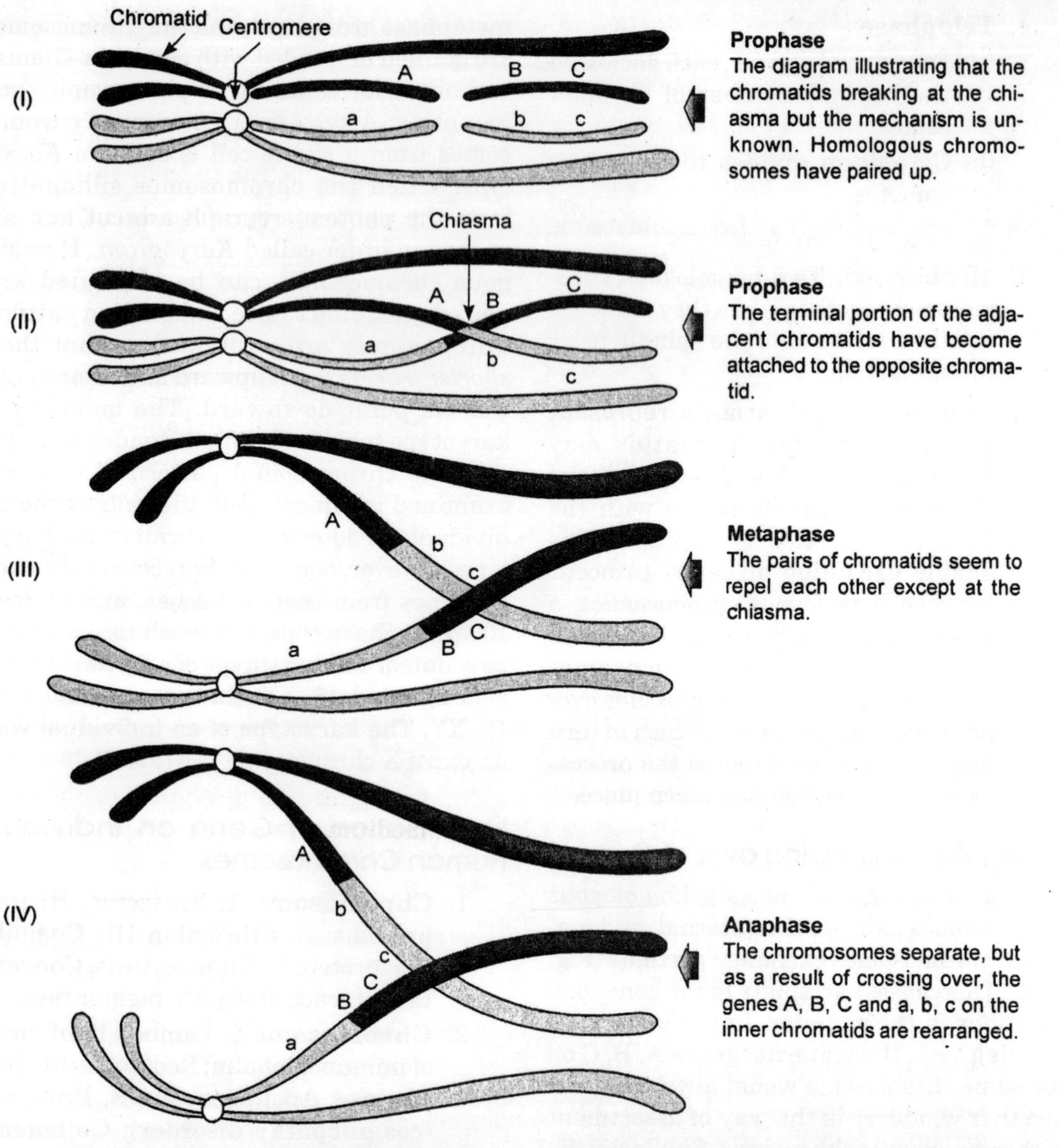


**Fig. 3.14:** Diagram illustrating the linkage and crossing over of genes.

6. **Chromosome 6:** HLA genes (major histocompatibility complex); Spinocerebellar ataxia locus I; Congenital adrenal hyperplasia (21-hydroxylase).
7. **Chromosome 7:** Collagen type I, alpha-1 chain (osteogenesis imperfecta); Cystic fibrosis regulator.
8. **Chromosome 8:** Retinitis pigmentosa locus I; Congenital adrenal hyperplasia (II-beta hydroxylase).
9. **Chromosome 9:** Galactosaemia; Friedreich's ataxia; Nail-patella syndrome; Complement factor 5; ABO blood group.
10. **Chromosome 10:** Multiple endocrine neoplasia syndrome (type 2); Congenital erythropoietic porphyria.
11. **Chromosome 11:** Beta globin genes; Insulin genes on short arm; Neimann-Pick disease; Wilm's tumour 1 gene;

Multiple endocrine neoplasia syndrome (type 1); Ataxia telangiectasia; Acute intermittent porphyria.

12. **Chromosome 12:** Von Willebrand disease; Collagen type 2; Phenyl-ketonuria.
13. **Chromosome 13:** Retinoblastoma; Wilson's disease; Coagulation protein VII; Coagulation protein X.
14. **Chromosome 14:** Lambda light chain of immunoglobulin; Spherocytosis type 1.
15. **Chromosome 15:** A deletion of band 15q. II; if inherited from father produces Prader-Willi syndrome, and if deletion in the same locus from mother produces Angelman syndrome; Fibrillin (Marfan syndrome); Hexosaminidase A (Tay-Sachs disease).
16. **Chromosome 16:** Alpha globin genes (Alpha-thalassaemia), Adult polycystic kidney disease-locus.
17. **Chromosome 17:** Growth hormone genes in the long arm; Charcot-Marie tooth disease; Neurofibromatosis type 1 (Von Recklinghausen's disease).
18. **Chromosome 18:** Specific genes not identified.
19. **Chromosome 19:** Myotonic dystrophy; Apolipoprotein E.
20. **Chromosome 20:** Proto-oncogene (Rous sarcoma virus).
21. **Chromosome 21:** Superoxide dismutase (familial motor neurone disease); Down syndrome; Alzheimer's disease.
22. **Chromosome 22:** Heavy chain of immunoglobulin; Neurofibromatosis type 2 (bilateral acoustic neuroma).

### • Chromosome X

In *short arm*: Xg blood group; Duchenne and Becker muscular dystrophy; Retinitis pigmentosa locus 2.

In *long arm*: Lesch-Nyhan syndrome; Fragile X-syndrome; coagulating factor VII (Haemophilia A); Coagulation factor IX (Christmas disease); Glucose-6-phosphate dehydrogenase.

### • Chromosome Y

SRY (Testis-determining) gene in the *short arm;* H-Y antigen in the *long arm.*

### • Classification of Human Chromosomes

*Denver-London system of classification (1960)* in order of decreasing length and arranged into seven groups:

1. *Group A*: Pairs of 1, 2, 3 chromosomes; Each of them is long and metacentric; Chromosome 2 is submetacentric.
2. *Group B*: Pairs of 4 and 5 chromosomes, which are fairly long with submetacentric centromeres.
3. *Group C*: Pairs of 6 to 12 chromosomes, X chromosome. Most of them are of medium size and sub-metacentric. Banding pattern helps identification of individual chromosomes.
4. *Group D*: 13 to 15 chromosomes pairs belong to this group. All of them are medium sized and acrocentric. A *satellite body* is attached to the free end of the short arm of each chromosome.
5. *Group E*: Chromosome numbers 16 to 18 and they are fairly short submetacentric chromosomes.
6. *Group F*: Pairs of 19 and 20 chromosomes belong to this group. Each of them is short and metacentric.
7. *Group G*: 21 and 22 pairs of chromosomes plus Y chromosome. Each of them is very short and acrocentric, 21 and 22 chromosomes present *satellite bodies* on their short arms. The distal ends of long arms of Y chromosome present *fluorescent* bodies after staining with a fluorochrome dye.

### • Type of Chromosomes

Each chromosome in a genome can be distinguished on the basis of the position of centromere and it divides the chromosome into two arms having different lengths. The part of the chromosomes on either side of the centromere is called the *arm of chromosomes* which may be *equal* or *unequal*, i.e. one arm of chromosome is longer than the other. The long arm is called q arm and short arm is p arm. Based on the *position of centromere,* chromosomes are grouped into 4 categories.

I. **Metacentric chromosomes:** The centromere is at the centre hence two arms of the chromosomes are equal or almost equal and thus are "V" shaped during anaphasic movement.

II. **Submetacentric chromosomes:** Centromere is located on one side of the central point, i.e. centromere is sub-median and divides the chromosome into two unequal arms and look "L" shaped in anaphase.

III. **Acrocentric chromosomes:** The centromere is located almost near at one end of chromosome, i.e. centromere is sub-terminal in position and it gives to two arms, one exceptionally short and the other long. The acrocentric chromosomes look "J" shaped in anaphase.

IV. **Telocentric chromosomes:** Telocentric chromosomes are made through a transverse division of the centromere. This type of division is known as *misdivision of centromere* a process which leads to the formation of isochromosome (Fig. 3.15). According to the number of centromeres the chromosomes are divided into:

1. *Acentric*: The chromosome is without any centromere.
2. *Monocentric*: Having only one centromere.
3. *Dicentric*: Having two centromeres.
4. *Polycentric*: Having more than two centromeres.

## Human Chromosomes

### Method of Study

1. **Tissue culture:** Human cells are grown in tissue culture, treated with the drug Colchicine, which arrests mitosis at the metaphase; exposed to a hypotonic solution that makes the chromosomes swell and disperse, and then "squashed" onto slides. *Fluorescent* and other staining techniques make it possible to identify the individual chromosomes and study.

2. **Electron microscopy:** Electron microscopy revealed little information regarding submicroscopic morphology of the chromosomes because:
   (a) Osmium tetroxide, normally used to fix the material, does not produce much electron density;
   (b) The chromosomes are not sharply delimited by membrane as are the cytoplasmic organelles;
   (c) E/M sections are so thin that only a very little of the total length of each chromosome will appear in any one section.

3. **Aldehyde fixatives:** This technique has overcome the first of these difficulties and has shown the chromosomes to be composed of what appear to be numerous granules and fine filaments.

4. **Isolation technique:** More information may be obtained by the use of isolation technique. Chromosomal material is isolated from the nucleus, spread on a carbon-coated E/M grid and stained with heavy metal atoms; under these conditions the central filamentous "core" is easily studied.

5. **Fluorescent dyes:** Recent technique such as staining of fluorescent dyes have helped to identify the individual chromosomes in a metaphase preperation. The arrangement in order of the chromosomes from photographs of the metaphase stage of cell division is called karyotype which is useful in medicine with the realization that some clinical conditions are due to chromosomal abnormalities. A typical

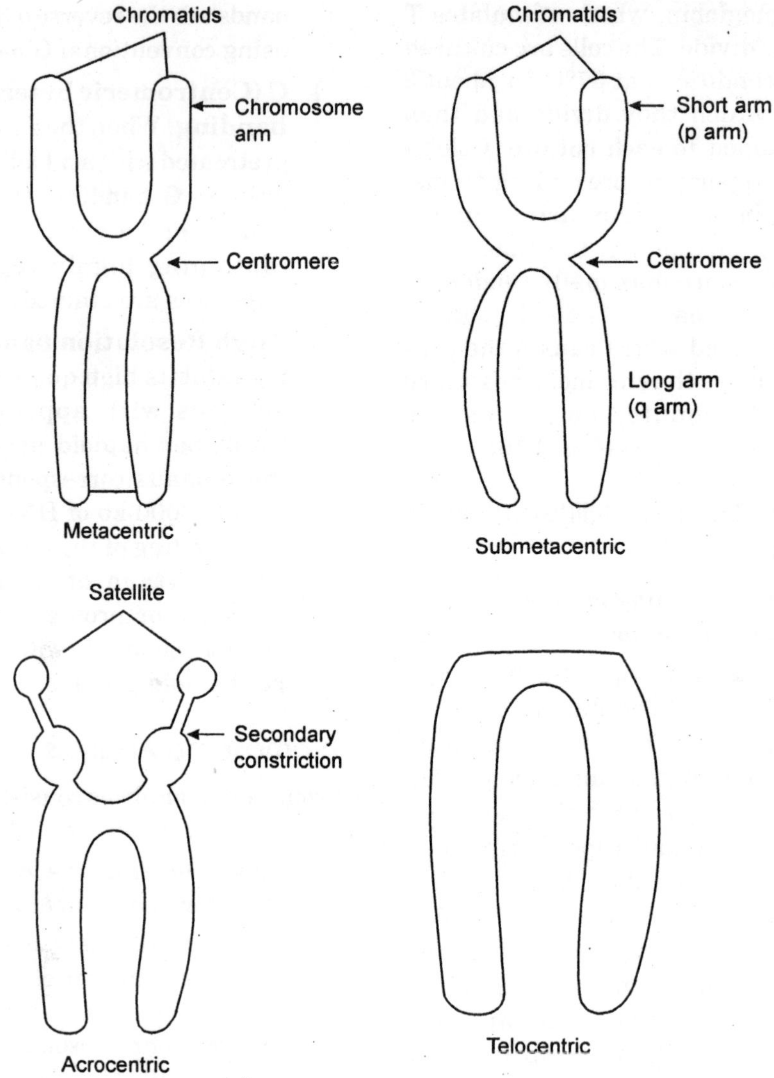


**Fig. 3.15:** Classification of chromosomes according to location of the centromere.

example is the presence of an extra chromosome number 21 to give Down's syndrome (Mongolism).

- **Observation**

Electron Micrograph of human chromosome has revealed that human chromosomes contain about 15 per cent DNA, 10 per cent RNA, and 75 per cent protein. They consist of long chromatin fibres about 20–30 nm in diameter and about 700–800 μm long. Each chromatin fibre is believed to contain on *DNA duplex* about *4 cm long* in *supercoiled* arrangement, around which the protein is wound.

## Chromosome Preparation

Any tissue with living nucleated cells which undergo cell division can be employed for the study of human chromosomes. In general circulating lymphocytes from peripheral, i.e. venous blood are studied although samples for chromosomal analysis can be prepared easily using skin, bone marrow, chorionic villi or cells from amniotic fluid, i.e. amniocytes.

In the study with venous blood, the lymphocytes are separated off and added to a small volume of nutrient medium containing phytohaemoglobin, which stimulates T lymphocytes to divide. The cells are cultured under sterile conditions at 37°C for about 3 days, during which they divide and then colchicine is added to each culture. Colchicine has the property of preventing formation of the spindle thereby arresting cell division during metaphase, the time when the chromosomes are maximally condensed and therefore most easily visible. Hypotonic saline is then added which causes the cells to swell and release the chromosomes which are then fixed, mounted on a slide and stained ready for analysis (Fig. 3.16).

## Chromosome Banding (different staining method)

There are several staining methods to identify individual chromosome.

1. **G Giemsa banding:** This method is used commonly. The chromosomes are treated gently with trypsin to denature their protein content and then with Giemsa which gives each chromosome a characteristic and reproductive pattern of light and dark band. By this technique approximately 350–550 bands per haploid set are observed; each band represents approximately 5–20 × $10^6$ bp of DNA. Because one gene may range from $10^3$ bp (α-globin) to more than 2 ×$10^6$ bp (e.g. dystrophin), and because it is estimated that there is approximately one gene every 30,000 bp, each band may represent several to hundreds of individual genes.
2. **Q (Quinacrine) banding:** This method gives a banding pattern similar to that of Giemsa. It requires examination of the chromosomes with the ultraviolet fluorescent microscope.
3. **R (Reverse) banding:** When the chromosomes are heated before staining with Giemsa, the light and dark bands are the reverse of those obtained using conventional G banding.
4. **C (Centromeric heterochromatin) banding:** When the chromosomes are pretreated with acid followed by alkali prior to G banding, the centromeres and other heterochromatic regions containing highly repetitive DNA sequences are stained.
5. **High Resolution banding:** G banding exhibits high quality chromosome analysis with approximately 400 bands per haploid genome. Each of these bands corresponds to approximately 8000 kb of DNA. High resolution banding of the chromosome at an earlier stage of mitosis such as prophase or prometaphase provides greater sensitivity with upto 800 bands per haploid genome.

## Chromosome Analysis

Chromosomal analysis consists of the following:

1. First *counting the number* of cells often termed as *metaphase spreads.*
2. Next careful analysis of the banding (staining) pattern of each individual chromosome in those cells. Generally the total chromosome count is determined in 10–15 cells, but if mosaicism is suspected, then 30 or more cell counts will be done.

   More detailed analysis of the banding pattern of the individual chromosomes is carried out on both members of each pair of homologous in approximately 3–5 metaphase spreads which show high quality banding. The pattern of banding of each chromosome is specific and can be demonstrated in the form of a stylized ideal karyotype termed as *idiogram.* The cytogeneticist analyses each pair of homologous chromosomes, either whilst observing through the

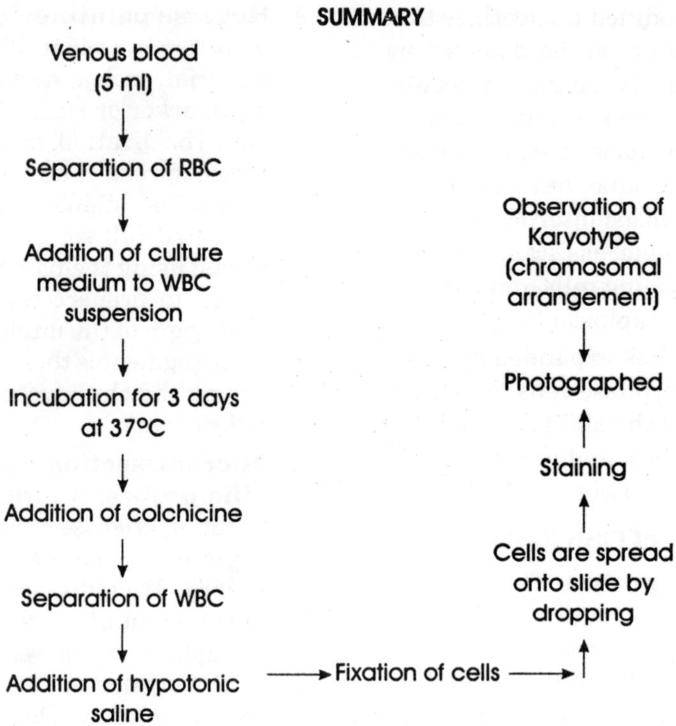


**Fig. 3.16:** Schematic representation of preparation of chromosome.

microscope or less commonly, on a photograph of the metaphase spread which can now be produced electronically. After the advent of banding, chromosomes could be classified only based on their overall morphology and a formally presented karyotype or *karyogram* will show each chromosome pair in descending order of size.

## Molecular Cytogenetics, Fluorescence in situ Hybridization (FISH)

In recent year most important advance in cytogenetics is the development of *fluorescence in situ hybridization (FISH)* technology. This is a new diagnostic tool combines conventional cytogenetics with molecular genetic technology. It is based on the unique ability of a portion of single-stranded DNA, i.e. a probe to anneal with its complementary target sequence wherever it is located on a metaphase spread.

**Use**

1. Identification and definition of both numerical and structural chromosome abnormalities in metaphase spreads associated with birth defects and cancer.
2. Gene mapping.
3. Study of chromosomes in cells which are resting between cell division in the interval known as *interphase* hence study at this period known as *interphase cytogenetics*.
4. Clinical evaluation of constitutional chromosome disorders and of certain leukemias, lymphomas and solid tumours as well as prenatal diagnosis.
5. Can also be applied to haploid cells such as sperm to obtain information on the frequency of chromosome abnormalities in gametes.

- **Technique**

The technique involves preparation of specific DNA probes, labelled by incorporation of chemically modified nucleotides that fluorescence directly or can be detected by binding a fluorescently tagged molecule. These single-stranded probes may be hybridized to metaphase chromosomes, as in standard cytogenetic technique, but can also be hybridized by immobilised interphase nuclei of nondividing cells, whereas standard cytogenetics requires dividing cells and chromosomes arrested in metaphase.

FISH technology has expanded chromosome analysis to interphase cells. Major advances have included the ability to generate a wide variety of probes, and improvements in digital imaging microscope.

- **Different Types of FISH Probes**

1. **Centromeric probes:** These consist of repetitive DNA sequences observed in and around the centromere of a specific chromosome. They are useful for making a rapid diagnosis of one of the common aneuploidy syndrome (trisomy, 13, 18, 21) using cells in interphase obtained from a prenatal diagnostic sample of chorionic villi.
2. **Chromosome-specific unique sequence probes:** These are specific for a particular single locus. They are particularly useful for identifying tiny sub-microscopic deletions and duplications. The group of disorders referred to as the *microdeletion syndromes.*
3. **Whole chromosome paint probes:** These consist of a cocktail of probes obtained from different parts of a particular chromosome. When this mixture of probes is used together in a single hybridization, the entire relevant chromosome fluorescence, i.e. is painted. Chromosome painting is extremely useful for characterizing complex rearrangements such as subtle translocations and for identifying the origin of additional chromosome material such as small supernumerary markers or rings.
4. **Reverse painting:** In this procedure a portion of unidentified chromosome material, such as a small supernumerary marker or ring is used as a paint for hybridization to a normal metaphase spread. The unidentified chromosome is obtained using a fluorescent activated cell sorter and is then amplified using the polymerase chain reaction to prepare the labelled paint. The origin of the unidentified chromosome segment is then revealed by identifying the chromosome(s) to which it hybridizes.
5. **Microdissection and Region-specific probes:** A further development is the microdissection of chromosome regions to produce region-specific probes. Microdissection of a specific chromosomal region from single metaphase chromosomes, followed by PCR (Polymerase Chain Reaction) amplification of the DNA, allows generation of region-specific probes without the time-consuming cloning procedures that would otherwise be necessary.

- **Latest Development in FISH Technology**

1. **Multicolour Spectral Karyotyping:** The development in FISH technology utilizes a pool of whole human chromosome paint probes to provide a multicolour human karyotype in which each pair of homologous chromosomes can be identified on the basis of their unique colour. Each whole chromosome paint probe is prepared in such a way that it gives a different spectral signal when bound to a fluorescent label and analysed using computer software. This approach should be extremely useful for detecting subtle chromosome rearrangements, both constitutional as in patients who show developmental abnormalities and acquired as in cancer.
2. **Fibre FISH:** The another development in FISH technology includes

high-resolution *Fibre FISH* which is useful in gene mapping and comparative genome hybridization, which involves combined hybridization to a normal metaphase spread of equal amounts of test and reference DNA labelled with two different colours. This technique is particularly useful in detecting and mapping deletions or amplifications in tumour cells.

## Chromosomal Abnormalities

**Types:** Different types of *abnormalities* which can occur are divided into the following orders.

### I. NUMERICAL ABNORMALITIES

- *Aneuploidy*: Numerical abnormalities involve the gain or loss of one or more chromosomes is known as aneuploidy.
- *Monosomy*: Loss of a single chromosome.
- *Trisomy*: Gain of a single homologous chromosome.
- *Tetrasomy*: Gain of two homologous chromosomes.
- *Polyploidy*: Addition of one or more complete haploid complements.

(a) **Aneuploidy:** *Aneuploidy* is the result of *nondisjunction* or failure of chromosomes to separate normally during cell division, and can occur during meiosis or mitosis. The causes of nondisjunction are not known but it has been observed that the risk of meiotic nondisjuction increases with increasing maternal age. When nondisjunction occurs after fertilization, i.e. mitotic nondisjunction, it is expected to the mosaicism, or two or more populations of chromosomally different cells.

1. **Monosomy:** The absence of a single chromosome is called monosomy and for an autosome is almost always incompatible with survival to term, with the possible exception of a few very rare reported cases of monosomy 21. Absence of an X or a

**Table 3.2:** Different types of chromosomal abnormalities

|  |  |
|---|---|
| **I. Numerical** | |
| (a) Aneuploidy → | 1. Monosomy<br>2. Trisomy<br>3. Tetrasomy |
| (b) Polyploidy → | 1. Triploidy<br>2. Tetraploidy |
| **II. Structural** | |
| (a) Translocations → | 1. Reciprocal<br>2. Robertsonian |
| (b) Deletion | |
| (c) Insertion | |
| (d) Inversion → | 1. Paracentric<br>2. Pericentric |
| (e) Rings | |
| (f) Isochromosome | |
| **III. Different cell lines** | |
| (a) Mosaicism | |
| (b) Chimaerism | |

Y chromosome results in a 45X karyotype which causes a condition called Turner's syndrome. Monosomy can result from nondisjunction in meiosis.

2. **Trisomy:** The presence of an extra chromosome is referred to as trisomy. Most cases of Down's syndrome are due to the presence of an additional number 21 chromosomes, hence Down's syndrome is often known as trisomy 21. Other autosomal trisomies which are compatible with survival to term are Patau's syndrome (trisomy 13) and Edwards syndrome (trisomy 18). Most other autosomal trisomes result in early pregnancy loss with trisomy 16 being a particularly common finding in first trimester spontaneous miscarriages. The presence of an additional sex chromosome (X or Y) has only mild phenotype effects.

   Trisomy 21 is usually caused by failure of separation of one of the pairs of homologous chromosomes during anaphase of maternal meiosis I. This failure of the bivalent to separate is called non-disjunction. Less often, trisomy can be caused by non-disjunction occurring during meiosis II when a pair of sister chromatids fail to separate. Either way the gametes receives two homologous chromosomes (disomy), and if subsequent fertilization occurs a trisomic conceptus results.

3. **Tetrasomy:** (2n + 2) have one chromosome in quadruplicate, while the others are in duplicate, (ABC, ABC, AAA).

(b) **Polyploidy:** Polyploid cells contain multiples of the haploid number of chromosomes such as 69, triploidy or 92, tertraploidy. In humans triploidy is found relatively often in material grown from spontaneous miscarriages but survival beyond mid-pregnancy is rare. Only a few triploid live births have been described and all have died soon after birth. Triploidy can be caused by failure of a maturation meiotic division in ovum or sperm leading, e.g. to retention of a polar body or to the formation of a diploid sperm. Alternatively it can be caused by fertilization of an ovum by two sperms, which is known as *dispermy.* When triploidy results from the presence of an additional set of paternal chromosomes, the placenta is usually swollen with what are known as hydatiform changes. In contrast when triploidy results from an additional set of maternal chromosomes, the placenta is usually small. Triploidy usually results in early spontaneous miscarriage.

**Euploid:** Refers to any exact multiple of N, the number of chromosomes in a normal haploid gamete. Euploid chromosome constitutions need not be normal, e.g. triploidy (69 chromosomes) is an euploid chromosome abnormality commonly found in spontaneous abortions.

**Diploid:** The number of chromosomes in a normal somatic cell is diploid or 2N.

## II. STRUCTURAL ABNORMALITIES

Structural chromosome rearrangements result from chromosome breakage with subsequent reunion in a different configuration. Structural abnormalities of chromosomes may involve either one or two or more chromosomes. When only one chromosome is involved, it may have a deletion or duplication of a portion of the chromosome, a pericentric (involving the centromere) or paracentric (not involving the centromere) inversion, or the formation of a ring chromosome or isochromosome. They can be:

(A) Balanced, (B) Unbalanced.

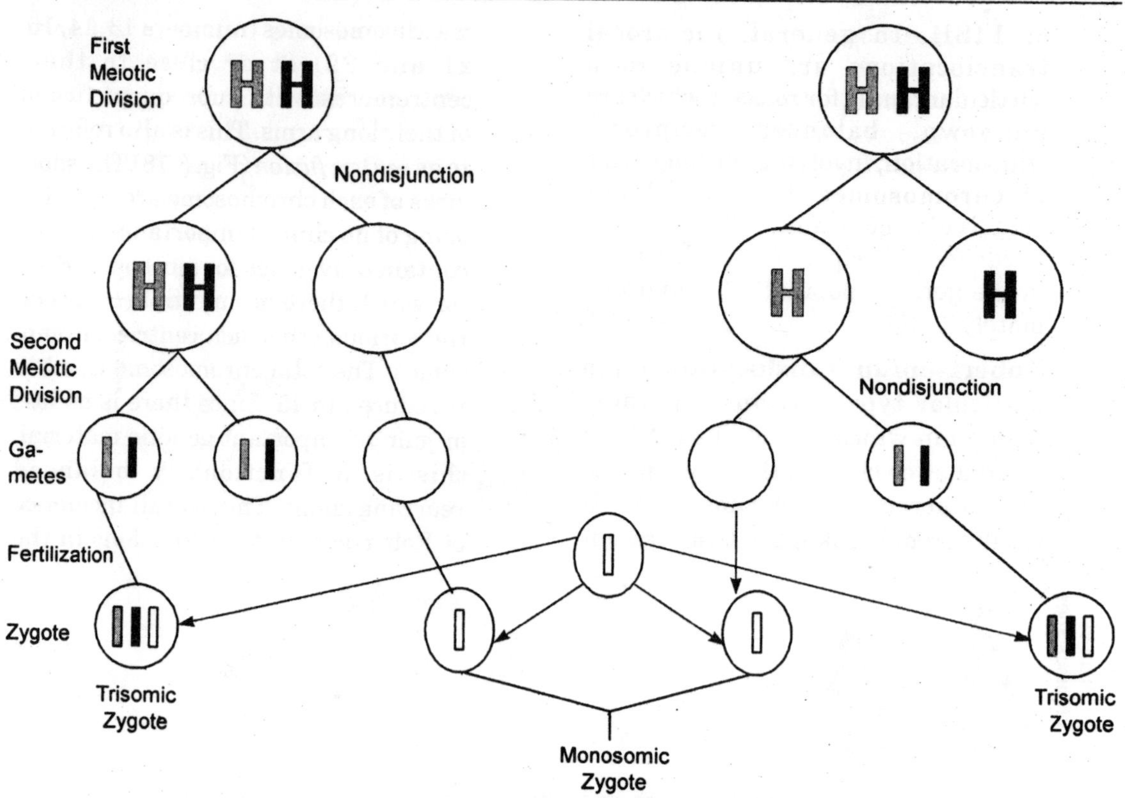


**Fig. 3.17:** Chromosomal abnormalities of nondisjunction occurring in the first or second meiotic division.

A. **Balanced:** In balanced rearrangements the chromosome complement is complete with no loss or gain of genetic material. Consequently, balanced rearrangements are generally harmless with the exception of rare cases in which one of the breakpoints damages an important functional gene. However, carriers of balanced rearrangements are often at risk of producing children with an unbalanced chromosomal complement.

B. **Unbalanced:** When a chromosome rearrangement is unbalanced the chromosomal complement contains an incorrect amount of chromosome material and the clinical effects are usually very rare.

- **Translocation**

A translocation refers to the transfer of genetic material from one chromosome to another. Structural abnormalities involving two or more chromosomes may result in insertions of material from one chromosome into another, or more commonly, translocation or exchange of chromosomal material between two or more chromosomes. Two major forms of chromosome translocation are recognized:

(A) *Reciprocal Translocation*
(B) *Robertsonian Translocation.*

A. **Reciprocal translocation:** It is formed when a break occurs in each of two chromosomes with the segments being exchanged to form two new derivative chromosomes. It involves breakage of at least two non-homologous chromosomes with exchange of the fragments. Usually the chromosome number remains at 46, and if the exchanged fragments are of roughly equal size, a reciprocal translocations can often only be identified with detailed chromosomal banding studies or FISH. In general, reciprocal translocations are unique to a particular family for reasons which are unknown, balanced reciprocal translocation, involving the long arms of chromosomes 11 and 22 are relatively common. The overall incidence of reciprocal translocations in the general population is approximately 1 : 500.

B. **Robertsonian translocation:** It is a particular type of reciprocal translocation in which the breakpoints are located at or close to the centromeres of two acrocentric chromosomes. It results from breakage of two acrocentric chromosomes (numbers 13, 14, 15, 21 and 22) at or close to their centromeres, with subsequent fusion of their long arms. This is also referred to as *centric fusion* (Fig. 3.18).The short arms of each chromosome are lost, this being of no clinical importance as they contain only genes for ribosomal RNA for which there are multiple copies on the various other acrocentric chromosomes. The total chromosome number is reduced to 45. Since there is no loss or gain of important genetic material, this is a functionally balanced rearrangement. The overall incidence of Robertsonian translocations in the

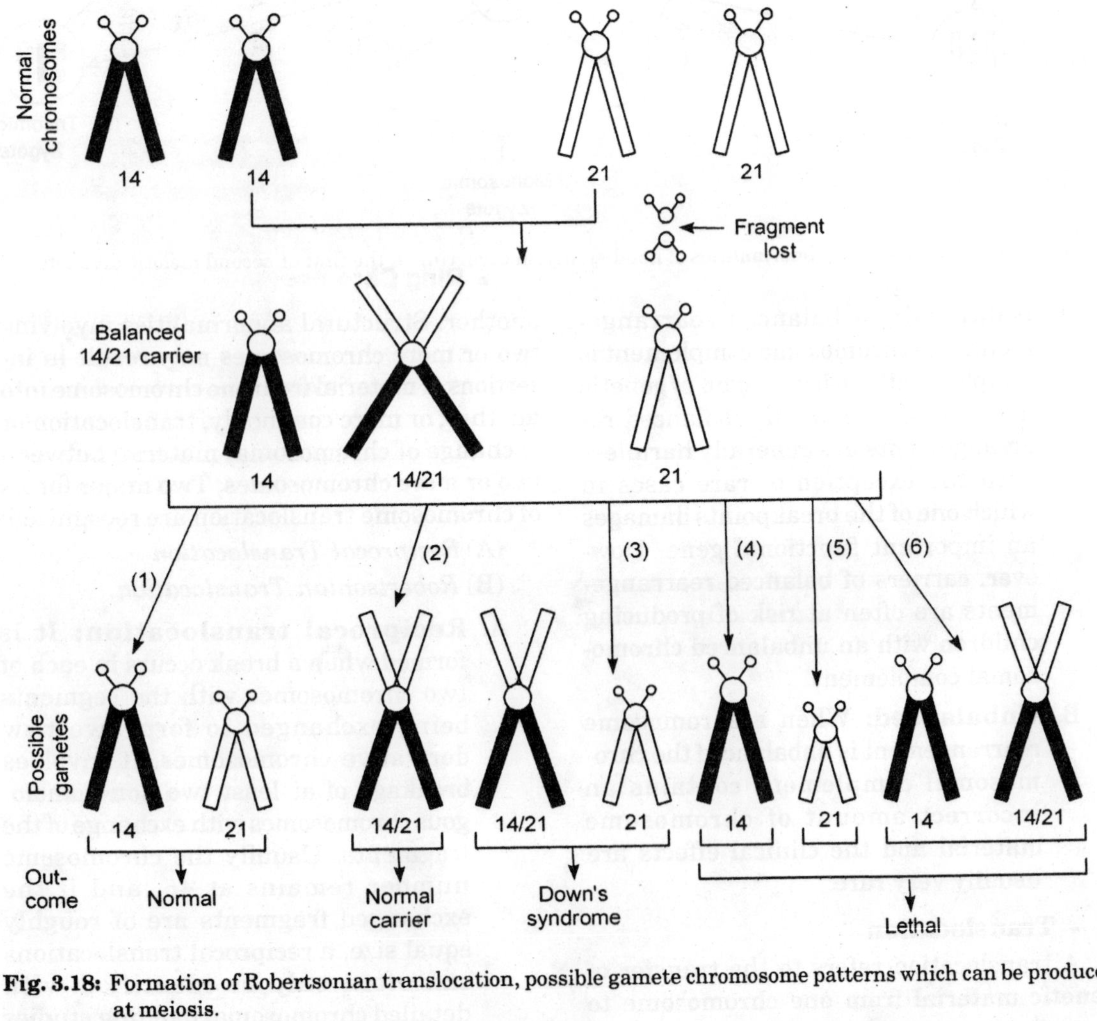


**Fig. 3.18:** Formation of Robertsonian translocation, possible gamete chromosome patterns which can be produced at meiosis.

general population is approximately 1 : 1000, with by far the most common being fusion of the long arms of chromosomes 13 and 14.

### • Deletion

A deletion involves loss of part of a chromosome and results in monosomy for the segment of the chromosome. Usually a very large deletion will be incompatible with survival to term and as a general rule any deletion resulting in a loss of more than 2 per cent of the total haploid genome will be lethal.

Deletion exists at two levels:

(a) **Microscopic or Chromosomal deletion:** It can be visualized under the microscope. Several deletion syndromes have been described such as:

1. Wolf-Hirschhon syndrome
2. Cri-Duchat syndrome

These involve loss of material from the short arms of chromosomes 4 and 5 respectively.

(b) **Submicroscopic Microdeletions:** More recently submicroscopic microdeletions have been observed with the help of high resolution prometaphase cytogenetics augmented by fluorescent in situ hybridization studies. It has been shown that the Prader-Willi and Angelman syndromes can be caused by microdeletions.

### • Insertions

It occurs when a segment of one chromosome becomes inserted into another chromosome. If the material is inserted from elsewhere in another chromosome, then the karyotype is balanced. Otherwise an insertion causes an unbalanced chromosome complement. Carriers of a balanced deletion-insertion rearrangement are at a 50 per cent risk of producing unbalanced gametes as random chromosome segregation at meiosis will result in 50 per cent of the gametes inheriting either the deletion or the insertion but not both.

### • Inversions

Inversion is a two-break rearrangement involving a single chromosome in which a segment is reversed in position, i.e. inverted. If the inversion segment involves the centromere it is termed *pericentric inversion* (Fig. 3.19). If it involves only one arm of the chromosome it is known as a paracentric inversion (Fig. 3.20).

### • Ring Chromosomes

A ring chromosome is formed when a break occurs on each arm of a chromosome leaving two 'sticky' ends on the central portion which reunite as a ring. Inevitably the two distal chromosomal fragments are lost so that, if the involved chromosome is an autosome, the effects are usually every serious. Ring chromosomes are often unable

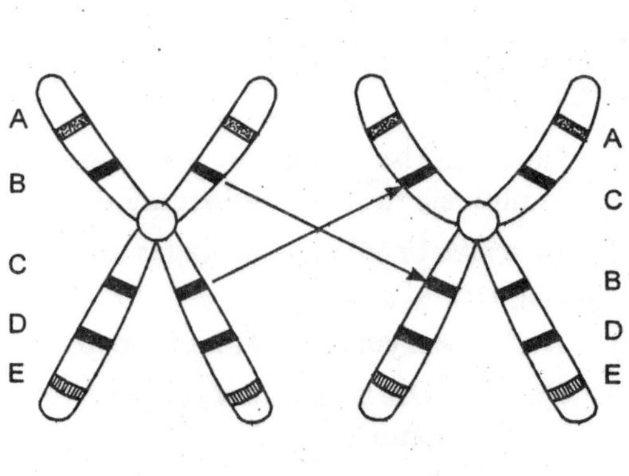


**Fig. 3.19:** A pericentric inversion.

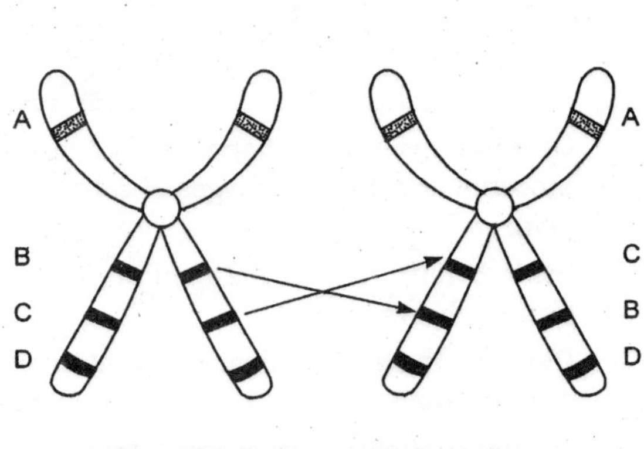


**Fig. 3.20:** A paracentric inversion.

to complete mitotic division so that it is not unusual to find a ring chromosome in only a proportion of cells. The other cells in the individual are often monosomic due to absence of the ring chromosome.

- **Isochromosomes**

The isochromosome exhibits loss of one arm with duplication of the other. The most probable explanation for the formation of an isochromosome is that the centromere has divided transversely rather than longitudinally. The most commonly encountered isochromosome is that made up of two long arms of the X which accounts for approximately 20 per cent of all cases of Turner's syndrome.

## III. DIFFERENT CELL LINES (MIXOPLOIDY)

(a) **Mosaicism:** Mosaicism is defined as the presence in an individual or in a tissue of two or more cell lines which differ in their genetic constitution but are derived from a single zygote. Usually it results from non-disjunction occurring in an early embryonic mitotic division with the persistence of more than one cell line. It can also exist at a molecular level if a new mutation arises in a somatic or early germ line cell division. The possibility of germ line or gonadal mosaicism is a particular concern when counselling the parents of a child who is an isolated case of a condition such as Duchenne muscular dystrophy.

(b) **Chimaerism:** Chimaerism can be defined as the presence in an individual of two or more genetically distinct cell lines derived from more than one zygote. In humans, chimaeras are of two types:

1. **Dispermic chimaeras:** These are the result of double fertilization whereby two genetically different sperms fertilize two ova and the resulting two zygotes fuse to form one embryo. If the two zygotes are of different sex the chimaeric embryo can develop into an individual with true *hermaphroditism* and an XX/XY karyotype.
2. **Blood chimaeras:** Blood chimaeras result from an exchange of cells via the placenta, between non-identical twins in utero. 90 per cent of one twin's cells can have an XY karyotype with RBC showing predominantly blood group B, whereas the other twin can have 90 per cent of cells showing an XX karyotype with RBC showing predominantly blood group A.

## QUESTION

1. Describe briefly the behaviour of chromosome at different stages of cell division.
2. State briefly the mechanism of migration of chromatids.
3. Describe the biochemical composition of chromosome.
4. Define replication of chromosomes.
5. Describe the behaviour of chromosome in meiosis in gamete-forming cell.
6. Describe briefly the linkages and crossing over of genes.
7. Define karyotype and karyogram.
8. Describe the methods of study of human chromosome.

# 4

# SEX DETERMINATION AND SEX DIFFERENTIATION

1. Fertilization, 2. Sex determination and sex differentiation, 3. Recombination of genes in the zygote, 4. Sex linkage.

## Fertilization

The union between male and female gametes is called fertilization. In most organisms sexual reproduction involves the union of gametes, a process known as *"syngamy"*. Generally the gametes differ from each other in structure, size and behaviour, for which reason they are known as *"heterogametes"* ovum and sperm are heterogametes, syngamy takes place by a process of fertilization.

The *Aim of Fertilization*

1. To restore the original number of chromosomes.
2. To transmit the hereditary characters of the parents.
3. To determine sex.

*Perquisite of Fertilization*

1. Maturity of sperm and ovum which is fulfilled by gametogenesis.
2. Coming together of these two cells which is facilitated by the act of sexual reproduction.

*Physical and Chemical Events*

1. Activation of ovum.
2. Amphimixis, i.e. intermingling of parental heredity traits in the off-spring.

Half billion sperm deposited in the female genital tract after coitus, but only 1000 to 3000 succeed in traversing the fallopian tube to reach the proximity of the ovum. Sperm can remain *fertile* in the female genital tract for 24–72 hours but can remain highly fertile probably only for 12–24 hours. But it remains alive in the female genital tract for 7 days. A mature ovum also is fertile probably for only 12–24 hours. Therefore, if coitus occurs when these two periods of fertility overlap, fertilization can take place.

- **Penetration:** Only one sperm can penetrate the ovum for fertilization. It pierces the zona pellucida of the ovum after dispersal of the corona radiata and penetrates into the cytoplasm of the ovum. After a sperm has penetrated the ovum, the tail is discarded and the head containing chromosomes and the middle piece (or body)

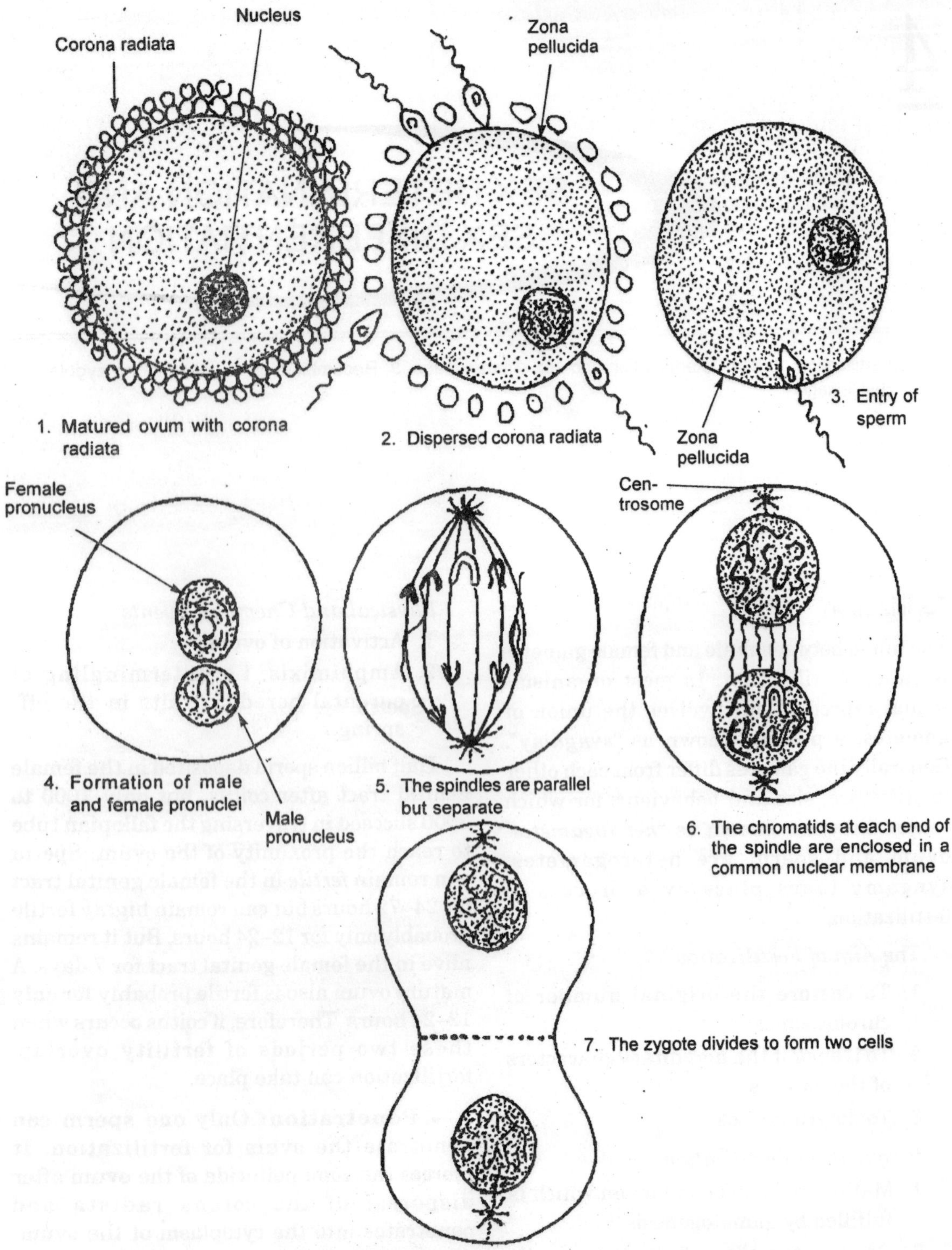


**Fig. 4.1:** Diagrammatic illustration of fertilization.

are drawn by force through the cytoplasm towards the nucleus.

The male and female pronuclei migrate towards each other, meet and fuse together near the centre of the ovum to form a single nucleus, called *"first segmentation nucleus"* from which all future developmental changes start. In the pronuclei the chromosomes were reduced to half the usual number; the mixed nucleus formed by the fusion of the two pronuclei contains the full number of 46 chromosomes (autosome) + 2 sex chromosomes, i.e. exactly the same number seen in the nucleus of a somatic cell. The resulting new structure is called a *"zygote"*.

## Segmentation

*Segmentation* or *cleavage* starts which consists of repeated division of the zygote, forming a mass of cells called *morula.* All these divisions are of the *homotype mitotic* form. The zygote divides at first into two cells and then to four, eight, sixteen, etc. and ultimately form a mass of cells.

In many cases at the time of zygote formation, there is no fusion of nuclear material at this stage. Each nucleus simultaneously undergoes a mitosis with the axes of the spindles parallel to each other, but at the telophase stage the adjacent chromatids, originally from parents, become enclosed in the same nuclear membrane, thus restoring the diploid number of chromosomes in all its cells (Fig. 4.1).

## Sex Determination and Sex Differentiation

Sexual development in the embryo involves two processes:

I. Sex Determination

II. Sex Differentiation

### I. Sex Determination

Sex determination is a genetic phenomenon and depends on the constitution of smallest sex chromosomes. What is determined is the *sex genotype* or *genetic sex* which almost invariably accords with the apparent sex of the fully-developed individual. Oogonia and spermatogonia each contain 46 chromosomes. The former contain 44 autosome + two identical sex chromosomes XX which carry *female-determining genes.* Spermatogonia contain 44 autosomes + XY sex chromosomes where Y is a smaller (not in all animals) chromosome carrying *male-determining genes.* When oocytes undergo reduction division halving of the chromosome number occurs so that each ovum finally contains 22 autosomes + an X sex chromosome and never Y sex chromosome hence called homogametic. When spermatocytes undergo reduction division (meiosis) there are two kinds of resulting secondary spermatocytes (which become spermatozoa). One type contains 22 autosomes type + X (female type) sex chromosomes and other 22 autosomes + Y (male type) sex chromosomes hence called heterogametes. The ovum may be fertilized by either kind of sperm with the following sex genotypes:

Sperm X + Ovum X = Offspring XX
= Female ♀

Sperm Y + Ovum X = Offspring XY
= Male ♂

The genetic sex is *determined exclusively by the sperm* and is quite *independent of the ovum.*

The nuclei of somatic cells in the female contain a visible chromatin mass about 1 μ in diameter which lies against the inner surface of the nuclear membrane. This *sex chromatin* (Barr body) is thought to result from apposition of certain regions of the XX chromosome pair.

The XY chromosome pair of the male forms no detectable mass. Thus the sex genotype can be identified by a cytological test, the most suitable cells being the epithelial cells of the *epidermal spinous layer, buccal mucosa* or *vagina* or the *blood leucocytes.* Though the X and Y chromosomes determine sex, it does not necessarily follow that male and female characteristics are determined by genes found only on the sex

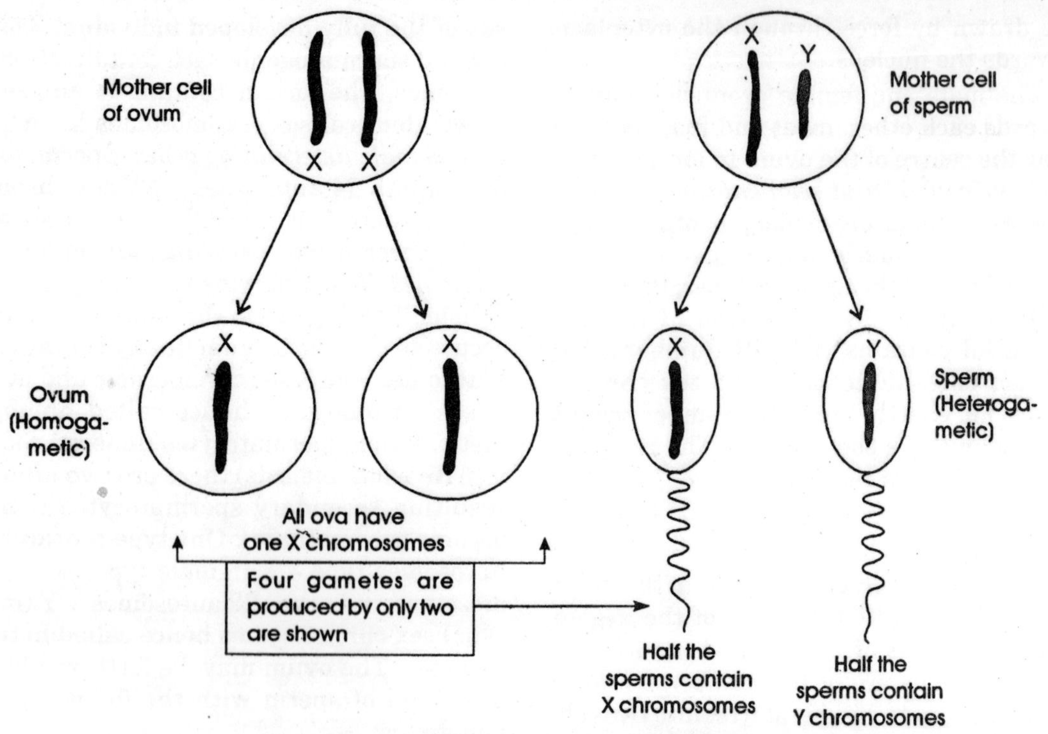


**Fig. 4.2:** Diagrammatic illustration of determination of sex.

chromosomes. In human, genes for *male and female characters* may be scattered fairly evenly throughout all the chromosomes, but the presence of the Y chromosomes in an XY zygote may tip the balance in favour of maleness. Femaleness results from the absence of the Y chromosomes, but this is not the case in all the animals studied, e.g. mouse but not for Drosophila.

After recombination of male and female pronuclei during fertilization, the *fertilized ovum* then *contains* 44 autosomal chromosomes and either 2X chromosomes in case of female child to develop or an X and a Y chromosome in case of a male child to develop.

The male is not invariably the heterogametic sex because in birds the male is XX and the female is XY. In some insects the female is XX and the male is XO, the Y chromosome being absent.

Sex chromosomes are transmitted by normal Mendelian process. In determining and individual's sex, the X and Y chromosomes are only important at a relatively early stage of development when the gonads are developing. Once the gonads have been formed, sex hormones take over. The potency of these hormone determining the sex is evident by frequent occurrence of phenotypically male or female, but are genotypically of the other sex (Fig. 4.2).

## II. Sex Differentiation

Sex differentiation in the embryo usually harmonizes with the genetic sex but environmental influences can lead to discordant development so that apparent sex may not accord with genetic sex. Sex differentiation takes place in 3 steps:

1. **Gonadogenesis:** *Primordial germ cells* migrate into the *genital ridge* where proliferation of both non-germinal and germinal cells leads to the formation of gonadal primordia which are identical in both sexes and later become divided into a peripheral cortex and a central medulla. In the

female the cortex becomes the ovary, and in the medulla becomes testis. The dominance of cortex or medulla is mainly determined genetically but can be influenced by hormonal and other environmental factors.

2. **The Genital ducts:** Before the primitive embryo has, as it were, chosen its sex, the *Mullerian ducts* and *Wolffian ducts* exist as a paired set of structures. If the primitive gonad becomes an ovary, the Mullerian ducts form the fallopian tubes, uterus and upper vagina, and the Wolffian ducts disappear. If the primitive gonad becomes a testis, the Wolffian ducts form the epididymis, vas deferens and seminal vesicle, the Mullerian ducts degenerating.
3. The urogenital sinus develops into the external genitalia, male or female.

- **Hormonal control of sex differentiation**

Removal of the foetal ovary in the rabbit or mouse does not prevent the differentiation of a female genital system, but removal of the foetal testis at an early critical period prevents the differentiation of male structures and results in entirely female development of ducts and urogenital sinus. Castration of male foetuses at a later stage does not prevent male sex differentiation. Thus the genital tract becomes female in type unless positively modified by the foetal testicular secretion formed by the Leydig cells during early gestation. The hormone is not testosterone or any other known androgen.

## Recombination of Genes in the zygote

In man there are 200 million sperms in a single ejaculation and about 300–400 ova produced in reproductive life. The genetic content of ovum differ from the sperms, e.g. white race wife with red curly hair and husband with black straight hair. If the chromosomes of the husband and wife combine, the zygote may contain genes for red, black, curly and straight hair. The genes for curly and black are dominant to those for straight and red, so that although the child may carry all four genes, his hair will be curly and black, a new combination of characters not represented in either parent. When this child grows up and his own gametes are formed, he could produce sperms with four alternative combinations of these genes, e.g. red and straight, black and straight, red and curly, black and curly.

## Sex Linkage

Certain genes present on the X chromosome are likely affect a male than a female. As for example, the gene or genes for a certain form of colour blindness in man are carried on the X chromosome.

Normal vision gene is *dominant* to *colour blindness* gene. If a colour-blind woman, who is homozygous for the character, marries a normal man, all the sons will be colour blind but not the daughter (Fig. 4.3). The explanation is that the Y chromosome is homologous with only a small section of the X chromosome and the non-homologous part of the X chromosome carries genes which are not represented on Y. It is assumed that the Y chromosome plays no part in the

**Table 4.1:** Shows colour linkages

| Parents | XN Xn (Carrier female) | | XN Y (Normal man) | |
|---|---|---|---|---|
| Gametes | XN | Xn | XN | Y |
| Possible combinations of gametes | XN XN Normal girl | XN Xn Girl carrier | XN Y Normal boy | Xn Y Colour-blind boy |

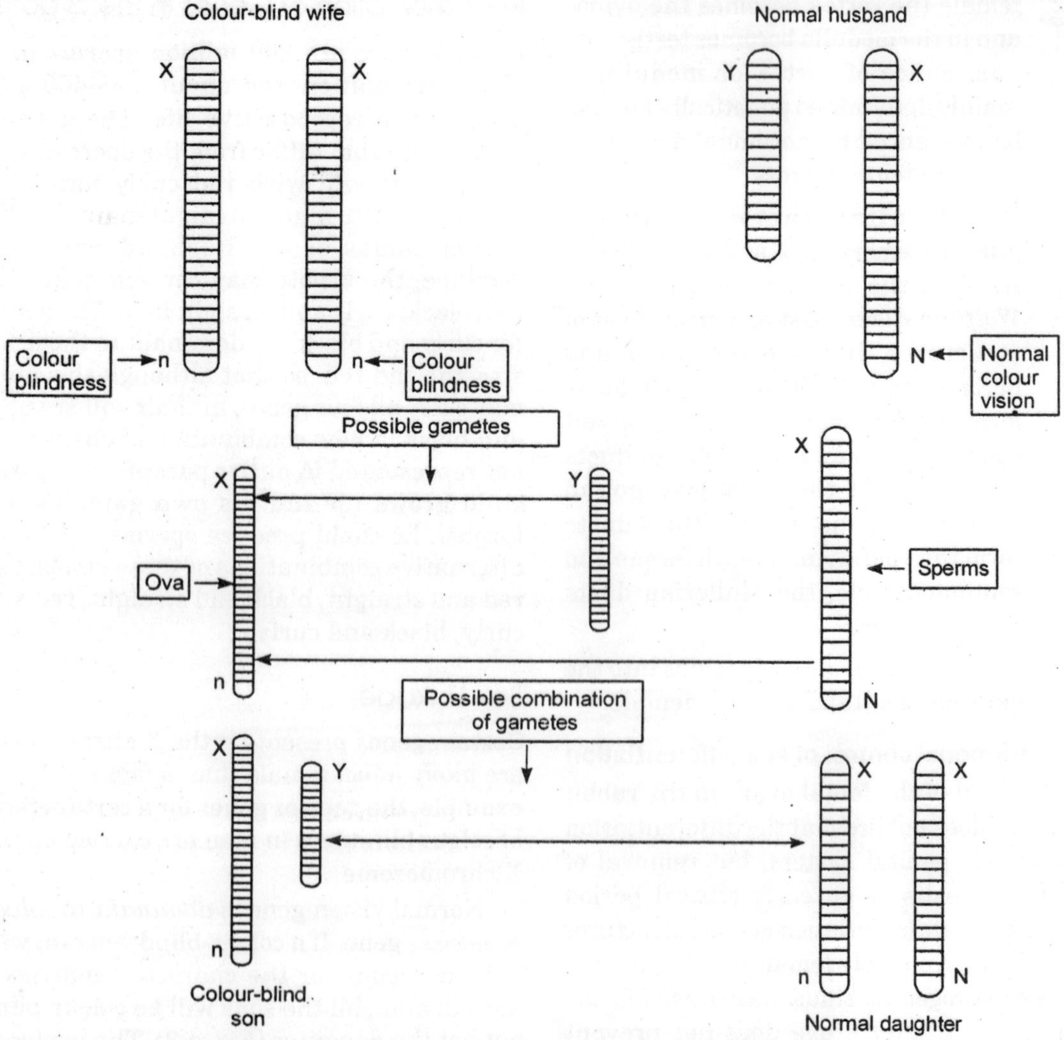


**Fig. 4.3:** Diagrammatic illustration of sex linkage showing distribution of X and Y chromosomes between the gametes and combination in the zygote.

determination of colour vision. The normal daughter is heterozygous for colour blindness and a *"carrier"* of *recessive gene.* If the daughter marries a normal man, the expected combinations of genes in the children would be:

The expected results are that all girls will be normal but half of them will be carriers, half the boys will be normal and half of them colour blind. The types of children expected from the marriage between a female carrier and a colour-blind man, or a normal female and a colour-blind man can be worked out in a similar way (Fig. 4.3).

The other X-linked diseases are *haemophilia* and *brown enamel on the teeth.* Haemophilia causes a delay in the clotting time of the blood. Though there are two kinds of sex-linked haemophilia, at least three other clotting disorders are known which are controlled by genes not on the sex chromosomes.

The sexual characteristics like change of voice, beard, masculine physique, etc. in males and development of breasts, wide pelvis are not the result of sex-linked genes but the different expressions of the same genes present in both sexes. Both sexes carry genes controlling the growth of mammary glands, hair, penis, etc. but in the physiological environment of maleness or femaleness they have different effects, with the result that, e.g. the mammary glands in males are small and functionless; the penis is represented in female by only a small organ called *clitoris*.

- **Globin Gene Mapping**

The arrangement of the various globin structural genes on the human chromosomes was provided by analysis of the Hb electrophoretic variant, Hb Lepore.

Trypsin digests of Hb Lepore with Hb from normal persons compared and revealed that the α-chains were normal. Non-α-chains appeared to consist of an amino-terminal δ-globin-like sequence and a carboxy-terminal β-globin-like sequence.

It is therefore proposed that Hb Lepore could represent a "fusion" globin chain which had arisen as a result of a cross-over coincidental with mis-pairing of the δ- and β-globin genes during meiosis as a result of the sequence similarity of the two genes and the close proximity of the δ- and β-globin genes on the same chromosomes. Further evidence at the protein level for the physical mapping of the human globin genes was provided by the report of another Hb electrophoretic variant, Hb Kenya.

Subsequent amino acid sequence analysis of the variant suggested that it was a γ–β fusion product with a cross-over having occurred somewhere between amino acids 81 and 86 in the two globin chains which suggests that the γ-globin structural gene must also be in close physical proximity to the β-globin gene.

- **Globin Gene Structure**

1. **DNA technique:** Immature RBC, reticulocytes provide a rich source of globin mRNA in the synthesis of cDNA because they synthesize little else! Use of β-globin cDNA for restriction mapping studies of DNA from normal persons has revealed that the non-α or β-like globin genes are located in a 50 kb stretch on the short arm of chromosome 11. The whole of this 50 kb stretch has been closed and the nucleotide sequence of each of the various globin structural genes is known. Of particular interest are regions with sequences similar to the globin structural genes but which are non-functional and do not produce an identifiable message or protein product, i.e. pseudogenes.
2. Similar studies of α-globin structural genes have revealed that there are, in fact, two α-globin structural genes, α1 and α2, located on the short arm of chromosome 16.
3. DNA sequence analysis has revealed differences between these two structural genes even though the α-globin chains produced have an identical amino acid sequence, evidence of the "degeneracy" of the genetic code.
4. Beside that, there are pseudo-α, pseudo-ζ- and ζ genes to the 5′ side of the α-globin genes as well as a recently discovered theta (θ) globin gene to the 3′ side of the α1-globin gene.

   This latter gene, whose function is not known, is of interest because, unlike the pseudoglobin genes, which are not expressed, its structure is compatible with expression. It has been suggested that it could be expressed in very early erythroid tissue such as the fetal liver and yolk sac.

## QUESTION

1. Describe the process of fertilization.
2. Describe the processes during sexual development.
3. Describe the recombination of genes in the zygote.
4. Write briefly on sex linkage.

# 5

# BIOLOGICAL FUNCTION OF SEX CHROMOSOMES

1. Biological function of sex chromosomes, 2. Y chromosome and sex linkage, 3. Sex chromatin (Nuclear sex), 4. Sex-linked diseases, 5. Chromosomal aberration, 6. Abnormalities of chromosomal aberration.

Biological function of the sex chromosomes has come to light from the examination of the patients with an abnormal number of sex chromosomes. In individuals with an XXY or an XXXXY complement of sex chromosomes, development of testes revealed that the Y chromosome carries strong male-determining genes which can induce fetal testicular development even in the presence of two or more X chromosomes. In absence of Y chromosome, testicular development does not usually occur. The *pericentric region* of the *Y chromosome* has been implicated as the source of genes directing testicular differentiation whereas *regions on the Yq long arm* have been associated with spermatogenesis. Apart from the genes associated with stature and spermatogenesis, there is no strong evidence in favour of the presence of other genes on the Y chromosome of man. The X chromosome has importance in spermatogenesis in that it must be inactivated prior to meiosis. Extensive studies on *H-Y antigen* have been done for its role as the determinant of testicular development in the fetus. H-Y antigen is a *protein* which lies on the surface of all tissues and initially it was described as a histocompatibility antigen since it was responsible for rejection. The relationship between the testis-determining gene on the Y chromosomes and the H-Y locus is not clear but it is believed to be closely linked. It has been observed that males with two Y chromosomes have more H-Y antigen than normal males which *supports* the *localisation of the H-Y gene* on the Y chromosome.

It is evident by the observations that H-Y antigen itself *plays a testis-determining role* because the protein is detected in most but not all cases where a testis is present, including sex reversed males. Some of these sex reversed males are well-masculinized having a testis and 46, XX karyotype. The discrepancy between gonadal and chromosomal sex is possibly due to translocation of the H-Y genetic material to one of the X chromosomes or to an autosome.

The problem arises in XY females without testes pertaining to the assignment of the gene for the H-Y antigen to the Y chromosome.

There are examples in which testes did not develop in the presence of H-Y antigen. This has been by postulating a defect in the H-Y receptor protein which rendered the target cells unable to respond.

The H-Y antigen might have been conserved during evolution because similar cross-reactive antigen has been observed in humans, mice, rats, amphibia, guinea pigs, rabbits and birds. In birds, females are heterogametic ZW. In amphibia, the male is the heterogametic sex in *Rana pipiens,* while it is the female in the *Xenopus laevis.*

In each case, the presence of a cell surface component, cross-reacting with antiserum to mouse H-Y antigen, is present in the heterogametic sex. From the mammalian species, the results obtained, it is reasonable to think that the *H-Y antigen* should more correctly be regarded as the *director of differentiation* of the *heterogametic gonad.*

The biological functions of X chromosomes are of the diverse nature. It is suggested that ovarian differentiation requires a *pair of X chromosomes* which is based on the fact that the ovaries do not develop with an *unpaired X chromosome* known as gonadal dysgenesis: 45, XO. But this observation does not hold:

1. For all species because functional ovaries develop in XO mice,
2. Deletion of portions of the X chromosome results gonadal dysgenesis,
3. The *short arm Xp contains* genes which escape inactivation, deletion of this region causes dysgenetic ovaries.

The germ cells in the ovary requires two active X chromosomes for normal development. The germ cell atresia, i.e. loss of germ cells may therefore result from *deletion of Xq* due to lack of genes required for germ cell development which results in gonadal dysgenesis and infertility.

The X chromosome contains genes which influence the differentiation of different types of extragonadal characteristics compared to Y chromosome which is mainly concerned with sex differentiation. The X chromosomes tends to be highly conserved in evolution, but not the Y chromosome. The function related to the X chromosome in one species tends to be X-linked in others while functions assigned to the Y chromosome may be distributed about the genomes of mammals.

The XO male Vole, *Ellobius lutescens,* is the extreme example of the widespread distribution of male-determining genes since this animal has no identifiable Y chromosome at all.

## Y Chromosome and Sex Linkage

The sex-linked genes contained in the X chromosome of Drosophila have no alleles in the Y chromosome; this is the reason why a male that carries a single dose of a recessive sex-linked gene shows its effects in the phenotype. The Y chromosome is, therefore, sex-linked genes are absent or inert, at least compared with the X chromosome. One gene, bobbed, however, has alleles both in the X and Y chromosomes. The recessive mutant allele of bobbed, when present in both X chromosomes of a female, causes the bristle on the body of the fly to be shorter and slenderer than normal. A male carrying bobbed in the X chromosome but a normal allele of that gene in the Y has normal bristles. If such a male is crossed with bobbed females, all the daughters are bobbed and all the sons have normal bristles; the normal bristles are transmitted from father to sons, just as the Y chromosome is several bobbed alleles in the Y chromosome are also known; and males that carry bobbed both in the X and in the Y chromosomes show bobbed bristles. The Y is also necessary for fertility. There is some evidence that in man also some genes are carried in the Y chromosome. The conclusion is drawn from the inheritance of sex-linked traits is that the sex-linked genes are absent from the Y chromosomes because it is evident from the experiment of *Drosophila* the Y chromosome is genetically empty but is difficult to conclude in case of man. If sex-linked trait is associated exclusively with the Y chromosome, it would be expected to show in men only but not in women. Some cases are the examples of sex-linkage of Y chromosome.

1. The possible Y-borne inheritance was the famous *"porcupine man"* of 18th century, England. It was manifested by patches of hard, spiny skin rather like porcupine and is alleged to have transmitted this condition to all his sons. Much doubt has been cast on the validity of the records and there is little evidence that it was associated with Y chromosome or with the sex chromosomes at all.
2. Another good documented is hairy ears. This trait, which shows as a tuft of hair sprouting from the pinna of the ear which is common in certain parts of India and exclusively in the male. But there is not conclusive proof that the Y chromosome is implicated.

**Biological function of Y chromosome** (See Biological function of sex chromosomes)

## Sex Chromatin (Nuclear Sex)

There is difference of *chromosomal composition* between male and female by only a single chromosome. There is also a *morphological difference in nuclei* from male and female animals. The difference between male and female is due to a special *mass of chromatin* called *sex chromatin* or *Barr body*. Soon after the cell division has started during embryonic development, one or the other of the two X chromosomes of the *somatic cells* in normal females becomes functionally inactive. Once inactivation has occurred, the same X chromosome is inactivated in all descendants of that cell. So one X chromosome is inactivated in approximately half of the cells and other X chromosome in the other half. The selection persists through subsequent divisions of these cells, and consequently some of the somatic cells in adult females contain an active X chromosome of paternal origin and some contain an active X chromosome of maternal origin.

The inactive X chromosome *condenses* and can be seen in various types of cells called *Barr body* or *sex chromatin*. The sex chromatin a Barr body has the following characteristics:

1. A plano-convex mass
2. Diameter is about 1 μm
3. Stains positively for DNA
4. Lies close to or in contact with the nuclear membrane.
5. Inactive chromosome is also visible as a small "drumstick" of chromatin protecting from the nuclei of 1-15% of the polymorphonuclear leukocytes in females but not in males
6. Chromatin–positive cells which contain two X chromosomes (46, XX and 47, XXY) have a single Barr body whereas those with more than two X chromosomes may have two Barr bodies. Therefore, it is the number of chromatin bodies in any diploid nucleus in one less than the total number of X-chromosomes.

**F body:** Another form of sex chromatin is the F body, *a portion of the Y chromosome* that can be *stained* with *fluorescence technique.* Thus, the number of Y chromosomes in a cell can be estimated by the number of F bodies.

## Sex-Linked Diseases

I. **Colour-blindness (Red-green):** Sex chromosome causes inheritance of red-green colour-blindness; the gene for colour-blindness is recessive and sex-linked. If a colour-blind man, marries a normal woman transmits his "colour-blind gene" to his daughters but since they are heterozygous, will not be colour-blind but are carriers of the defective gene. His sons will not have the defective gene because they receive their X chromosome from the mother. Now if one of the daughters marries a

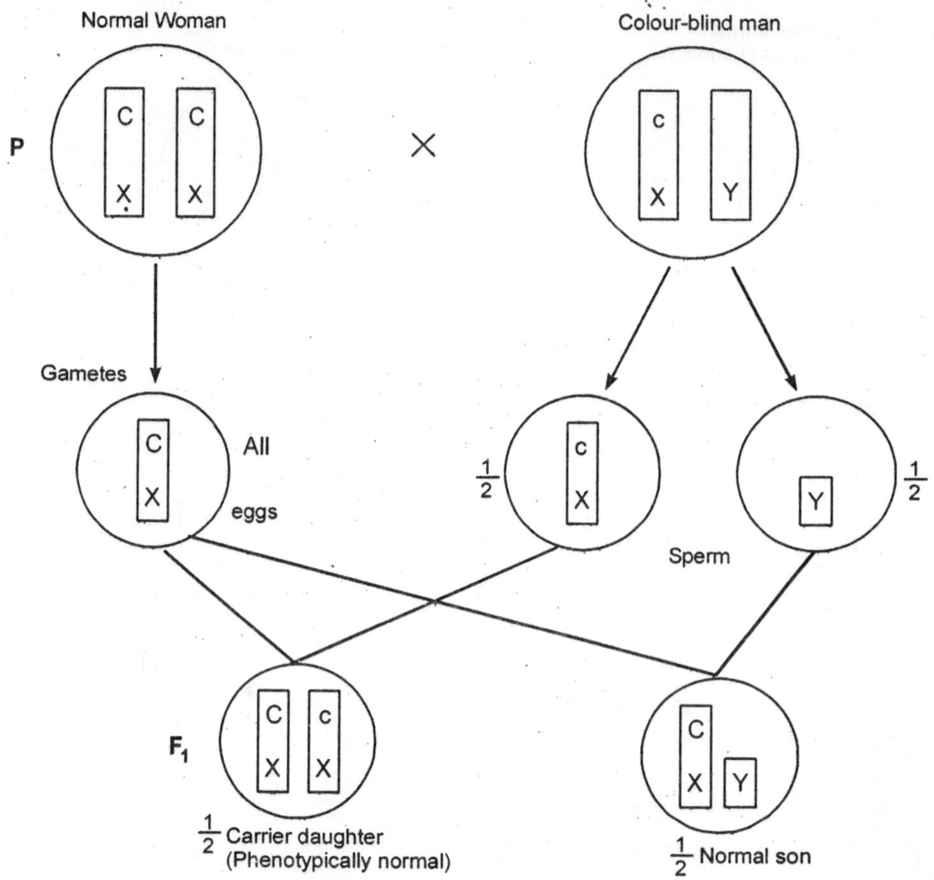


**Fig. 5.1:** Colour-blindness, sex linkage in man.

normal man and their son has a probability of $\frac{1}{2}$ that he will be colour-blind. A daughter will be phenotypically normal but there is a probability of $\frac{1}{2}$ that she will be carrier. Red-green colour-blindness is more common in man than in woman (Fig. 5.1).

II. **Haemophilia:** Another sex-linked trait, which is serious, known as *haemophilia* or "bleeder's disease". In this case the blood takes an abnormally long time to clot. It causes profuse and prolonged bleeding from even the slightest wound. Haemophilia, is caused by a recessive gene carried on the X chromosome. If a normal man marries a woman who carries the gene for haemophilia, there is probability of $\frac{1}{2}$ that if they produce a son he will be a haemophiliac, and if they produce a daughter, she will be a carrier. Both children may then transmit their defective genes to the grandchildren, and so on (Fig. 5.2). This is precisely what has happened in the Royal families of Europe during the last hundred years.

## Chromosomal Aberrations (Fig. 5.3)

The different cells of one's body and different individuals of the same species have identical numbers of chromosomes. In these chromosomes the genes are arranged in linear fashion which is seen in the linkage maps. But there is exception to this rule and study of these exceptions is very much useful for

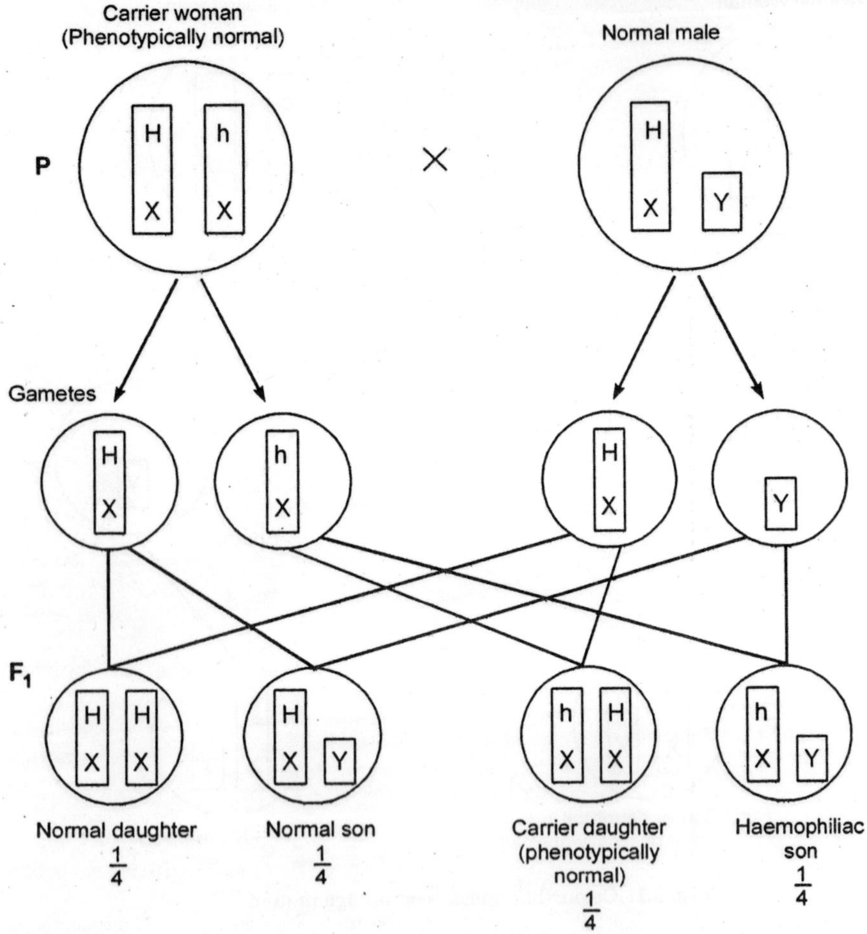


**Fig. 5.2:** Haemophilia, a sex-linked disease in man. Probability = $\frac{1}{4}$, haemophilia gene, h (recessive), normal gene, H (dominance).

the development of the genetics. Genes can reproduce themselves and can synthesize their perfect copies, similarly chromosome do the same. But genes occasionally mutate, and chromosomes produce abnormalities or aberrations. The different kinds of chromosomal aberrations are:

I. **Changes in the number of chromosomes**

A. **Monoploid:** Changes involving entire sets.

1. *Haploidy* (n): Each chromosome represented singly.
2. *Polypoidy*: Each chromosome represented by more than two homologues.

(a) Triploidy (3n)

(b) Tetraploidy (4n)

(c) Pentaploidy (5n) etc.

**Autopolyploid:** It is a form derived by chromosome multiplication from a single diploid, so that the homologues come from the same source as in pure strains or homozygotes.

**Allopolyploid:** It is a form derived from a hybrid between two diploids, so that the homologues come from different sources.

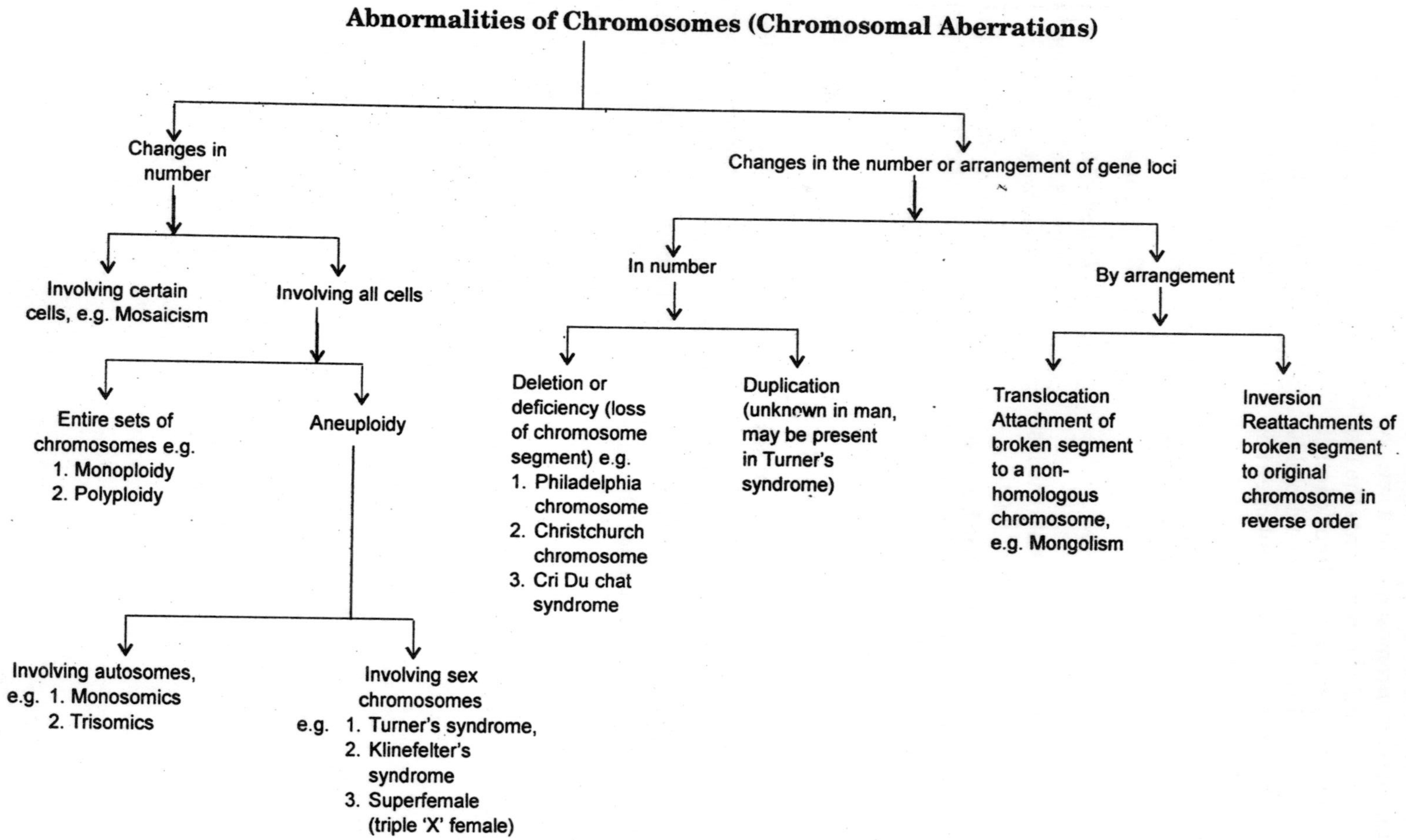


**Fig. 5.3:** Flow chart showing Chromosomal Aberration.

B. **Heteroploidy:** Changes involving the numbers of chromosomes in a set.

1. *Monosomics*: It represents the loss of one chromosome from one set. Where this occurs in the diploids, the chromosome complement is $2n-1$.
2. *Polysomics*: It represents the addition of one or more chromosomes to one set.

   Trisomic = $2n + 1$

   Tetrasomic = $2n + 2$
3. *Nullisomics*: It represents the loss of both chromosomes of a pair.

II. **Changes in the number or arrangement of gene loci within a chromosome**

A. **In number**

1. *Deficiency or deletion*: Loss of one or more genes.
2. *Duplication*: Addition of one or more genes, as a result of which the organism carries the same gene repeated in its haploid chromosome complement.

B. **By arrangement**

1. *Translocation*: It is due to exchange of parts between nonhomologous chromosomes.
2. *Inversion*: Within a chromosome, a block of genes may rotate by 180 degrees.

## QUESTION

1. Describe the biological functions of sex chromosome.
2. What do you know about Y chromosome and sex linkage?
3. What do you know about sex chromatin?
4. What is F-body?
5. What are the sex-linked diseases? Describe colour blindness and sex linkage in man.
6. Describe haemophilia and sex linkage and probability in man.
7. State briefly the chromosomal aberration.

# 6

# CHROMOSOMAL BANDS AND GENES

1. Chromosomal bands and genes, 2. Isolation of specific genes, 3. Chemical synthesis of genes and polydeoxyribonucleotides, 4. Gene function, 5. Mapping of chromosomal structure or locating genes on chromosomes.

Numerous data collected over many years from breeding experiments, *chromosome maps* show the positions of the various genes on the chromosomes in case of Drosophila and maize. The position occupied by a gene is called *gene locus.* In Drosophila, hundreds of gene loci have been established. In man, an effort has been made in mapping genes from the study of human pedigrees and the distribution of inherited characteristics in the population but not from breeding experiment.

But this indirect evidence to work out relative position of genes on chromosomes. At the time of chromosome maps, the chromosomes were visible under microscope as a uniform thread-like structure without any demarcation of any structures which can be identified as genes. The concensus of opinion was that the genes were not visible because it is hidden under the structure of chromosomes. A brilliant piece of work was done with the chromosome of the salivary glands of Drosophila larvae which had a great impact on the study of genes. They are about one times larger than the chromosomes present elsewhere in the body hence they are called *giant chromosomes.* These giant chromosomes are uniform in appearance, these giant chromosomes are traversed by numerous darkly-staining bands spaced at irregular intervals and of characteristic shape (Fig. 6.1). Some are narrow and rather difficult to detect, others

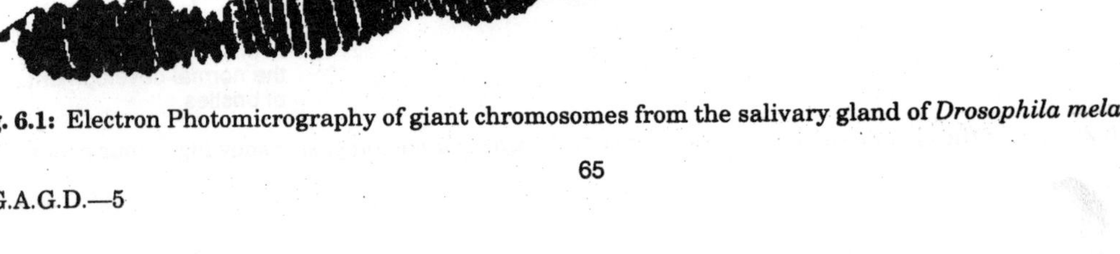

**Fig. 6.1:** Electron Photomicrography of giant chromosomes from the salivary gland of *Drosophila melanogaster.*

are thicker and clearer. Each type of chromosome has its own characteristic patterns of banding and this is constant in all organisms (Fig. 6.2). Now the question arises whether these bands represent the genes or not. Several attempts have been made in *Drosophila* to relate the bands to the chromosome maps. These were done mainly correlating the genetic abnormalities observed in the breeding experiments with structural abnormalities detected in the chromosomes. The normal arrangement of the bands is characteristic for each chromosome and this makes it possible to detect small changes in the structure of an individual chromosome. Sometimes a chromosome breaks and the detached part rotates through 180° and joins up again in the wrong way round. As a result of this, the normal pattern of banding for that particular portion of the chromosome is reversed causing alteration in the relative positions of the genes in the chromosome map. In certain cases this event is correlated with the genetic abnormalities detected during breeding experiments. This kind of procedure has helped to relate individual bands in the chromosomes to specific gene loci in the chromosome. But these bands do not represent individual genes. Detecting a band does not mean a single gene but it is smaller in several orders of magnitude.

The salivary chromosomes of *Drosophila* merely help to locate the positions of the genes than the ordinary chromosomes. We have example of how information obtained from breeding experiments which agrees with that derived from direct observation of cells, an agreement which supports the contention that *genes are carried in a linear sequence on the chromosomes.*

## Number of Genes

If all the DNA of the cell function as gene then there should be more than 2 million of genes. But DNA is responsible for other functions besides constituting genes, it has been calculated that there is probably a maximum of about 100,000 functional genes in the human cell. The number of genes is not known but it could be in the region of 1000 per chromosome.

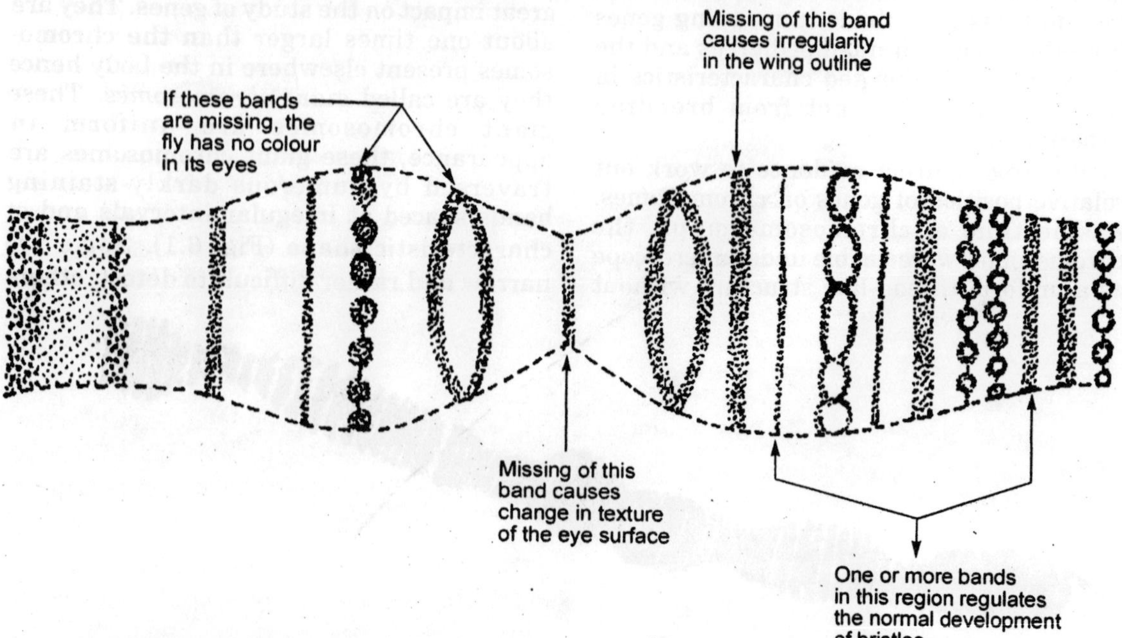


**Fig. 6.2:** Part of the chromosome of salivary gland of *Drosophila melanogaster* showing of four genes.

## Size of Gene and Mutational Units

Classical genetic study carried out by Benzer, S., valuable information on the *size of genes* and *mutational units* were achieved. Benzer examined a small region (r ll) of the T4 bacteriophage chromosome by the viral-recombination method. He concluded that each gene contains many mutable sites, which he was able to map in a linear sequence and discarded the old idea of each gene contained only one mutable site. From the known length of DNA molecule, in the T4 viral particle and the known small fraction of the genetic map occupied by the r ll region, the *r ll A gene of the T4 chromosome was calculated to consist of about 25,000 nucleotide pairs.* He found nearly *500 different mutants* of the *r ll A region,* indicating that the *mutation unit* is not greater than about *five nucleotides.* It was further concluded that mutation involves a chemical change in only a single nucleotide which is known as *point mutation.*

The conclusions given by the genetic experiments were confirmed by the subsequent biochemical investigations. The valuable information obtained from the discovery that each amino acid is coded for by a specific triplet of nucleosides. Thus, a gene specifying a single polypeptide chain of a protein may have anywhere from about 300–5500 or more nucleotide pairs, corresponding to polypeptide chain having from 100–1800 amino acids residues, the range of chain length of the polypeptides in most proteins.

Since a given protein or enzyme may contain two or more polypeptide chains, each coded by a different gene, the original hypothesis of one gene-one enzyme has been modified into the *one gene-one polypeptide chain* concept. The gene is also referred as *cistron.*

The genes specifying tRNA molecules have about 100 nucleotide units. The smaller genes are those specifying the sequence of amino acids in short polypeptides (e.g. ACTH) which has only 39 nucleotides.

## Isolation of Specific Genes

In some case single genes or sets of genes have been isolated from DNA specimens by special procedures. Beckwith, T. et al. isolated the set of three genes and the regulatory regions of the DNA concerned in the transport and metabolism of β-galactosides in *E.coli.* Brown, D.D., isolated genes coding for ribosomal RNA. These isolated DNA fragments are functional and capable of coding for characteristics gene products *in vitro.* The nucleotide sequence of isolated genes may be established indirectly and used as a guide for finding out direct methods for determining the nucleotide sequence of DNA.

## Chemical Synthesis of Genes and Polydeoxyribonucleotides

For the study of synthesis of genes, certain difficulties were encountered such as:

1. The direct analysis of base in a homologous DNA has some difficulties;
2. Even the smaller DNAs have larger number of residues than the RNAs and proteins that have been sequenced;
3. DNAs usually contain no minor bases to use as markers;
4. Deoxyribonucleases fail to cleave DNA into pieces of expected size. In spite of these difficulties, base sequences of DNA fragments having 100 or more residues have been determined. There is possibility to deduce the base sequences in certain genes, particularly those specifying RNAs of known sequence because the bases of the gene and those of the RNA it codes are complementary. The sequence of bases in the gene coding for yeast alanine tRNA, whose base sequence is known is readily deduced.

Khorana et al. synthesized the gene for yeast alanine tRNA by a combination of chemical and enzymatic methods, the first gene to be made by largely chemical methods. Generally short oligonucleotides of

5 to 10 residues are first separately. These are assembled to make longer chains. By these methods polydeoxyribonucleotide chains having up to 20 residues in known sequence have been synthesized. By joining 15 such chemically synthesized double-strand polydeoxyribonucleotide chains in the correct sequence by means of the DNA ligase reaction. Khorana *et al* succeeded to make the total synthesis of both strands of the gene for yeast alanine tRNA. They have also synthesized the gene for the precursor of tyrosine tRNA of *E. coli*. The synthetic DNA of the tRNA precursor contains 126 residues.

## Gene Function

1. *Carries hereditary character*: From the evidences it is clear that the genes carries the hereditary characters from parents to the offspring and arranged in linear sequence upon the chromosomes.
2. *Controls production of enzymes:* Gene controls the production of enzymes which in turn determine what functions go on in a cell and eventually in the organs and entire organism. If anything happens to a gene it will affect the organism, e.g. in mice there is a gene which determines that the coat will be coloured. If this gene is missing, the mouse will be without pigment and it will be white with pink eyes.

   These enzymes function for:
   (a) oxidative reaction for energy,
   (b) synthesis of lipids, glycogen, ATP, etc.
3. *Controls the formation of proteins*: Gene controls the formation of structural proteins for cells, e.g. organelles.
4. *Genes have more than one effect:* It is known that genes have more than one effect, and the characteristics by which they are recognized is not necessarily their most important function, e.g. the gene responsible for colour of flower also offer a resistance to any fungal disease because they determine the presence of certain chemical substances which act as a fungicide.
5. *Gene effect and physiology of the cell*:
   (a) During development of a particular cell, it seems that the way in which one of its genes will affect the cell depends not only on the gene itself but also on the physiology of the cell, which in turn is related to its particular position in the body.
   (b) The expression of gene depends to some extent on the physiology of the cell and the situation in which it finds itself is evident by the experiment with amphibian embryos. When a piece of tissue would normally become skin is taken from abdomen and grafted into a region overlying the developing eye, the graft will be incorporated into the eye as a lens. It has the same chromosomes and genes, but its new position has altered its fate. This effect is not seen in all animals and absent in insects because the fate of individual cells are determined at a very early stage of development and is not affected by shifting the cells to a new situation.
6. *Storage of genetic information*: Genetic information is stored in DNA, the informational macromolecule of the chromosome. The genetic information flows in the direction DNA → RNA → Protein. Purine and pyrimidine bases carry genetic information, whereas sugar 8-phosphate groups perform structural role.
7. *Amino acids sequence of polypeptide chain coded by DNA:* The sequence of amino acids in polypeptide chain of each type of protein is specified or coded by the sequence of nucleotide residue in DNA. The segment of DNA

molecule specifying one complete polypeptide chain is called *cistron* or *gene*.

8. *One amino acid is coded for by three successive nucleotide residue*. Each amino acid is coded for by three successive nucleotide residue.
9. *Interaction of genes*: Usually a characteristic is controlled by one pair of alleles. Sometimes, however, a single characteristic is controlled by two or more pairs of alleles interacting with one another, e.g. shape of comb in different breeds of poultry.
10. *Genes act in producing effects in the phenotype*: Genes act in producing their effects in the phenotype such as in case of sickle-cell anemia. Genes exert their effects, and express themselves in the phenotype, by influencing the organism's biochemical processes, sometimes in a very slight and subtle way.

## Different Types of Genes According to Mendelian Inheritance

1. Dominant gene
2. Recessive gene
3. Co-dominant genes
4. Carrier gene
5. Sex-linked genes
6. Sex-limited genes.

1. **Dominant gene:** Dominant gene expresses its physical or biochemical trait, when the allelic genes are either homozygous or heterozygous for that trait, e.g. tallness is caused by dominant gene. The genetic constitution of a tall person may be T : T or T : t (T for tallness, t for shortness). Most of dominant traits are expressed in heterozygote state. This can be studied from the pedigree record of the family.
2. **Recessive gene:** Recessive gene expresses the trait only in homozygote state, that means when both alleles are recessive for that trait, e.g. shortness is caused by a recessive gene, the genetic constitution of a short person is t : t where t for shortness.
3. **Co-dominant genes:** When both the allelic genes are dominant but of two different types, both traits may have concurrent expression. In ABO blood groups, A gene and B gene are both dominant, whereas O gene is recessive. When A gene and B gene occupy identical loci in homologous chromosomes, AB blood group is expressed.
4. **Carrier gene:** Heterozygous recessive gene acts as a carrier which may be expressed in subsequent generation. When both parents are heterozygous tall (T : t), the possibilities of height of the offspring may be such that out of four children three are tall and one short, in the proportion of 3 : 1. One tall child is homozygous, and the other two are heterozygous.
5. **Sex-linked genes:** The genes located on the X and Y chromosome are known as the sex-linked genes. X-linked inheritance is more common, and is mostly expressed as recessive traits, when the genes are mutant.
6. **Sex-limited genes:** These genes are borne by the autosomes, but the trait is expressed preferentially in one sex. Gout and baldness are found predominantly in males.

## Types of Genes According to Gene Action

According to gene action, in prokaryotes the genes are mostly of Mendelian types, i.e. Dominant and Recessive. But in eukaryotes, there are various types of genes affecting the phenotypes. The types of genes are:

1. **Lethal genes:** The genes which control the manifestation of phenotype trait and affect the viability of the organism by producing harmful effects.

2. **Complementary genes:** Two or more dominant genes present on separate gene loci which interact to produce a particular phenotype trait. But neither of them produces the phenotypic trait in the absence of other.
3. **Supplementary gene:** Two or more independent pairs of dominant genes which interact together in such that the dominant gene will produce its effect whether the other is present or not. The second dominant gene, when added, produces an altogether changed expression, but only in the presence of the first one.
4. **Inhibiting gene:** Genes present on two separate loci, i.e. two non-allelic genes interacting together, in which one gene suppresses the expression of the other gene.
5. **Duplicate gene:** When two or more genes situated on different chromosomes affecting the single trait called duplicate gene.
6. **Cumulative genes:** Two or more independent pairs of genes which affects the same characteristics in an additive way.
7. **Collaborative genes:** Two different genes on separate loci influencing the same trait but interact to produce a new trait phenotype which cannot be produced by the individual genes.

## Origin of New Gene and Phenocopies

At the time of DNA duplication new genes may be produced which may or may not exist gene ratio after generation and existence will ultimately depends upon the natural selection.

### • Variable Effects of Gene and Phenocopies

The phenotypic effects of genes are variable for the following reasons:

(a) Regulator gene effects on the expression of many genes.

(b) The complicated pathways by which some genes exert their effects on phenotype.

(c) The occasional complex interactions among environmental factors, regulator genes and pathways of gene on phenotype expression.

### • Position Effect of Gene

When a gene is in normal position in the chromosome, it produces its phenotypic expression properly, but if the position of the same gene is altered by any means from its normal position then its phenotypic expression will also vary and this is known as *position effect of gene.*

## Cistron (or gene)

Mutation in one functional region, e.g. A or Bs, are complementary only to mutations in other region and complementation is detectable by cis-trans effect.

Each functional region is responsible for the production of a given polypeptide chain. Benzer defined the functional unit as *cistron* and conformed operationally more closely to what we commonly think as gene. This cistron may be considered as gene at the functional level. There may be over a hundred points within a functional unit where in a mutation can take place and cause a detectable phenotypic effect.

A cistron has over hundred nucleotide pairs in length and there is some evidence that some cistron may be as long as 30,000 nucleotide pairs. Actually each cistron represents part of gene which is responsible for coding of only one polypeptide chain of an enzyme that has two or more different polypeptide chains in its complete enzymatic unit (Fig. 6.3).

## The Muton

The muton is the smallest unit of DNA which, when altered, can give rise to a mutation. Study of the genetic code makes it clear that an alteration of a single nucleotide pair in DNA may result in a missense codon in tran-

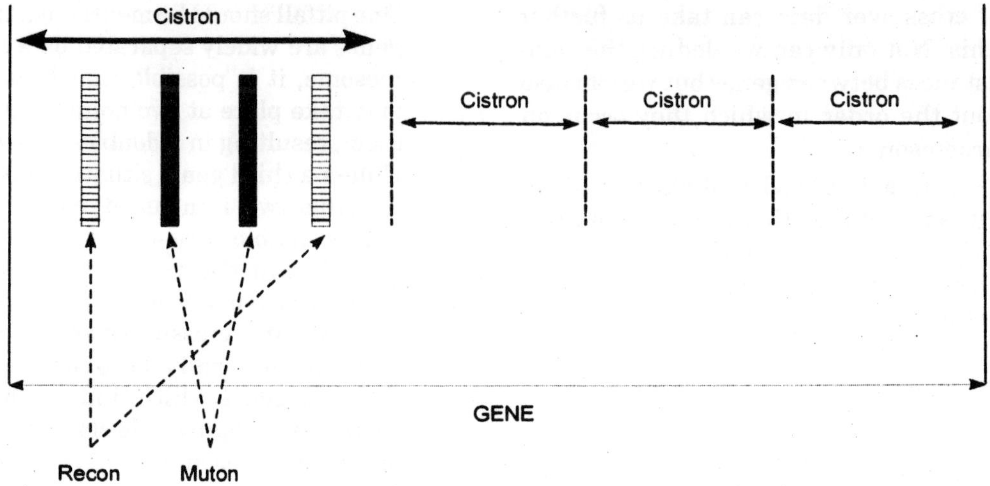


**Fig. 6.3:** Schematic representation of fine structure of gene.

scribed mRNA, e.g. AGC $\rightarrow$ AGA or nonsense, e.g. UGC $\rightarrow$ UGA. So a cistron may be expected to consist of many mutable units or mutons and this term was coined by Benzer (Fig. 6.3).

## The Recon

The Recon is the smallest part of DNA which is interchangeable through crossing over and recombination. Extremely delicate studies of recombination in *E. coli* indicates that a recon consists of not more than two parts of nucleotides, may be only one.

## Mapping of Chromosomal Structure or Locating Genes on Chromosomes

- **Introduction**

To a biologist crossing-over has a special significance for it enables to work out the relative positions of genes on a chromosome. Genetic mapping of the relative positions of different genes in bacterial chromosomes is made possible by the fact that under special circumstances a segment of the chromosomes of one type of cell, containing one or more genes may become detached and transferred to the chromosome of a second type of cell, into which it becomes covalently incorporated. The identity of the recombined gene or genes can be deduced from changes in the phenotype of the acceptor cell and its progeny. Such recombination of genes is a common event in eukaryotic organisms, which usually reproduce by sexual conjugation, but it is relatively infrequent among prokaryotic organisms.

- **Discussion with Example**

To work out the relative portion of genes on chromosome, let us imagine two hypothetical linked genes A and B. The further apart A and B are on their chromosome, the more likely it is that the chromosome will break and rejoin at some point between them. Conversely, the closer they are, the less likely it is that breakage and crossing-over will occur between them. So if in a breeding experiment we find that a relatively large percentage of the offspring show separation of A and B, we conclude that these genes are relatively for apart on the chromosome. On the other hand if a low percentage show separation, we conclude that the genes are relatively close together. The percentage offspring which show separation of the genes represents the "*cross-over frequency or cross-over value (COV)*" and for convenience a COV of 1 per cent is taken to represent a distance of 1 unit on the chromosome.

But cross-over data can take us further than this. Not only can we deduce the relative distances between genes but we can also work out the order in which they occur on the chromosome.

1. Imagine that in a breeding experiment it is found that the COV of A and B is 4 per cent. This means that A and B are 4 units apart on their chromosome (Fig. 6.3.1).

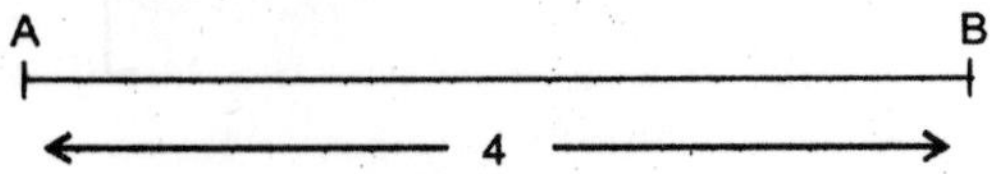


**Fig. 6.3.1**

2. We then go on to investigate the transmission of each of these genes (A and B) with a third linked gene C. If we find that A and C give a COV of 10 per cent whereas B and C give only 6 per cent, we conclude that B is located somewhere between A and C (Fig. 6.3.2).

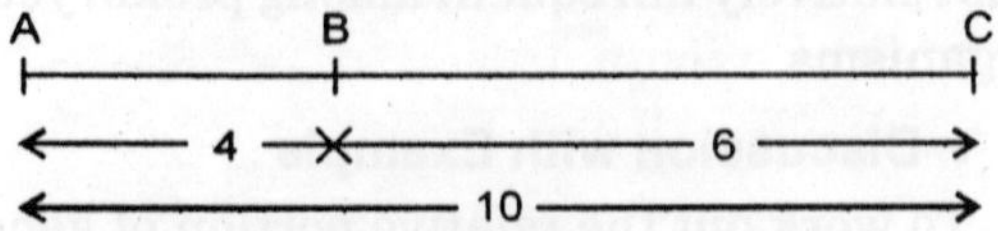


**Fig. 6.3.2**

3. If on the other hand, A and C give a COV of 2 per cent, and B and C 6 per cent, we conclude that C lies to one side of A (Fig. 6.3.3).

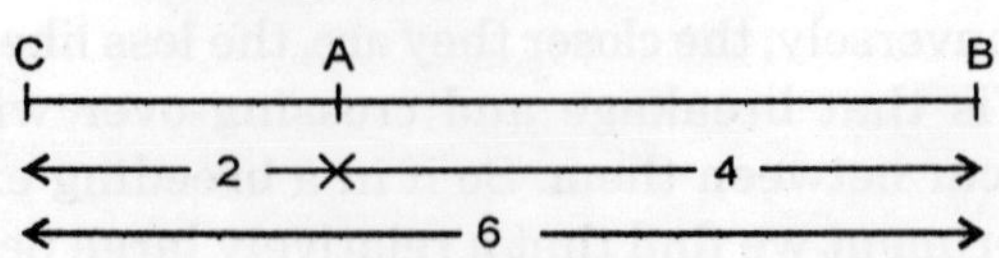


**Fig. 6.3.3**

4. Another possibility might be that A and C, and B and C both give COV's of 2 per cent. From this we would conclude that C lies midway between A and B (Fig. 6.3.4).

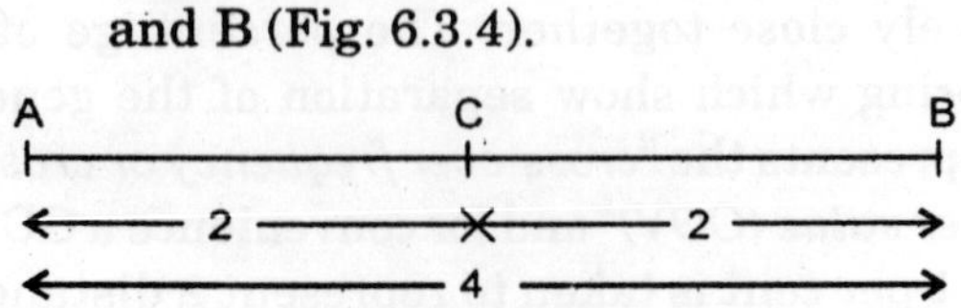


**Fig. 6.3.4**

One pitfall should be mentioned. If two genes are widely separated on a chromosome, it is possible that breakage may take place at two points between them, resulting in a double cross-over. Unless a third gene, situated between the other two, is included in the analysis, the double cross-over will go undetected, and the number of recombination will not be a true reflection of the amount of crossing-over. The problem can generally be overcome by choosing genes which, judging by recombination figures, lie close together on the chromosome.

From data collected over many years from numerous breeding experiments "*chromosome maps*" showing the positions of the various genes on the appropriate chromosomes, have been built up for several organisms notably Drosophila and maize. The position occupied by a gene is known as the "*gene locus*". In Drosophila literally hundreds of genes loci have been established in this way. Even in man a start has been made in mapping genes though the data on which this is based are obtained not from breeding experiments but from the study of human pedigrees and the distribution of inherited characteristics in large populations.

- **Method of Study**

Experimental conditions can be chosen that greatly increase the frequency of recombination processes in bacteria, thus yielding a powerful tool for mapping the relative positions of genes in bacterial chromosomes. Four types of genetic recombination process are useful in mapping becterial and viral chromosomes:

1. Sexual conjugation
2. Transformation
3. Viral transduction
4. Viral recombination

1. **Sexual conjugation:** This process consists of the transfer of some or all

of the chromosomes of a "male" or (+), cell into a "female" or (–) cell. Conjugating *E. coli* cells require about 90 min for complete transfer of the (+) chromosome. The conjugation process can be interrupted at any time by agitating the cell suspension in a blender. In this way (–) cells carrying varying amounts of male chromosomal material can be recovered. From the type and number of phenotypic traits specified by the acquired segment of male chromosome in the asexual progeny of the (–) cells, the relative portion of the genes specifying these traits in the inserted male chromosome can be deduced. This experiment also genes conclusive evidence that *E. coli* consist of an endless, closed loop. The genetic map of bacterial chromosomes is thus circular.

2. **Transformation:** A recipient bacterial cell can be genetically transformed by addition of exogenous DNA derived from another strain. This effect is due to the incorporation of one or more genes from the donor DNA into the chromosome of the recepient cell. By determining the frequency with which two or more phenotypes from the donor cells appear together in the progeny of the recipient cell it is possible to deduce the relative positions of the two corresponding genes in the donor chromosomes. Pairs of genes incorporated into the recipient cells with high frequency are closer together in the chromosome than pairs of genes that enter together infrequently.
3. **Viral transduction:** A small portion of the chromosome of a bacterial cell infected with a virus becomes incorporated into the DNA molecules of the viral progeny. When the progeny viral particles infect another cell, the genes derived from the original host cell may be incorporated into the chromosome of the new host cell. In effect, the transducing viral particle carries a portion of the chromosome of one bacterial cell to the chromosome of another. Transduction is of great importance in fine-structure genetic mapping, since the transducing phage usually carries only a very small portion of a bacterial chromosome, of the order of 1 per cent. Thus transduction is particularly suitable for mapping very small segments of the chromosome rather than its gross structure.
4. **Viral recombination:** Fine structure genetic mapping may be carried out on viral chromosomes. Bacterial viruses undergo spontaneous or induced mutations. When the chromosomes of two viral particles (one normal and the other a mutant) enter the same host cell, they can fragment and then recombine in the course of viral replication to form hybrid viral chromosomes. Such hybrid chromosomes can then direct the formation of progeny which contain genes derived from both parent viruses recognizable by their phenotypic features. From such viral recombination experiments it has been deduced the genetic maps of $T_4$ and other bacterio-phages of *E.coli* are also circular. The viral recombination experiments by S. Benzer has given valuable information on the size of genes and mutational units. He concluded that each gene must contain many mutable sites which he was able to map in a linear sequence. This discovery was important because it had previously been thought that the gene itself was the unit of mutation, i.e. that each gene contained only one mutable site.

## QUESTION

1. Describe the chromosomal bands.
2. State briefly about size of gene and mutational units.
3. Write short notes on synthesis of genes, isolation of genes.
4. State briefly gene function.
5. Describe briefly the mapping of chromosomal structures.
6. What are the types of genes?

# 7

# GENERAL CHARACTERISTICS OF GENE

1. Gene and theory, 2. Interaction of gene, 3. Lethal genes, 4. Lethal hereditary diseases in man, 5. Mutations, 6. Mutagens, 7. Crossing-over of genes, 8. Linkage of genes, 9. Chiasmata formation, 10. Recombination, 11. One-gene one-enzyme hypothesis.

## What is Gene?

A cell having nucleus contains chromosomes. The chromosomes-contain band like structure called gene. A gene is an element that serves as a specific transmitter of a single hereditary character. At the molecular level, a gene is a portion of a DNA molecule arranged in a fixed position in a linear fashion along the DNA chain and which functions by controlling the synthesis of a specific polypeptide chain. Thus *each* gene controls the synthesis of a single protein, regardless of the functional role of that protein in the cell in particular or the body in general.

## Theory of Gene

The theory of gene can be summed up as follows:

1. Genes are the heritable living but nonvisible units varying in chemical composition in different species determining the characters of the offspring. They occur in large numbers, e.g. 5000 genes are estimated in a Drosophila fly.
2. They are serially arranged in linear rows and distributed among the chromosomes.
3. They are identified by the chromosomes they inhabit, and the chromosomes maintain their individuality unaffected by metabolism of the cell in which they occur and are independent of environmental effects. The chromosomes occur in like pairs and show constancy in number.
4. The characters of an individual are referable to paired genes.
5. Each pair of genes separates at the maturation of germ cells so that each gamete contains one set.
6. Genes assort independently.
7. Interchange or crossing-over of genes may occur between homologous pair of chromosomes.
8. Frequency of crossing-over furnishes evidence of linear arrangement and indicates position of genes in the chromosomes. Chromosome maps can be out accordingly.

9. Translocation of genes may occur or the genes may be lost on account of wrong synapsis.
10. A single character is usually the effect of one gene, but may also be the combined effect of several genes.
11. The relative position of genes has an effect on other genes in determining the character.
12. Genes affect the rate of developmental processes of the body accounting for malformation, origin of vestigial organs and death.
13. Sex is determined by genetic balance.
14. The fact that the genes and their arrangement normally remain constant, accounts for the resemblance between the parent and the offspring.
15. The recombination of genes during zygote formation producing new combinations of old genes and crossing-over, accounts for the differences between the parents and the offspring or the variations between members of a family.
16. Mutations are caused by some change in genes for the character concerned.
17. Gene mutation, though a rare event as for instance, occurring in certain insects at a rate of 1 in $10^5$ individuals, leads to the foundation for species formation (or speciation) by throwing up new genes (or gene complex).
18. Genes may have their expression altered by modifiers (as set up in the internal environment on account of interaction with the environment) so as to change the selective value in the evolutionary process.
19. Genes contain the genetic material which is chemically DNA.
20. DNA of genes has double strands, double helical in nature running parallel in opposite direction and genetic property is limited in one strand.
21. Gene can replicate, i.e. it participates in its own synthesis.
22. Gene regulates not only genetic character but also day-to-day functions of cell, i.e. protein synthesis, enzyme synthesis, etc.
23. Genes contain genetic code which is transcribed to mRNA and is sent to cytoplasm the to ribosomes for translation for protein synthesis.
24. Gene replicates once and only once.
25. If, due to some accident in replication, one or other of the bands is lost, there is a corresponding malformation in the organisms.
26. DNA molecule in the gene is having bases whose sequence is the important factor in heredity.
27. DNA of gene is the molecule of heredity. The purine and pyrimidine bases of DNA carry genetic information, whereas the sugar and phosphate groups perform a structural role.
28. Every gene has a specific position in a specific chromosomes and the position is called *locus*.
29. A single gene may occur in several forms or in several functional states. All the forms of a normal gene are known as allele.
30. Genes can duplicate themselves very accurately.
31. Genes express themselves by producing proteins.
32. One gene is a segment of DNA which contains the information for one protein (enzyme).

## Interaction of Genes

Mendel's laws of segregation and independent assortment was confirmed by subsequent experiments and it was certain that these laws were not applicable universally to all heredity in all organisms. Though Mendelian inheritance might be an expectation but most heredity might be of the blending type, in which the heredities of the parents mix or blend in the offspring. The classical

Mendelian ratios, such as 3 : 1, 1 : 9, 3 : 1, etc. do not occur in all cases. It is known that a characteristic is controlled by one pair of alleles. In some cases, a characteristic is controlled by two or more pairs of alleles by interaction with one another which is seen in case of shape of comb in different breeds of poultry. Each comb type is produced by the interaction of two genes, each of which is transmitted independently in a Mendelian fashion. Depending upon the form of interaction, phenotype ratios are modified in various ways, although the fundamental laws of the transmission of heredity remain the same.

- **Experimental Results**

If a true-breeding *pea-combed* fowl is crossed with a true-breeding rose-combed fowl, all the $F_1$ offspring would be a different kind of comb known as *walnut.* If two of these walnut are crossed, the $F_2$ produces fowl with all three types of combs, i.e. *pea, rose, walnut* and a fourth type called as the *single comb* (Fig. 7.1).

- **Discussion**

It is observed that a totally new comb is produced in both $F_1$ and $F_2$ generations. If we assume that these results of four types of combs are produced by the interaction of two pairs of alleles designated as *Pp* and *Rr*.

1. A *pea-comb* fowl develops when *P* alone is present without *R*;
2. A *rose-comb* develops when *R* are present without *P*;
3. A *walnut-comb* develops when *P* and *R* are present together;
4. A *single-comb* develops when *p* and *r* are present without *P* and *R*. The genotypes and phenotypes of the resulting. The zygotes for the different combs are given below (Fig. 7.2). The genotypes are:

Pea : *PPrr* or *Pprr* (3 : 16)
Rose : *ppRR* or *ppRr* (3 : 16)
Walnut : *PPRR, PPRr, PpRR,* or *PpRr* (9 : 16)
Single : *pprr* (1 : 16)

1. Pea comb

2. Rose comb

3. Walnut comb

4. Single comb

**Fig. 7.1:** Different types of comb in poultry.

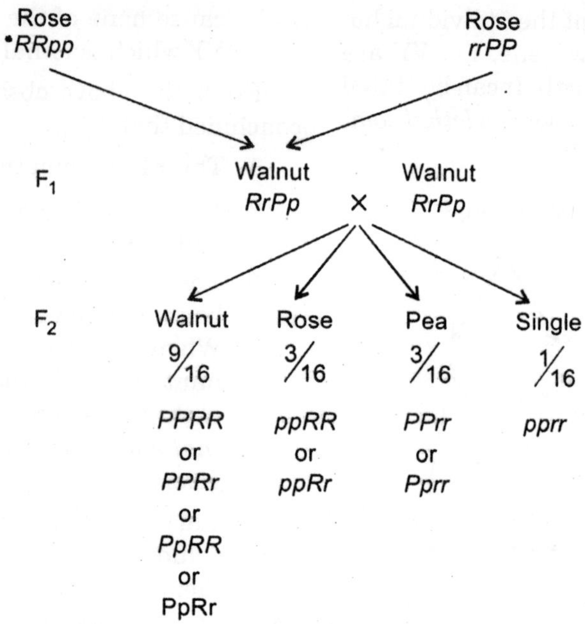


**Fig. 7.2:** The interaction of two independent pairs of alleles affecting comb form in fowls. The sixteen possible combinations of $F_1$ gametes, and the genotypes and phenotypes of the resulting zygotes in $F_2$ generation.

The two pairs of alleles are transmitted in a normal Mendelian manner. They show free assortment and give the same ratios as in ordinary dihybrid crosses. The mode of inheritance of the genes for rose and pea does not differ at all from the usual Mendelian fashion. The differences that distinguish this and similar cases from ordinary dihybrid inheritance are:

1. $F_1$ resembles neither parent, and
2. Apparently different characters appear in $F_2$ generation.

   One of these new characters, i.e. walnut comb evidently results from an interaction between two independently inherited dominant genes, and the other, i.e. single comb results from the interaction of their two recessive alleles. These peculiarities are due not to a new process of inheritance but simply to the circumstance that both genes involved happen to express themselves in the same part of the organism, e.g. the comb, of the fowl.

## Lethal Genes

Carriers of the same genotype may have different phenotypes. But it may be difficult to tell to which genotype an individual belongs by observing the appearance. The irregularities in the external expression of genes are particularly frequent among gene mutations which produce hereditary diseases in the organisms and investigation of them is of considerable importance. From genetical studies we get information about genes from abnormal ratios than from normal ones.

- **Experimental Results**

We can study the inheritance of fur colour in mice of which one mice has yellow fur (*Y*) is dominant to mice having grey fur (*y*). If a pair of yellow mice (*Yy*) are mated, the result will be the same, i.e. 2/3 of the offspring will be yellow, $\frac{1}{3}$ grey. Yellow to grey ratio is 2 : 1. The offspring are of both types of fur colour, i.e. both parents must be heterozygous (*Yy*). If this happens, then we should expect a ratio of 3 : 1 in the $F_1$ generation. It can be explained that that the individual homozygous for the yellow gene, i.e. *YY* are enviable and die before birth. It can be stated that the genotype *YY* represents a *lethal combination of genes* (Fig. 7.3).

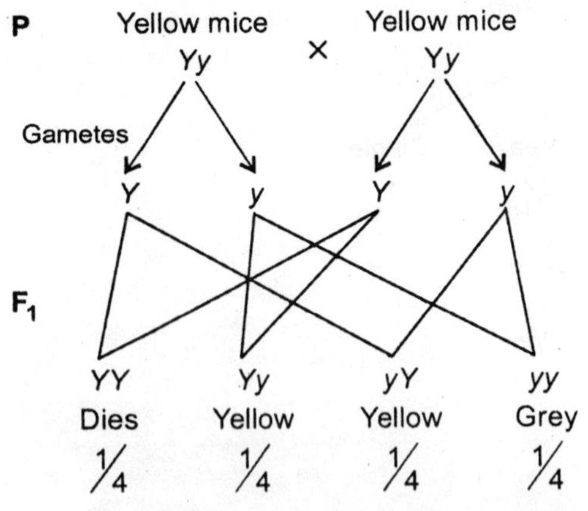


**Fig. 7.3:** Inheritance of fur colour in mice. The *Y* is dominant to *y*. The *YY* combination is lethal. The normal 3 : 1 ratio is disturbed.

The death of such individuals in the embryonic state would have the effect of removing a quarter of the offspring from the litter. The dead embryo would represent a third of the potential yellow offspring thus reducing the proportions of yellow offspring from what would have been $\frac{3}{4}$ to $\frac{2}{3}$. This hypothesis of that the zygotes homozygous for yellow has been supported by the following observations:

1. Crossing yellow with yellow, never produces exclusively yellow offspring, i.e. there is impossibility to breed true for a pair of yellow mice which means yellow mice are heterozygous and a living homozygous yellow mouse is an impossibility.
2. A quarter of the total offspring are found dead in the embryonic stage in the uteri of yellow mothers who have been mated with yellow males. No such dead embryos are found in the uteri of grey mothers or of yellow mothers mated with grey males because none of the offspring would be *YY* which is lethal.

From the above observations it can be concluded that:

1. The *YY* combination is lethal,
2. *Y* gene not only affects colour but also viability,
3. The gene is dominant or recessive can be considered depending upon action. When it controls the colour it is dominant but as a lethal gene affecting viability it is recessive, exerting its *lethal effects only when in the homozygous state.*

Lethal genes exist in wide range of organisms including man. It is difficult to observe the homozygous lethal because it is perished in the embryonic stage. But its existence can be deduced from:

1. The unusual ratios of living offspring,
2. The total number of offspring will be lower than expected.

The existence of a lethal gene can be observed from the heterozygous condition, though not lethal has an observable effect on the individual's phenotype.

In such cases the lethal gene is not completely recessive but exerts partial dominance so that it expresses itself to a slight extent in the heterozygote.

## Semilethal and Subvital Genes in Man

Sickle-cell anemia is a possible exception but the genes we have seen in case of mice and men are lethal which kill all or nearly all homozygous individuals before they reach the sexual maturity. These genes are complete lethals. But there are genes which kill some of the affected persons but allow others to survive long enough to have families. These types of genes are called *semilethal* or *subvital.*

The line of demarcation between lethal and semilethal is not sharp because the genes that are lethal under some conditions may permit survival under certain conditions, e.g. with special medical treatment. Usually, genes that kill fewer than 100 per cent but more than 50 per cent of their victims are called *semilethals,* whereas those that cause less than 50 per cent mortality are known as *subvital genes.*

**EXAMPLE**

1. **Epiloia:** It is a rare disease in man and a *dominant semilethal,* heterozygous carriers of a gene causes a combination of morbid trait, including severe mental deficiency, abnormal growths on the skin, and tumors in internal organs, most persons die during childhood but some manage to survive, marry, and beget children.
2. **Retinoblastoma:** It is due to another *dominant human gene.* Causing malignant tumors in the eyes and almost leads to death in early childhood. The gene acts as an almost *complete lethal.*

## Lethal Hereditary Diseases in Man

1. **Juvenile amaurotic idiocy:** Infants homozygous for this gene appear to be normal at birth, eyesight begins to fail between 4–7 years of age leading to blindness. Deterioration of mental and physical powers leads to death before or during adolescence.
2. **Tay-Sachs Disease:** Another recessive lethal gene produces, when homozygous, the condition is known as Tay-Sachs disease or infantile amaurotic idiocy which results in early death of the homozygotes. It is due to deficiency of the enzyme *hexosaminidase-A,* in tissues. The characteristic clinical symptoms are manifested by the increased accumulation of Gangliosides ($GM_2$) in the brain and spleen, demyelination, cerebral degeneration mental retardation, severe motor deterioration, blindness, cherry-red spot in the eye muscles within first year of life; the circumference of the head becomes 50 per cent greater than normal at about two years, repeated respiratory infection and the patient dies between 3–4 years.
3. **Sickle-Cell Anemia Molecular Disease:** The persons with this (genotype $Hb_1^S\ Hb_1^S$) have no normal haemoglobin (HbA) in their RBC and instead have (1) an abnormal haemoglobin ($Hb_1^S$) having lower negative charge than $Hb_1^A$ and (2) that the $Hb_1^A\ Hb_1^S$ heterozygous carriers of the disease have both $Hb_1^A$ and $Hb_1^S$ in their RBC leads to mild sickling but not to anemia. According to Pauling this is to be a gene-controlled molecular disease.

**Normal** :
(1) Valine — (2) Histidine — (3) Leucine — (4) Threonine —
(5) Proline (6) [Glutamic acid] (7) Glutamine (8) Lysine —

**Sickle cell** :
(1) Valine — (2) Histidine — (3) leucine — (4) Threonine —
(5) Proline — (6) [Valine] — (7) Glutamine — (8) Lysine —

**Fig. 7.3.1:** Normal (HbA) carries normal α and β chain having all amino acids in order. Sickle cell (HbS) carries normal α chain but β chain contains valine in place of glutamic acid in number six and other amino acids are in order.

**Molecular Basis:** Sickle-cell anemia is caused by mutation of autosomal gene ($HbS_1$) with a lethal effect when homozygous and with only a slight but discernable effect when heterozygous. In this case normal haemoglobin in the RBC is entirely replaced by the abnormal haemoglobin S(HbS). HbS carries normal $\alpha$ chain and abnormal $\beta$ chain—the $\beta$ chains have valine instead of glutamate at the 6th position amino acid residue along the chain (Fig. 7.3.1).

**Characteristics of HbS:** The HbS is much less soluble, deoxygenated (when oxygen concentration falls in the capillaries and tissues). with their abnormal Hb, the sickled red cells are inefficient to carry oxygen. Haemoglobin of sickle-cell anemia differs from normal haemoglobin (HbA) electrophoretically. Normal Hb having glutamic acid in position 6 has a negative charge and migrates to the positive pole, whereas HbS with valine replacing glutamic acid has no net charge and does not migrate in the electric field.

The amino acid sequences of both HbA and HbS were studied by "finger-printing" technique (Fig. 7.3.1).

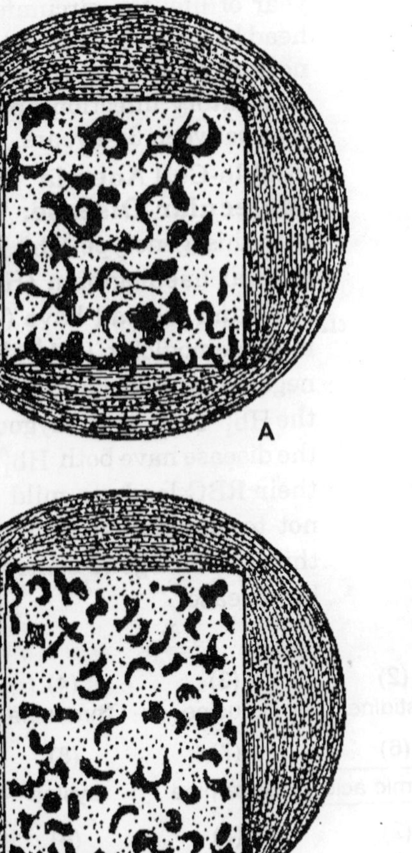


**Fig. 7.4:** Sickle cells in man
A. Red blood cells from a patient with sickle-cell anemia (homozygous).
B. Red cells from a person with sickle-cell trait (heterozygous).

**Clinical Manifestations:** The sickling leads to increased fragility of RBC causes severe haemolytic anemia which results is generally total. There is oxygen shortage, weakness, emaciation, kidney and heart failure, hepatosplenomegaly, rupture of spleen, bone-marrow hyperplasia, haematuria jaundice Occurrence—the condition is not very common, about 0.2 per cent of American Negroes suffer from it. It is relatively frequent among Western Indian tribals, many Africans (20%) in certain racial groups in Arab races.

HbS is precipitated as polymer crystals in RBC, changing the latter into highly fragile crescentic or sickle cells which are easily haemolyzed (Fig. 7.4A).

**Sickle-Cell Trait:** A related condition which is common is known as *sickle-cell trait.* In this case abnormal HbS is present only to the extent of about 30-40 per cent, the remaining haemoglobin being normal and the RBC are of the normal shape. The anemia is slight and generally not sufficient to interfere with normal function (Fig. 7.4B). The relationship between sickle-cell anaemia and sickle-cell trait is that the gene responsible for normal blood is only partially dominant to its alleles producing the sickle-cell condition. A person *homozygous* for the defective gene suffers from *sickle-cell anemia* whereas *heterozygous* individual, however, only develop the milder *sickle-cell trait.* Because of the partial dominance of the normal allele, the sickle-cell gene is prevented from expressing itself fully in the heterozygote.

**Gene Action:** The disease is inherited in the Mendelian fashion. The defective HbS is due to a change in as few as two nucleotides in the DNA responsible. It is not reasonable to think that bad genes are always quickly needed out of a population by natural selection. Sometimes a gene which is disadvantageous in homozygous state but advantageous in heterozygous state as we see in case of sickle-cell anemia. This favours its survival and such genes may show a high frequency in a population. Much more common is the milder sickle-cell trait when the gene is in heterozygous and high incidence of this case caused by heterozygote being more resistant to *malaria.* This ensures the survival of the sickle-cell gene. Due to heterozygote condition, individuals and population always harbour a certain number of bad genes which is referred as the *genetic load.*

- **Genetic Code and Sickle-cell Anemia**

Studies of cell free systems reveal that one of the DNA triplets which codes for glutamic acid is CTC. This CTC triplet would code for the complementary GAG in the mRNA transcript substitution of an A for T(CAC) in DNA for $Hb_1^S$ would result in GUG in mRNA and mRNA triplet codes valine.

| $Hb_1^A$ (Normal) | | $Hb_1^S$ | |
|---|---|---|---|
| *DNA* | *mRNA* | *DNA* | *mRNA* |
| C | G | C | G |
| T | A | A | U |
| C | G | C | G |

*The One Cistron-One Polypeptide Hypothesis*: From different studies it is evident that α and β chains are determined by different genes, one is synthesized separately and both are then assembled to give haemoglobin by random association of α and β subunits according to one cistron-one polypeptide hypothesis.

## Mutations

- **Introduction**

Genetic variation by reshuffling of genes (i.e. by independent segregation, crossing-over, chiasmata or fertilization) in cellular mechanisms provide the basis of continuous variation in a population but play only a limited role in evolution. The reason is that independent segregation of chromosomes and crossing-over may establish a new combination of genes in one generation but the same processes can just as readily 'undo' it in later generations. Thus genetic change is non-persistent and plays a limited role in long-term evolutionary change. A notable exception is seen when two genes are brought into very close proximity as a result of crossing-over. This lessens the chance of a chiasma being formed between them in later generations with the result that they may remain together indefinitely. But for the most part more reshuffling of genes produces only short-term changes. Much more profound initiators of genetic variety from the evolutionary point of view are *mutations.*

- **Definition**

Every now and again a natural population of animals or plants throw up an individual with some characteristics strikingly different from the rest of the population, e.g. haemophilia and cystic fibrosis in man, white eyes and vestigeal wings in Drosophila, resistance of bacteria to penicillin or flies to DDT, etc. It can be shown from studies of inheritance that these conditions are generally recessive and are transmitted in a Mendelian fashion. The genetical mechanism producing the change is known as a *mutation.* Any descendants of an individual with a mutation, showing the new characteristic, are called *mutants.* Mutations are the basis of discontinuous variation in population.

### • Mutation Rate

Mutations are comparatively rare, at least in terms of the rate at which gametes are produced. In most organisms the mutation rate varies between 1 and 30 per million gametes. The rarity of mutations can also be seen in their occurrence in population, known as *mutation frequency.* At any one gene locus at each generation a mutation occurs in only one individual out of approximately 500,000. The exact number is variable since the genes at different loci have different mutation rates. Expressed in this way the mutation rate certainly seems to be extremely low. However, it must be remembered that each individual has a very large number of gene loci at which it can happen. The overall mutation rate is therefore much greater than is realized by considering a single locus.

**Occurrence:**

1. Spontaneous mutations—occur spontaneously or,
2. Induced mutations—under the influence of external factors, e.g. chemical, radiation.

## Agents Causing Mutations (Mutagenic Agents)

The factors responsible for mutations known as *mutagenic agents.* Such as

1. Irradiation with X-rays
2. Gamma rays
3. Ultraviolet rays
4. Mustard gas
5. Base analogs, e.g.
   (a) 5-bromouracil } Transitional mutants
   (b) 2-Aminopurine }
6. Nitrous acid
7. Acridine derivative, e.g. acridine (insertion mutant)
8. Proflavin (deletion mutant)
9. Lysergic acid diamide (LSD)
10. Caffeine (has been reported as mutant)

These have made it possible to study the development and transmission of mutations in some detail.

### Salient Facts about Mutations

1. *First,* mutations arise spontaneously and are undirected by the environment. Environment can affect mutation rate but cannot induce mutation. The direct part played by environment it to select mutants.
2. *Second,* mutations, though not necessarily permanent, are relatively persistent. A mutant gene may be transmitted through many generations without further change, after which it may mutate again producing yet another novel characteristic, or it may change back to its original condition. But the fact that it can persist unaltered for at least several generations makes it possible for natural selection to get to work on it.
3. *Third,* the vast majority of mutations confer disadvantages on the organisms that inherit them.

## Types of Mutations

There are two types of mutations:

A. Chromosomal mutations (changes in the supermolecular structures),

B. Gene mutations or point mutations or replacement (changes in the nucleotide sequence in DNA and, accordingly, in mRNA).

A. **Chromosome Mutations:** It can occur in 6 ways during meiosis:

1. **Deletion:** A chromosome may break in two places and the section in between may drop out, taking all its genes with it. The two ends then join up, giving a shorter chromosome with a chunk missing in the middle. This is called *deletion.* Deletions are lethal even very shortest one.
2. **Inversion:** If a chromosome breaks in two places and the middle piece then turns round and joins up again so that the normal sequence of genes is reversed. This is called *inversion.*

3. **Translocation:** Sometimes a section of one chromosome breaks off and becomes attached to another chromosome. This is rather like crossing-over, except that it occurs between non-homologous chromosomes; it is known as *translocation*.
4. **Duplication:** When a section of chromosome replicates so that a set of genes is repeated: This is *duplication*.
5. **Non-disjunction:** Caused by the addition or loss of one or more whole chromosome. On some occasions in meiosis, the two homologous chromosomes, instead of separating, go off into the same gamete and known as non-disjunction. The defects are seen

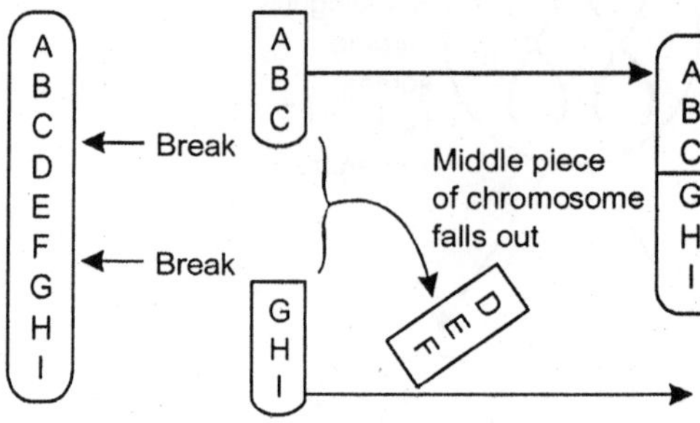


1 DELETION

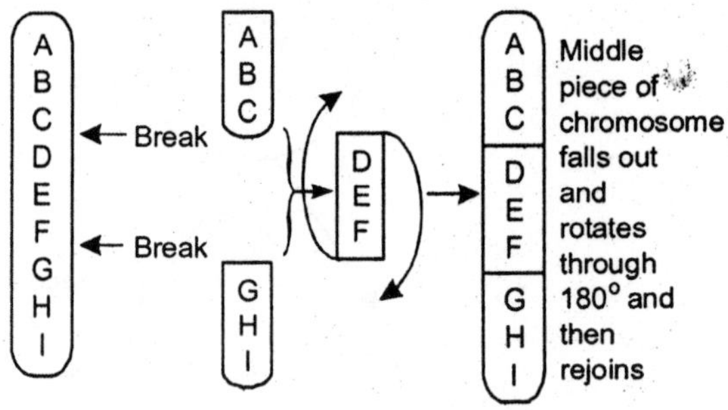


2 INVERSION

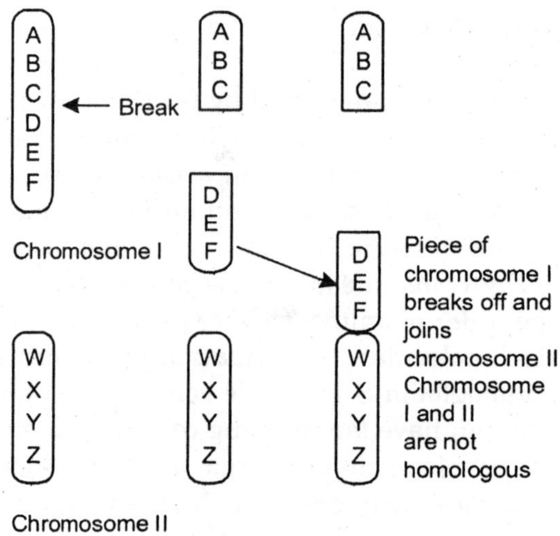


3 TRANSLOCATION

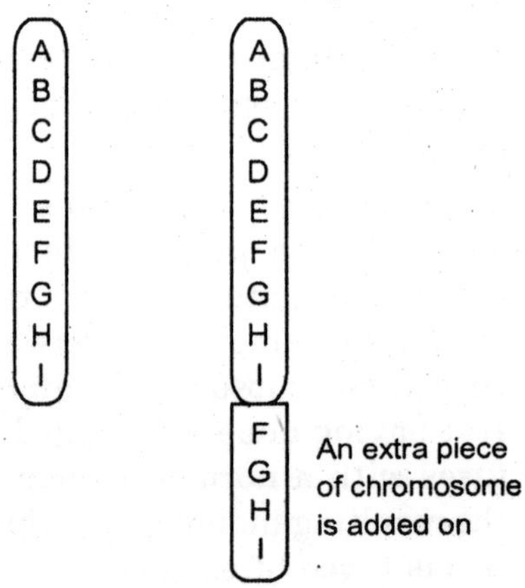


4 DUPLICATION

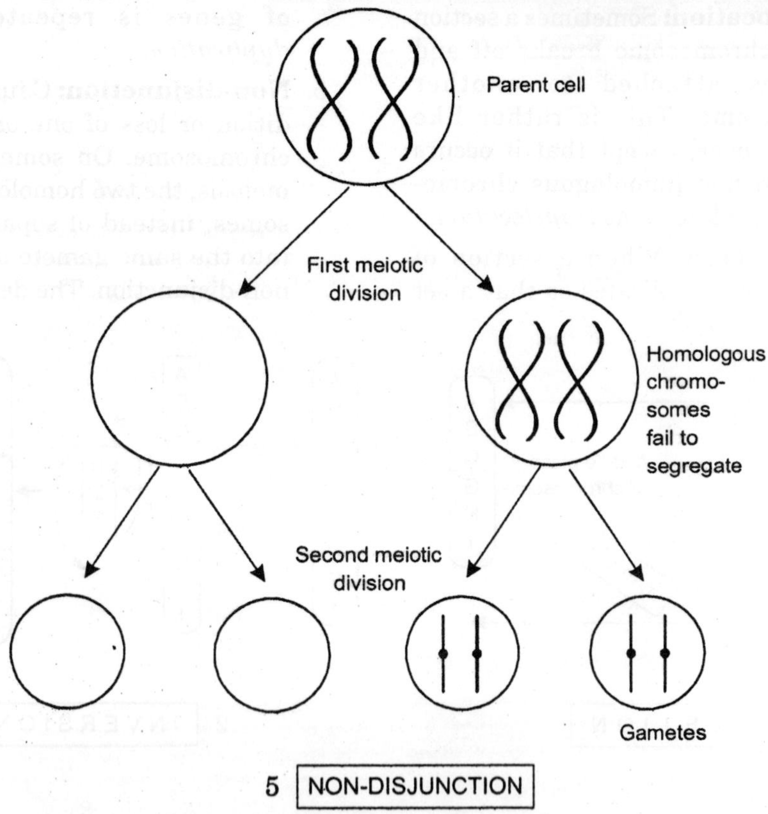


**Fig. 7.5:** Types of Chromosome Mutation.

in mongolism (Down's syndrome), Klinefelter's syndrome, Turner syndrome (Fig. 7.5).

6. **Polyploidy:** sometimes all the chromosome pairs undergo non-disjunction simultaneously, resulting in half the gamete having two of each type of chromosome, i.e. diploid, the rest having none. If a diploid gamete fuses with a normal haploid gamete the resulting individual is triploid, i.e. it has three of each type of chromosome. If two diploid gámetes fuse, a tetraploid individual results. It is thus possible for an organism to acquire one or more complete extra sets of chromosomes, a phenomenon called *Polyploidy*.

B. **Gene Mutations (Point Mutations or Replacement):** Gene mutations are thought to have played by far the most important part in generating evolutionary change. A gene mutation arises as a result of a *chemical change in an individual gene.* An *alteration* in the *sequence of nucleotides of DNA molecule* in a single gene will change the order of amino acids sequence in a protein molecule (e.g. formation of abnormal haemoglobin-S in Sickle-cell anemia) and this can have far-reaching consequences on the development of an organism. Sometimes it causes early death of the organism, the genes are described as *lethal genes.*

Although the initial change may not occur in the sense strand of the double-strand spontaneous gene mutations are determined by errors in DNA replication that arise of the thermal motion of atoms and molecules. These errors of transcription and translation are not inherited.

DNA molecule of that gene but after replication the daughter DNA molecules with mutation in the sense strand will segregate and appear in the population of organisms.

Various types of chemical changes in DNA can lead to mutant gene. By transduction method it was shown that mutation may involve chemical change in only a single nucleotide known as *point mutations,* or *single-point mutations. Point mutations* can be divided into following types:

| | | |
|---|---|---|
| 1. Transitional | Silent mutations | Point mutation |
| 2. Transversional | | |
| 3. Insertion | Frame-shift mutations | |
| 4. Deletion | | |

5. Substitution
6. Inversion

### • Silent Mutations

There are replacements of in codon leasing to degeneration of the codon together with original one i.e., which do not change the amino acid residue in the protein. These do not manifest themselves phenotypically, but they are important for evolution.

1. **Transitional:** A given pyrimidine is changed to the other pyrimidine or a given purine is changed to the other purine.

   Thymine (Pyrimidine) ⇄ Cytosine (Pyrimidine)

   Adenine (Purine) ⇄ Guanine (Purine)

2. **Transversional:** From a purine to either of the 2 pyrimidines or the change of a pyrimidine into either of the 2 purines

   Thymine → Adenine, Guanine

   Cytosine → Adenine, Guanine

   Adenine → Thymine, Cytosine

   Guanine → Thymine, Cytosine

If the nucleotide sequence of the gene containing the mutation is transcribed into an mRNA molecule, then the mRNA molecule will possess a complementary base change at this corresponding locus. Single base changes in the mRNA molecules may have one of several effects when translated into protein:

(a) *No detectable effect*: There may be no detectable effect because of the degeneracy of the code.

(b) *Missense effect*: It causes replacements of nucleotides in codons and changing the sense of codons, i.e., a different amino acid is incorporated at the corresponding site in the protein molecule. This mistaken amino acid or missense, depending upon its location in the specific protein, might be
   1. Acceptable
   2. Partially acceptable
   3. Unacceptable

   to the function of that protein molecule.

(c) *Nonsense codon formation*: A nonsense codon may appear that will result in the premature termination of amino acid incorporation into protein chain and production of only a fragment of the intended protein molecule.

3. **Insertion:** The third type of point mutation called insertion which involves extra base bair. The result is one type of *frame-shift mutations,* which result in a deletion or an insertion so called because the normal reading frame relationship of nucleotide mRNA triplets codons is shifted by the mutation. Insertion can be induced by *acridine.*
4. **Deletion:** The fourth type of mutation results from a deletion of one or more bases from DNA. Deletion can be induced by *proflavin.*

Not all point mutations in DNA are necessarily lethal. Transactions and transversional are relatively benign, because they result in replacement of only one amino acid in the peptide chain coded; very often the defective protein is still functional. Such a mutation is called a *silent mutation.* However, insertion and deletion mutants cause all the DNA beyond the point of mutation to be misread; such mutations often are lethal.

5. **Substitution:** Gene mutation that causes sickle-cell anemia is called a substitution because an incorrect nucleotide is substituted for the correct one.
6. **Inversion:** This is due to placing of DNA in inverted way.

- **Mutational Units**

Benzer by viral-recombination method had shown *that each gene must contain many mutable sites,* which he was able to map in a linear sequence. This discovery was important because it had previously been thought that the gene itself was the unit of mutation, i.e. each gene contained only one mutable site. From the known length of DNA molecule and the known small fraction of genetic map was calculated that gene consist of 2500 nucleotide pairs and found nearly 500 mutants, indicating that the unit of *mutation is no greater than about 5 nucleotides.* Benzer and other investigators concluded that mutation may involve a *chemical change in only a single nucleotide* and such mutation called *point mutation.*

**The Spread of Mutations through a Population:**

1. If there are beneficial mutant genes, and *dominant,* it can spread quickly through population.
2. If they are *recessive,* it will show its effect in organism homozygous for it. The chances of two such genes being brought together into the homozygous state are slight, and many generations may pass without it happening ultimately the gene may be lost from the population.
3. The whole situation can be altered if the mutant gene, recessive at first, becomes dominant in the course of time. Selection will then favour the heterozygote, and the gene will spread rapidly through the population.

**The Importance of Mutations in Evolution:** Although chromosome mutations such as deletion and translocations are important in the genetics of individuals, but from evolutionary point of view the significant result is obtained if there is increase in the amount of genetic material. Although the addition of whole chromosome may have been disastrous but the addition of new genes within individual chromosomes, i.e. insertion has been extremely important in promoting evolutionary novelty.

## Mutagens

- **Introduction**

The induction of mutations by mutagenic agents was first shown by Muller in 1927 applying X-rays to the fruit-fly Drosophila. With bacteria, which do not require the penetrating power of X-rays, the more convenient ultraviolet irradiation is equally effective, and powerful chemical mutagens were subsequently discovered. In addition, through the selection of rare mutants from large populations of bacteria, a slight mutagenic action was later demonstrated for many chemicals (e.g. formaldehyde, caffeine, $Mn^{2+}$) and even for elevated temperature. Bacteria thus provide sensitive test systems for detecting potential environmental mutagens for man.

- **Definition**

The agents cause induction of mutation are called *mutagens* or *mutagenic agents* and the changes occur called *mutagenesis.*

- **Type**

I. **Physical** Radiations → Ultraviolet radiation, X-rays, γ-rays

II. **Chemicals**

(a) *Base analogs*: 5-Bromouracil, 5-BudR, 2-Aminopurine

(b) *Deaminating agents*: Nitrous acid

(c) *Alkylating agents*: Nitrosoguanidine EES (Ethane sulfonate)

(d) *Acridine derivative*:

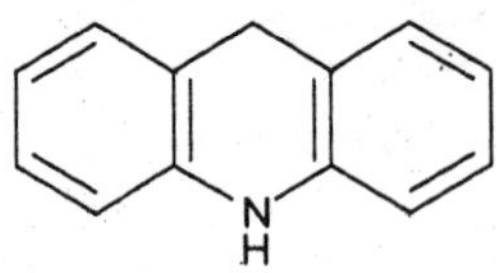

III. **Carcinogens**

4-Nitroquinoline -1-oxide

IV. **Miscellaneous**

Formaldehyde, Caffeine, $Mn^{2+}$, Proflavin.

- **Mechanism of Action**

Various types of chemical change in DNA that can lead to mutant gene products have been identified. *Single-point mutations* can be divided into *4 major classes* on the basis of the *change produced in the DNA by the mutagens.*

1. **Transitional Mutation:** One purine-pyrimidine base pair is replaced by another which can brought by base analog *5-bromouracil* and *2-aminopurine and nitrous acid, 5-BUdR, Hydroxylamine.*
2. **Transversional Mutation:** A purine-pyrimidine pair is replaced by pyrimidine-purine pair induced by carcinogens like *4-nitroquinoline-1-oxide, $Mn^{2+}$*.
3. **Insertion:** It involves insertion of an extra base pair. The result is one type of *frame-shift mutation,* so called because the normal reading frame relationship for readup of nucleotide triplets is put out of register by the mutation. This can be induced by *acridine derivatives.*
4. **Deletion:** It involves of deletion of one or more bases from the DNA; it too a frame-shift mutation. It is induced by covalent cross-linking reagents, alkylating agents, deaminating agents and *proflavin.*

## Crossing-over of Genes

- **Introduction**

Linkage is often considered as an exception of Mendel's law of independent assortment and crossing-over is in the same way an exception to linkage, i.e. in linkage the linked gene always stay together without seperation but during inheritance of linked gene it is unusual to find complete linkage. A certain proportion of the offspring show new combination as in independent assortment.

- **Definition**

Crossing-over means breaks in the linkage of genes within the chromosomes and bodily transhipment of genes from one chro-

**Fig. 7.6:** Showing transipment of genes from one chromosome to other during crossing-over.

mosomes to the corresponding position in the mate. The phenomenon of crossing-over closely resembles independent assortment of Mendel but it is a different thing. Independent assortment is concerned with the whole chromosome while crossing-over (Fig. 7.6) involves parts of chromosome. It is a sort of shuffling of genes between homologous pairs of chromosomes which always bring forth new combination. The gametes containing the new combination are known as *cross-over* or *recombinant gametes*. The gametes in which the linked genes remain in their original combinations are called *non-cross-over gametes*.

- **EXAMPLE**

If a maize plant homozygous for kernels which are coloured (C) and smooth (S) is

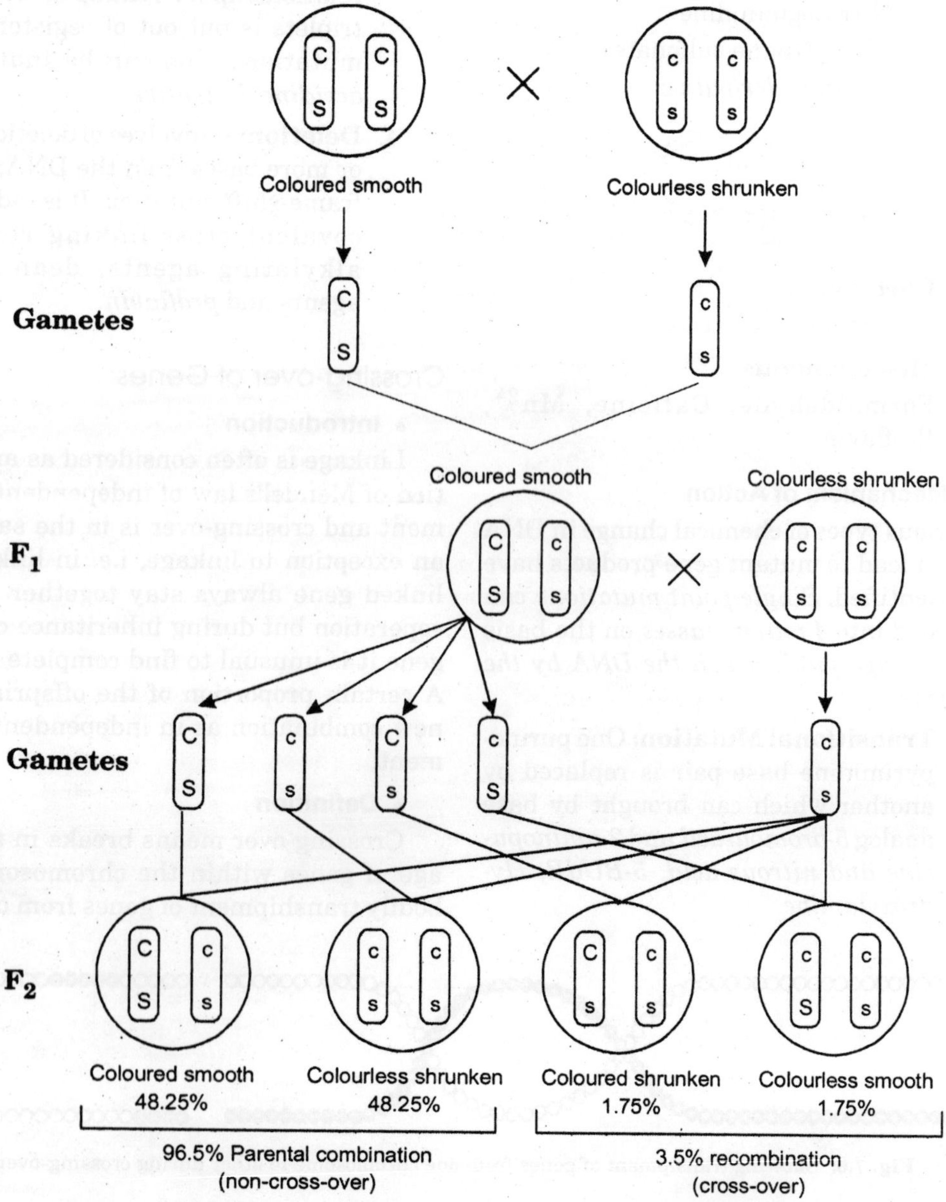


**Fig. 7.7:** Representation of example of cross-over of genes.

crossed with one having colourless (c) and shrunken (s) kernel, the $F_1$ generation produces nothing but the coloured smooth variety. If coloured smooth variety is crossed with colourless shrunken, the $F_2$ generation gives all four combinations of characteristics but not in the proportions, i.e. 96.5% parental combination (non-cross-over) and 3.4% as recombination (cross-over) (Fig. 7.7).

- **Inference**

That the genes are in fact situated on the same chromosomes and for the most part they are transmitted together through the gametes to the offspring. However, in the formation of a small proportion of the gametes the genes, instead of staying on their own chromosome, change places: This is called "crossing-over" and it has the effect of separating genes that were previously linked. The result is that, whilst the majority of offspring have the same combination of characteristics as the parents, i.e. *parental combination* (non-cross-over), a few show new combination i.e. *recombination* (cross-over).

- **Explanation of Crossing-over (Cytological Basis)**

Two basic facts must be remembered:

1. In meiosis (the type of cell division occurs in the formation of gametes), homologous chromosomes become intimately wrapped round one another; and
2. Allelic genes normally occupy the same relative positions on their respective chromosomes and therefore lie alongside each other when the chromosomes come together.

During prophase of the first meiotic division when homologous chromosomes become intertwined, the chromatids of homologous chromosomes are seen to be in contact with each other at certain points along their length. At these points, known as *chiasmata* (Fig. 7.8) (singular : chiasma), the chromatids break and rejoin. The result is that portions of the chromatids belonging to the homologous chromosome change places, taking their genes with them. These chromatids, with their new complements of genes are termed "cross-overs". Eventually they finish up in separate gametes and after fertilization, give rise to new combinations of genes in the offspring.

The number of chiasmata formed in a bivalent during meiosis, and therefore, the amount of crossing-over, varies from one pair of homologous the amount of crossing-over, varies from one pair of homologous chromosomes to another. One or several chiasmata may be formed, and of course the longer the chromosomes the greater is the number of chiasmata likely to be. The number of possible combinations of genes in the gametes will depend on the number and positioning of the chiasmata relative to the sequence of genes. The percentage of crossing-over varies between different genes but for each pair of genes the percentage remains constant. The cross-over percentage, according to Morgan, is related to the length of chromosomes, i.e. occurs more frequently in long chromosomes than in short one and between distant points on the same chromosome.

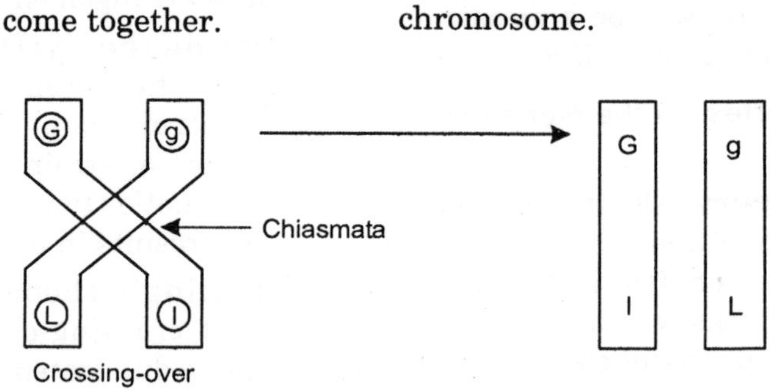


**Fig. 7.8:** Chromatids of homologous chromosomes are in contact at certain points.

The importance of crossing-over is that by establishing new gene combinations, it is an instigator of genetic variation by reshuffling of genes.

### • Interference and Coincidence

Crossing-over at a particular region of a chromosome tries to prevent another crossing-over close to it which is called "interference". Due to interference there are no or few double cross-over within a section of chromosome. The amount of interference becomes less if the distance is greater between the genes.

The double cross-overs are nothing but coming together or "Coincidence" of two single cross-overs.

### • Double cross-over

Crossing-over just once is known as single cross-over and the resultant gametes are called single cross-overs. But sometimes crossing over occurs at two points in the same chromosome pair which is known as double cross-over and the gametes so formed are called double cross-overs.

### • Somatic cross-over

Pairing of chromosomes is restricted to the germ cell and it takes place during the first maturation division. Somatic cross-over is a rare phenomenon. In Drosophila such rare instance of somatic cross-over has been shown by Stern. Such somatic cross-overs seen in one or two cells during the course development of the fly. But these cells give rise to cluster of cells resulting in formation of a patch or spot on the body with the cross-over cells. Thus the fly will be a mosaic of cross-over and non-cross-over tissues.

### • Different Theories on the Mechanism of Cross-over

1. **Classical theory:** During the early prophase stage of meiosis the chromosome split up longitudinally. Each chromosome forms two sister chromatids. The two non-sister chromatids of the homologous pairs of chromosomes coil round each other. At their points of contact the chromatids break first and cross.
2. **Chiasma type theory:** Breaking up of chromatids occurs at pachytene stage. After breaking up the chromatids unite again and form a chiasma. This theory states chiasma is the result of crossing-over.
3. **Copy choice theory:** Synthetic activity and duplication of chromosomes and intimately associated with recombination. In the mechanism, the sister chromatids duplicate their genetic parts and non-sister chromatids develop fibres at random. The entire recombinants are formed from newly formed sections. This theory states that duplication occurs in late meiotic prophase but it has been established that DNA replication occurs long before synapsis.

### • Theories that Explain the Happenings during Crossing-over

1. **Contact first theory:** The chromatids destined to undergo crossing-over, touch each other first and then crossing-over give rise to chiasma. After this, breakage takes place at the point of contact and new attachment of chromatid parts takes place.
2. **Breakage first theory:** Muller advocated the theory that the chromosomes destined to cross-over first break into two segments and then reunion occurs between non-sister chromatids to give new arrangement.
3. **Strain theory:** Darlington advocated that the chromosomes break as a result of strain of the time of pairing. A sort of strain develops, when two chromatids twist round each other and this results in breakage and reunion.
4. **Belling's theory:** Belling believes that crossing-over occurs between newly duplicated genes and that there is no breakage or reunion during crossing-over.

- **Significance of Crossing-over**

1. Crossing-over supports the fact that genes are arranged in a linear fashion on chromosomes.
2. Crossing-over provides opportunities for reshuffling of genes and thus bridges variations which play major role in the process of evolution.
3. By calculating the cross-over frequency it is possible to plot the genes on the chromosomes (i.e., gene locus) and chromosome map can be shown.

## Linkage of Genes

- **Introduction**

There are several examples of dihybrid inheritance in which genes segregate and assort independently, giving all possible combinations in predictable proportions in the $F_2$ generation. However, this does not always happen. Sometimes the genes fail to assort freely or do so in unexpected proportions.

The seven pairs of characters in garden peas with which Mendel worked were located on separate chromosomes. But genes for different characters stay together in the same chromosomes, a very confusing phenomenon occurs. Genes of different characters in the same chromosomes do not abide by the principles of independent assortment. Thus, Mendel's law of independent assortment is not universal but is limited to genes situated in different chromosomes.

Apart from their intrinsic interest these situations are important because they tell us much about the relationship between chromosomes and genes.

- **Definition**

Genes of different characters located in the same chromosomes, tend to remain linked together while passing from one generation to another. Such genes are called *"linked genes"* and the process is known as *"linkage"*.

### I. Bateson and Punnet's (1906) Experiment:

The exception to Mendel's law was first noticed by Bateson and Punnet.

They breed a dihybrid strain of sweet peas in which one pair of characters was Purple (P) and red (p) for the colour of the flowers and the other pair of character Long (L) and round (l) for the shape of pollen grains. The $F_1$ dihybrid were all purple long when a cross was made between purple long and red round peas because purple and long are dominant characters. But when these $F_1$ dihybrids producing PL, pl, pL, and pl gametes were crossed together to form $F_2$ generation. The four expected phenotypic kinds in the ratio of 9 : 3 : 3 : 1 never turned up.

- **Interpretation**

They put forward that when two dominant characters go in together from the same parent to form the dihybrid offspring there is a tendency for the same pair of characters to enter the same gamete and to be transmitted together. This phenomenon, they called *"coupling"* occurs when two dominant or two recessive factors enter the $F_1$ dihybrid from each parent instead of two dominant factors from one parent and two recessive factors from the other.

There occurs always a tendency for two dominant factors entering separately and likewise for two recessive factors to avoid or repel each other. This tendency of unlike pairs (one dominant and one recessive) to stay together avoiding union with their own dominant or recessive sort is termed *"repulsion"*.

### II. Morgan's Experiment

*Morgan* (1910) observed the similar result in Drosophila.

A wild type Drosophila having grey body (G) and long wings (L) is crossed with a fly having two recessive mutations of black body (g) and vestigial wing (l). The members of $F_1$ generation were all like wild type, that is grey bodied and long winged GL, gl as grey body colour and long wings are dominant over black body and vestigial wings.

Now a male fly from this dihybrid is back-crossed to a double recessive black vestigial female. Had there been independent assortments, the kinds of flies would have

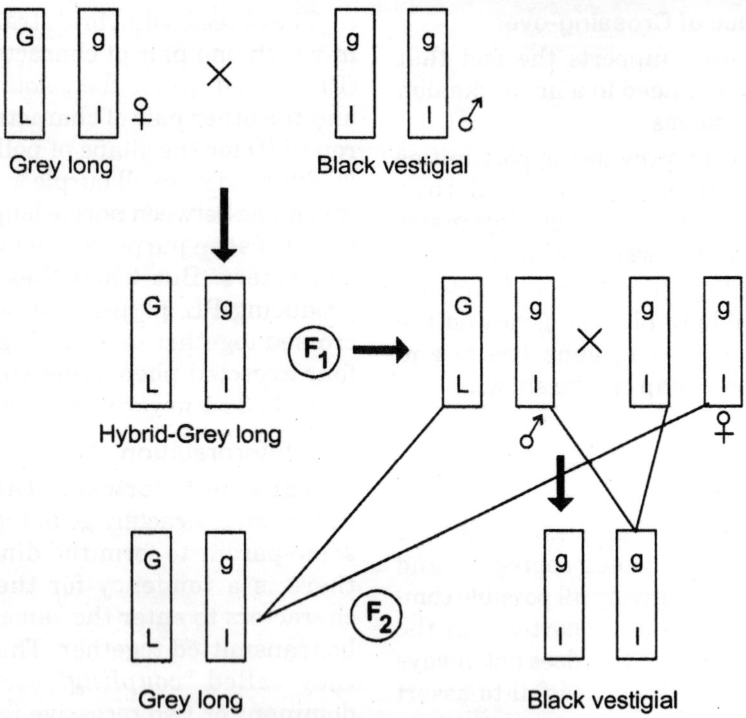


**Fig. 7.9:** Morgan's Experiment.

been formed in the following way as shown in the Fig. 7.9. But from actual experiments only two classes of offsprings, namely, grey long and black vestigial like two grandparents were obtained. This shows that grey body and long wings entering the dihybrid cross from one parent stay linked together.

- **Interpretation**

He put forward a satisfactory explanation of coupling and repulsion which are two aspects of a single phenomenon called *linkage*. He further stated that this tendency of linked genes to remain in their original combination was due to their location on the same chromosome.

- **Characteristics of Linkage**

1. Linkage is an exception to Mendel's principle of independent assortment.
2. Linked genes are housed in the same pair of chromosomes.
3. Lesser the distance between the linked genes stronger is the linkage bond. Genes lying further apart show less linkage.
4. Linked genes tend to be transmitted together from one generation to the other.
5. Separation of linked genes occurs very seldom.

- **Linkage Groups**

From a long series of experiments, it has been established that *"Drosophila melanogaster"* has four linkage groups. The genes in any one linkage group are transmitted together but independently of the genes in the other linkage groups. The interesting thing is that there are also four pairs of chromosomes, a powerful evidence that genes are located on the chromosomes (Fig. 7.10).

In general the number of linkage groups found in a particular species corresponds to the number of different types of chromosomes (the haploid number) characteristic of the species. Thus in the different species of Drosophila the haploid number varies from

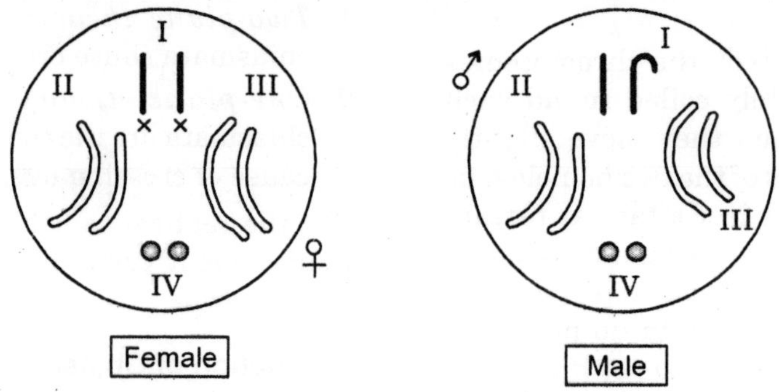


**Fig. 7.10:** *Drosophila melanogaster* fly has only four pairs of chromosomes which can be distinguished by their size. The black pair are the sex chromosomes, the Y chromosomes being hooked.

3 to 6 and in each case the number of linkage groups in the same. In men it is difficult to establish the number of linkage groups but it is expected that there are 23 linkage groups corresponding to the 23 pairs of chromosomes.

(a) Study of linkage groups predicts not only the number of chromosomes in a particular species, but their relative sizes as well.

(b) If a linkage group contains a large number of genes, we conclude that the chromosome concerned is correspondingly large.

(c) If a linkage group contains only a few genes, the chromosomes is presumably comparatively small.

- **Sex Linkage**

All the genes carried on the sex chromosomes are transmitted along with those determining sex, that is they are "*sex-linked*". The principle of sex linkage seen in Drosophila apply equally well to other organisms including man. The male is heterogametic (XY) and the female is homogametic (XX), and Y chromosome is genetically empty. As a male always receive his Y chromosomes from his father he cannot inherit any of his father's sex-linked traits. A female, however, receives X chromosomes from her father which she may subsequently transmit to her children. The latter may therefore show their grandfather's sex-linked traits. To put it in general terms: a male transmits his sex-linked traits to his grand children via his daughters, he cannot transmit them to or through his sons.

The way of this work in man can be seen in inheritance of

(a) colour blindness,

(b) haemophilia.

## Chiasmata Formation

- **Introduction**

In meiosis, however, the number of chromosomes is halved, the daughter cells receiving only one of each type of chromosome instead of two and the daughter cells are said to be in haploid state (Greek word "haploos", meaning single).

The halving of the chromosomes number is seen by behaviour of chromosomes during prophase and metaphase. Meiosis consists of two successive divisions: the parent cell splits into two (first meiotic division) and the products then divide again (second meiotic division) giving a total of four daughter cells.

In the first division homologous chromosomes get separated from each other and go into different cells. The second division is concerned with separating the chromatids.

In prophase I of meiosis the homologous chromosomes come to lie side by side a process known as "*synapsis*". In this condition each pair of homologous chromosomes constitutes a "*bivalent*".

- **Definition**

As prophase proceeds the chromosomes may become intimately coiled round each other, and later, when they move slightly apart, the chromatids of the two homologous chromosomes remain in contact at certain points called "*chiasmata*" (singular chiasma) and the meiosis is called "*chiasmate meiosis* or *division*" and if chiasmata do not occur known as "*achiasmate meiosis* or *division*".

- **Process**

Chiasma formation was first correctly understood by a Belgian cytologist, Janssens (1909).

Chiasma formation is very varied. Every bivalent forms at least one chiasma, but in fact as many as eight may be present, and these can be formed between any two of the four chromatids.

Except at the chiasmata the homologous chromosomes in a bivalent become increasingly separated from one another. This results in the bivalents adopting characteristic shapes, which vary according to the number and position of the chiasmata. Such as:

(a) **Cross-shaped:** A bivalent with only one chiasma assumes a cross-shaped configuration.
(b) **Ring-shaped:** The formation of two chiasmata results in a ring.
(c) **Interconnected loops:** Where three or more are present a series of interconnected loops.

However, not all the loops need necessarily be visible because consecutive one may be at right angles to each other like links in a chain. Moreover, the chromosomes continue to contract at this stage and this, coincident with the separation of homologous chromosomes, has the effect of pushing the chiasmata towards the end of the bivalent, a process called "*terminalization*".

- **Theories of Chiasmata**

Many speculations have been made for cause of breakage of chromatids:

1. *Two-plane theory*: advocates that chiasmata cause crossing over.
2. *One-plane theory*: claims that chiasmata are the consequence not the cause of crossing-over.

But the real cause behind the breakage and subsequent rejoining of chromatids is still not known.

- **Function of Chiasmata**

Chiasmata have two functions: one mechanical, the other genetic.

A. **Mechanical:** The mechanical function is to hold the two homologous chromosomes together while they manoeuvre themselves on to the spindle prior to segregation.
B. **Genetic function:** The chiasmata represent places where the chromatids may break and rejoin, a portion of one chromatid changing places with the equivalent portion of another. This enables exchange of genes to occur between homologous chromosomes, a process known as crossing-over. Since the two homologous chromosomes contain different genes, crossing-over promotes genetic variety, which in turn plays an important part in the process by which evolution takes place.

## Recombination

(Exchange of genes between chromosomes or directed incorporation of foreign genes into chromosomes)

- **Definition**

It is exchange of genes between chromosomes recombination is (1) a vital genetic process by which new combinations of existing genes may be generated and (2) ensures the accuracy of chromosome segregation during meiosis and (3) conservative from the standpoint of evolution.

- **Occurrence**

1. In the transformation of bacterial cells.
2. In viral transduction.

3. In the exchange or crossing-over of genes during meiosis of eukaryotic germ cells.

- **Description**

Recombination could be described in simple terms as *reciprocal exchange* between two homologous chromosomes occurring with equal probability at the junction between any two gene loci. It may be recalled the *chiasmata* are usually formed between homologous chromosomes after they have paired during meiosis. A chiasma in some cases may be proved to be a consequence of physical exchange or *crossing-over* between homologous chromosomes. This results in exchange of parts between homologous chromosomes. It may be pointed out here that homologous chromosomes are in some cases held together by means other than chiasmata and as a result crossing-over does not take place.

- **Source of Recombination Studies**

1. One main source is the close study of events when all the products of meiosis are recovered and analysed in detail. A few species of fungi offer such an opportunity.
2. Other source comes from the bacteria and the phage where mutations affecting recombination are traced back. Behind all such studies lie that each chromatid is a double-stranded structure, a molecule of DNA. Therefore, purely mechanical break and rejoin or even copy choice theories are not sufficient to explain the mechanism. Recombination must have to be studied in terms of a biological system in which molecular events are catalysed by enzymes.

- **Mechanism of recombination**

Many possible models for the mechanism of recombination have been proposed that employ enzymes such as:

1. Endonucleases
2. DNA polymerase I

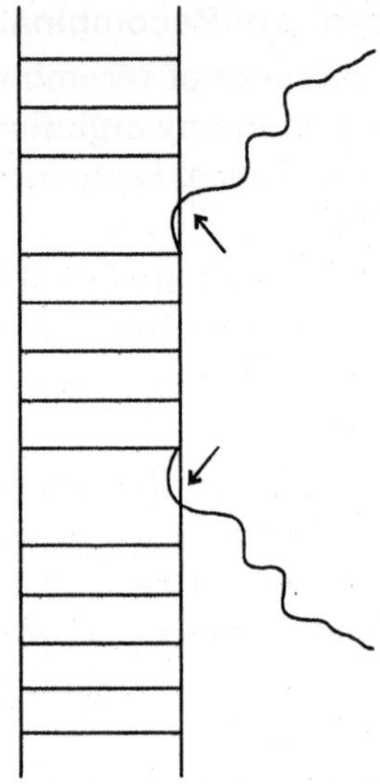

**Fig. 7.11:** Insertion of genes into duplex DNA.

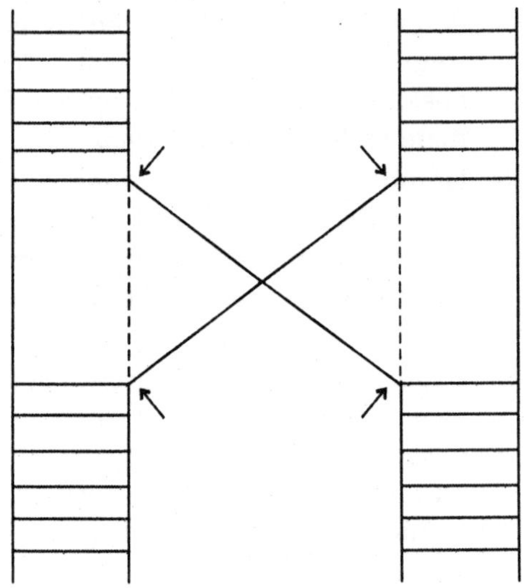

**Fig. 7.12:** Exchange of DNA segments.

3. DNA ligase

Two possible models of recombination is shown below:

I. Insertion of gene into duplex DNA (Fig. 7.11).

II. Exchange of DNA segments (Fig. 7.12). It occurs in 3 stages:
  (a) Endonuclease cleavage takes place at the points,
  (b) Filling of gaps by DNA polymerase I,
  (c) Sealing by DNA ligase.

- **Significance of Recombination**

1. The occurrence of recombination and segregation in any organism provides evidence of sexual reproduction in that organism.
2. Recombination between linked genes occurs when homologous chromosomes come together and pair intimately during meiosis.
3. Measurement of frequency of recombination is expressed as genetic map in which the successive loci remain ordered at distances approximately in proportion to the recombination between them.
4. Recombination is an orderly sequence of events conducted by enzymes, determined and regulated by genes. Regulation extends to restriction of possible recombination by a mechanism called "*Synaptinemal Complex*".
5. Homologous chromosomes or DNA molecules if subjected to appropriate biological environment will interact to produce recombinant version of themselves.
6. Experiments on transference of genes also raise hopes that gene deficiencies in the chromosomes of higher animals and man may one day be repaired by transplantation of "good" genes from another organism.

## "One-gene One-enzyme" or "One-gene One-protein" Hypothesis

- **Introduction**

Inborn error of metabolism was first recorded by Garrod in 1909 during study of albinism and alkaptonuria. He hinted at the functioning of genes and was able to prove that these traits behave genetically as recessive alleles and become expressive in persons with homozygous recessive genotype. From his studies Garrod postulated hypothesis of gene action which states that genes regulate and control the synthesis and action of enzymes. Garrod's views were not accepted in his time but present-day knowledge has supported the view that the genes regulate the synthesis and action of enzymes for cellular metabolic reactions.

- **One-gene One-enzyme Hypothesis**

The study of albinism provided the answer to the question—how the genes function to elaborate a phenotype? The last supported hypothesis is "*one-gene one-enzyme hypothesis*" which states that gene controls the synthesis and activity of enzymes for metabolic reactions.

In the elaboration of a particular phenotype, a number of metabolic reactions occur in stages. Each of these stages is regulated by a specific enzyme which is produced under the direction of specific genes. Thus when the gene becomes defective due to mutation or any other cause, the enzyme under its control also becomes defective. The phenotypic manifestations in albinism result out of a "*genetic block*" involving a step between the amino acid tyrosine and the pigment melanin. The reduction in enzyme activity affected by a mutation is called "*genetic block*".

There are three different ways by which genetically determined enzymic block may be produced. They are:

1. **Accumulation:** Production and accumulation of substances in excess just proximal to the block, e.g. alkaptonuria.
2. **Complete Block:** Complete absence of the product for enzyme action, e.g. albinism.
3. **Incomplete Block or Leaky Block or Partial Block:** The formation of altered enzyme that has some limited degree of activity. In such cases a particular product is produced in minor quantities for normal cellular metabolism, e.g. phenylketonuria or

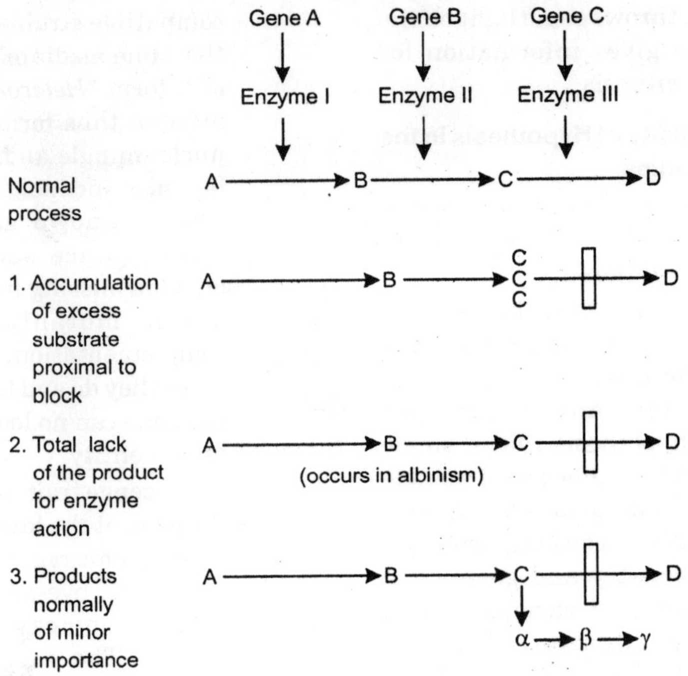


**Fig. 7.13:** Phenotypic abnormalities due to genetic block.

PKU where the defect is enzyme-involved in the conversion of phenylalanine to tyrosine (Fig. 7.13).

- **Evidence**

1. **Beadle and Tatum's experiment** on the "*Neurospora crassa*" fungus (see the protein biosynthesis).
2. **Garrod's study** in case of albinism and alkaptonuria.
3. **Study of sickle-cell anemia**
   (a) *Pauling and Itano* and others have done work on "sickle- cell trait"—a disease occurring among Africans. It is manifested by peculiar sickle or crescent shape of RBC, anemia, kidney damage, spleen enlargement, skin lesions and early death. In the sickle-cell trait, blood of the genotype is heterozygous having both normal $Hb^A$ and mutant $Hb^S$ alleles and the haemoglobin consists of a mixture of both kinds. The sickle-cell $Hb^S$ comes out of the solution when $O_2$ content is low and it behaves as positive ion in an electric field whereas normal $Hb^A$ behaves as negative ion.
   (b) Ingram studied the structural differences between normal and sickle-cell haemoglobin. He showed that normal Hb contains glutamic acid at certain position whereas a different amino acid valine was found in the same position in case of sickle-cell haemoglobin. Thus normal Hb is converted to an abnormal Hb by substitution of an amino acid which causes sickling of RBC It is known that amino acids form chain—the polypeptide and a protein molecule consists of few polypeptides. The change in a gene at the sickle-cell locus results in a very specific change in a protein by the substitution of one amino acid for the other. Thus this study is significant in correlating chemical changes with gene changes and

therefore throws some light on the way gene gives information for protein synthesis.

- **Position or Validity of Hypothesis in the Light of Recent Studies**

Various investigations are being carried out to find out the relations between protein differences and gene differences:

A. On the assumption that the inability to synthesise the amino acid was caused by the absence of a specific enzyme, Beadle and Tatum put forward the hypothesis that a single gene controls the production of a single enzyme, the *"one-gene one-enzyme hypothesis"*. Numerous experiments on the biochemistry and genetics of micro-organism and various human disorders, e.g. albinism, phenylketonuria known to be due to single genes have been shown to be caused by absence of one specific enzyme.

B. But from studies on the genetic control of complex molecules like haemoglobin, e.g. Hb (foetal and adult), Hb of sickle-cell anemia, we find that genes not only control the production of enzymes but also other proteins. Moreover a single protein may contain several polypeptides each controlled by a different gene. In view of this it would be more accurate to restate Beadle and Tatum's hypothesis as the *"one-gene one-protein hypothesis"* or even more accurately as the *"one-gene one-polypeptide hypothesis"*. The following findings have further modified the concept.

C. *One-cistron one-polypeptide hypothesis or one-cistron one-subunit concept*:

1. Neurospora is normally able to synthesize adenine but various mutants can grow only if adenine is added to the culture medium. If two of these "non-allelic" adenine requiring mutants (which are of compatible strains) are cultured in the same medium their hypae fuse and form *"Heterokaryons"*. In the mycelia thus formed two kinds of nuclei mingle and the new growth becomes independent on adenine. This is known as *"Complementation"* since each supplies an enzyme missing in the other. Most allelic mutants do not show complementation, but in a few rare cases they do and this suggests that the gene can no longer be regarded as one entity.

2. The lozenge trait in Drosophila that the gene at the lozenge locus on the chromosome regulates the eye-size, eye-facets, eye-pigment and fertility in homozygous recessive female. The several effects of a single gene on different traits known as *"Pleiotropism"*. Thus it is more likely that a gene consists of smaller subunits of function and the subunits have been termed *"cistron"*. It is likely that a cistron determines one polypeptide chain and a reformulation of the "one-gene one-enzyme" or "one-gene one-protein" hypothesis has been made. The present reformulated concept is *"one-cistron one-polypeptide hypothesis"*.

3. **Antibody or Immunoglobulins Diversity:** There are five major classes of immunoglobulin in human plasma (IgG, IgA, IgM, IgD, IgE) of which the IgG globulins are by far the most abundant and the best understood. The blood serum of a normal person contains a mixture of many different IgG immunoglobulins, all having the basic structure. Each individual type of immunoglobulin is specific for combining with a certain antigen. From such a mixture it is very difficult to

obtain a single molecular species of an immunoglobulin for analysis. Now question arises: how can such an immensely large number of different antibody molecules be genetically specified? Does each individual human being possess all the genes necessary to specify the million or more different antibodies for million of antigens which can theoretically be formed?

It has been found that each molecule of antibody possesses four polypeptide chains:

1. Two identical "heavy" (H) chains, having about 430 amino acid residues.
2. Two identical "light" (L) chains having 214 amino acid residues.

   These chains are linked by disulfide bonds into Y-shaped but flexible structure.

   Each immunogolobulin *light chain* is the product of at least *3 separate structural genes:*

   (a) A variable region ($V_L$) gene
   (b) A joining region (J) gene
   (c) A constant region ($C_L$) gene

Each heavy chain is the product of at least 4 different genes:

(a) A variable region ($V_H$) gene
(b) A joining region (J) gene
(c) A constant region ($C_H$) gene
(d) A diversity region (D) gene

Each person is capable of generating antibodies directed against perhaps 1 million different antigens. The generation of such immense antibody diversity appears to depend upon combined action of a small number of genes contributing to the formation of each immunoglobulin chain.

*Thus the "one-gene one-protein or one-enzyme" concept is invalid.* The cistron is the genetic unit coding for the structure of the subunit of a protein molecule, i.e. smallest unit of genetic expression. Thus the hypothesis can be regarded as a *"one-cistron, one-subunit concept"*.

## QUESTION

1. What is gene? Explain the theory of gene.
2. What do you know about interaction ot gene?
3. What do you mean by lethal genes? Name the lethal hereditary diseases in man.
4. What is mutation?
5. What are mutagens? Describe different types of mutagens and mode of action.
6. Describe the crossing-over of genes.
7. Describe the linkage of genes.
8. Describe briefly the formation of chiasmata.
9. Describe the process of recombination of genes.
10. Justify the statement one-gene one-enzyme hypothesis.

**PART-II**

# GENETICS

# 8

# GENETICS

1. Introduction, 2. Application of genetics for human, 3. Consanguinity, 4. Discontinuous and continuous variation.

## Introduction

An individual of a particular species inherits the characteristics of the species and resembles its parents in a general way. As for example, human being inherits highly developed cerebrum, vocal cords and the nervous coordination necessary for speech, arrangement of teeth, ability to maintain posture. Beside that, the offspring inherits certain characteristics peculiar to his parents but not common to species as a whole, e.g., eye and hair colour, facial appearance and blood group. The passage of characters from one generation to another is called *inheritance* or *heredity*. Heredity can be defined as the occurrence of qualities in the living organisms derived from their ancestors. Heredity includes:

1. *Physical qualities*,
2. *Physiological qualities*.

These qualities may be either:

1. *Expressed* or
2. *Latent*.

Though the individual belongs to the same species in major respects but there is variations among the members of the species in minor respects and then it is called *variations*. These may be:

1. *Phenotypic* or *Fluctuating variations*: Depends upon environmental factors, not heritable.
2. *Genotypic* or *Germinal variations*: Not influenced by the environment, handed down to the progeny from one generation to another. Since a new individual is derived from the gametes of the parents, the heriditary characters or information must lie within the gametes. And it has been proved that this information is present in the nucleus of the gamete and located on the chromosomes.

The branch of science which deals with the study of the method of inheritance of the resemblance and difference between the young and its ancestors is known as *Genetics*.

1. Genetics explains us the physical basis of heredity, distribution of genes during mitotic division and maturation of sex cells.

2. It helps us to understand the conception of 'carrier', i.e. hidden recessive defective gene.
3. It gives information regarding causation of disease by interaction between heredity and environment.
4. It helps to study of blood groups necessary for blood transfusion, medico-legal cases like disputed paternity. Rh factor in blood and its relation to fetal death and abnormalities and adoption of methods for its prevention.
5. Mutations of genes and its significance, factors causing mutation and its applied knowledge for production better crops and fruits and effects of ionising radiation.
6. It gives knowledge about genetics of cancer.
7. It gives knowledge of genesis of histocompatibility.
8. It helps to study inborn error of metabolism.
9. It helps in the diagnosis of genetic diseases.
10. It helps in the pedigree analysis.
11. It helps to study sex-linked genetic diseases.

## Application of Genetics for Human

12. **Eugenics:** It is the study of human genetics for the following:
    (a) *Encouraging* breeding for good characteristics for improvement of human shock,
    (b) *Eliminating* or reducing harmful characteristics.

    It seems that if all the congenitally (hereditary condition) insanes are sterilized or prevented from breeding by any means, then the numbers of congenital idiots would be greatly reduced or the condition completely eliminated. But this is not the case if we consider the condition of *recessive albinism.* Only the person having two recessive genes aa, will show these characteristics and whose number about one in 20,000 of the population. If there is marriage between the two albinos, then it will give rise to all albino children but it occurs unlikely. It has been shown that 1 in 70 of the population must be carrier for albinism having genetic constitution Aa. Such persons are indistinguishable from normal AA individuals but a marriage between two such people could produce some albino children aa. Out of all albinos, 99 per cent arise from these types of marriages. But there should be no sterilization or birth control until the first albino baby was born and the genetic constitution of the parents are revealed. It would take 22 generations for reduction of frequency of harmful gene to 0.1 per cent, if no homozygous recessive person, aa, were to breed. If every congenitally mental defective were sterilized, the incidence would fall to 8 per cent only. The elimination of a harmful dominant trait, Dd or DD, can be accomplished theoretically in one generation provided that the condition appears before reproductive age.

13. **Genetic Advice:** If sterilization or prevention of breeding, etc. are not adopted, genetic advice sometimes avoid unwanted breeding for which in many countries have *Genetic Counselling Services* to give information about inborn defects. Suppose, a person having congenital juvenile cataract caused by a dominant gene would be told that if he marries a normal person, half their children might inherit the disease; this can be observed by considering the $F_1$ offspring from the mating of Dd (affected) and dd (normal). An albino (aa) marrying a normal person who might be either AA or Aa would find that the chance of any one of his children being an albino is 1 : 40, while the normal brother of an albino, who might be a carrier, Aa, has

a chance of 1 : 420 begetting an albino child if he marries a normal person. The chances for other relations like aunts and uncles, can also be calculated.

Genetic advice would be accurate if heterozygous (affected) people is distinguished from homozygous (normal). As for example, Heterozygotes having the disease *Phenylketonuria* can be detected by determining the blood level of *Phenylalanine*.

I4. **Consanguinity:** It is the study of human genetics to predict the chances of recombination of two harmful genes in the children of marriages between first cousins. Suppose, two cousins are Martin and Rita. Martin is assumed to be heterozygous for a rare recessive gene, Nn, if one of his parents was nn. Martin could have inherited this gene from his grand parents A or B. There is 1 : 2 chance of presence of the gene from grand-parents B and in this case there is also a 1 : 4 chance that Rita has inherited the gene. There is thus a chance of $1:8\left(\frac{1}{2}\times\frac{1}{4}\right)$ that cousin Rita is also Nn, in which the chance of an effected child from their marriage is 1 : 4. The overall chances of an affected child are 1 : 32 $\left(\frac{1}{8}\times\frac{1}{4}=\frac{1}{32}\right)$ (Fig. 8.1). If the gene is fairly rare in the general population, e.g. occurs once in 100 individuals, the chances of Martin marrying an Nn person from the general population are 1 : 100. The overall affected child would be $\frac{1}{100}\times\frac{1}{4}$ or $1:400$.

For this, cousin marriages are discouraged and marriages between brother and sister are prohibited by law. Though by this means, there is no reduction of homozygous recessives number in population but definitely reduce the chances of occurring in a particular family (Fig. 8.1).

## Discontinuous and Continuous Variation

The individuals within a species are alike in all major respects which determines their species character but it may differ in many minor aspects from other individual of the same species. A goat may be black, white or brown, the size of its tail, ears may vary but it is still recognizable as a goat.

The variations may be:

A. *Discontinuous variation*

B. *Continuous variation.*

### A. Discontinuous Variation

1. **Coat Colour:** The variations of *coat colour* are the great examples of

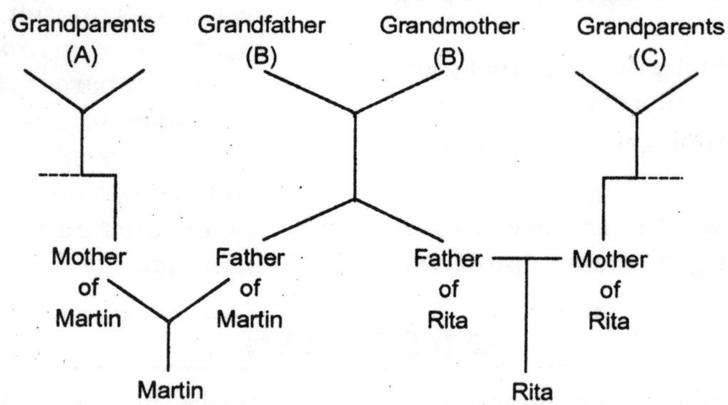


**Fig. 8.1:** Lineage of first cousins.

discontinuous variation because there is no intermediate of the coat colour. There is no possibility to arrange the animals in a continuous series of colours ranging from brown to black with almost imperceptible differences of colours between the members of the series.

2. **Sex Inheritance:** Inheritance of sex is also another example of discontinuous variation. There will be either male or female and there is no intermediate except few abnormalities.
3. **Blood Groups:** There are four major blood groups designated A, B, AB and O. A person must be one or other but no intermediate between A and O.
4. **Eye Colour:** Eye colour in white races is inherited in a discontinuous manner; one has blue eyes or pigmented eyes, but there are some individuals would be difficult to classify.
5. **Achondroplastic Dwarf:** Good examples of discontinuous variation occur among the more serious variants such as one is either achondroplastic dwarf or one is not but there is no intermediate.

- **Features of Discontinuous Variation**

1. The features are genetically determined and cannot be altered during lifetime of the individual, e.g. eye colour cannot be changed by changing the diet.
2. An achondroplastic dwarf cannot grow to full height by more diet.
3. An albino cannot get dark colour by exposure to sun rays.

These variations are under the control of genes. The dominant gene makes one achondroplastic dwarf and absence of one gene which makes pigment cause albinism.

**B. Continuous Variation**

1. It is difficult to classify the individuals according to height and weight. Because apart from tall and short, there are intermediate sizes differing from each other but they do not represent discontinuous variations for such little difference between individuals.
2. Though continuous variations are not genetically controlled but they are likely to be under the influence of several genes. Suppose, height might be controlled by 30 genes, each gene influences a few centimetres to the stature. If one person inherits all 30 genes would be tall whereas a person who inherits only 10 genes would be short. But it is known that height is at least partially genetically determined because tall parents have, on an average tall children and vice versa but how many genes are related is not known.
3. Continuous variations are also likely to be influenced by the environment. Though a person inherits the genes for tallness but due to lack of nutrition during the growth period, he will not grow tall as he might unless he gets proper nourishment. Most continuous variations are the result of interaction of the genotype and the environment. A person gets a disease or not when he exposes himself to a bacteria of the disease is an environmental factor but yet it is apparent that he may inherit resistance or susceptibility to a disease. A person with inherited susceptibility to an infectious disease should not expose himself to the same type of infection.

## QUESTION

1. Describe the application of genetics for human.
2. What do you mean by consanguinity?
3. Describe the discontinuous and continuous variations.

## PART-III

# MOLECULAR GENETICS

# 9

# NUCLEIC ACIDS—BIOCHEMISTRY OF GENETIC MATERIALS



## Introduction

The molecular genetics deals with the problems related to the structure and properties of nucleic acids up to protein biosynthesis. This includes the following:

1. The secondary and tertiary structures of nucleic acids.
2. Relation between structure of nucleic acids and biological properties.
3. The genetic code which determines the relationship between the nucleotide sequence in genetic DNA, RNA and in messenger RNA (mRNA) and the amino acids sequence to the protein chain.
4. The genetic code which arises in the atomic molecular treatment of the processes of protein synthesis is physical in nature and the problem performing to it has been solved.
5. Regarding the physical meaning and origin of the genetic code, the code is not accidental and originated in the course of prebiologic evolution. Physics has been applied to disclose the internal relations of the code.
6. Regarding the conformation properties of nucleic acids, the behaviour of nucleic acids, like that of proteins, is monitored by electronic conformational interactions.
7. Regarding internucleic and nucleic proteins molecular recognition, the nucleic acid-protein interaction has greatest significance for the regulation of protein biosynthesis and of the action of the genes.

## Nucleic Acids

- **Definition**

Nucleic acids are the non-protein nitrogenous substances and polynucleotides, i.e. it contains several nucleotides or polymers of mononucleotides.

- **Occurrence**

A. Animal: Nucleoli, chromatin, cytoplasm, ribosomes, mitochondria.

B. Microbiological: Bacteria (Staphylococcus, bacteriophage and disease-producing viruses of plants and animals.

C. Plant: Chloroplast.

- **Types**

1. DNA or Deoxyribonucleic acid
2. RNA or Ribonucleic acid.

- **Distribution of Nucleic Acids in the Cell**

By chemical test and histochemical staining, e.g. Fuelgen reaction, it has been shown that *DNA* is confined to the *nucleus* and *in the genes* of chromosomes and *RNA* is found mainly in the *cytoplasm* and very minute amount may be present in the nucleus.

- **Constituents of Nucleic Acids**

1. DNA and RNA are long chain molecules like protein and the chains are much longer than those of proteins and the building units are more complicated than amino acids.
2. Nucleic acids are non-protein nitrogenous substances.
3. They are the *polynucleotides,* i.e. polymers of *nucleotides.* Nucleotide is a *monomeric unit,* i.e. *mononucleotide.*
4. A *nucleotide* is composed of sugar, a pentose (Ribose or Deoxyribose), a nitrogenous base (Purine-Adenine or Guanine or Pyrimidines—cytosine, thymine or uracil) + phosphoric acid.
5. The number of nucleotides in nucleic acids molecule varies from 74 in tRNA to about 1600–9000 in DNA.
6. Nucleic acids often remain noncovalently bound to basic proteins like protamines or histones by ionic bonds to form nucleoproteins; basic polyamines like spermine and spermidine also remain combined with nucleic acids.

**Nucleic Acids (Polynucleotide)**

↓

**DNA**
(Deoxyribonucleic acid, a polynucleotide having several monomeric unit, i.e. mononucleotide)

↓

Mononucleotide

If contains the following:

1. One *Purine base* or nitrogenous *base*: Adenine or Guanine or one *pyrimidine base*: Cytosine or thymine or uracil
2. One *Pentose* sugar: Deoxyribose
3. One Phosphoric acid

The Mononucleotides of DNA are:

A. Deoxyadenylate = Adenine + Deoxyribose + Phosphoric acid
B. Deoxyguanylate = Guanine + Deoxyribose + Phosphoric acid
C. Deoxycytidylate = Cytosine + Deoxyribose + Phosphoric acid
D. Deoxythymidylate = Thymine + Deoxyribose + Phosphoric acid

**RNA**
(Ribonucleic acid, a polynucleotide having several monomeric unit, i.e. mononucleotide)

↓

Mononucleotide

It contains the following:

1. Purine or Pyrimidine bases: like DNA it contains all purines and pyrimidines *except thymine* which is *replaced by uracil*
2. Pentose sugar: Ribose
3. Phosphoric Acid

The mononucleotides of RNA are:

A. Riboadenylate = Adenine + Ribose + Phosphoric acid
B. Riboguanylate = Guanine + Ribose + Phosphoric acid
C. Ribocytidylate: Cytosine + Ribose + Phosphoric acid
D. Ribouracidylate = Uracil + Ribose + Phosphoric acid

**Fig. I:** Types of nucleotides.

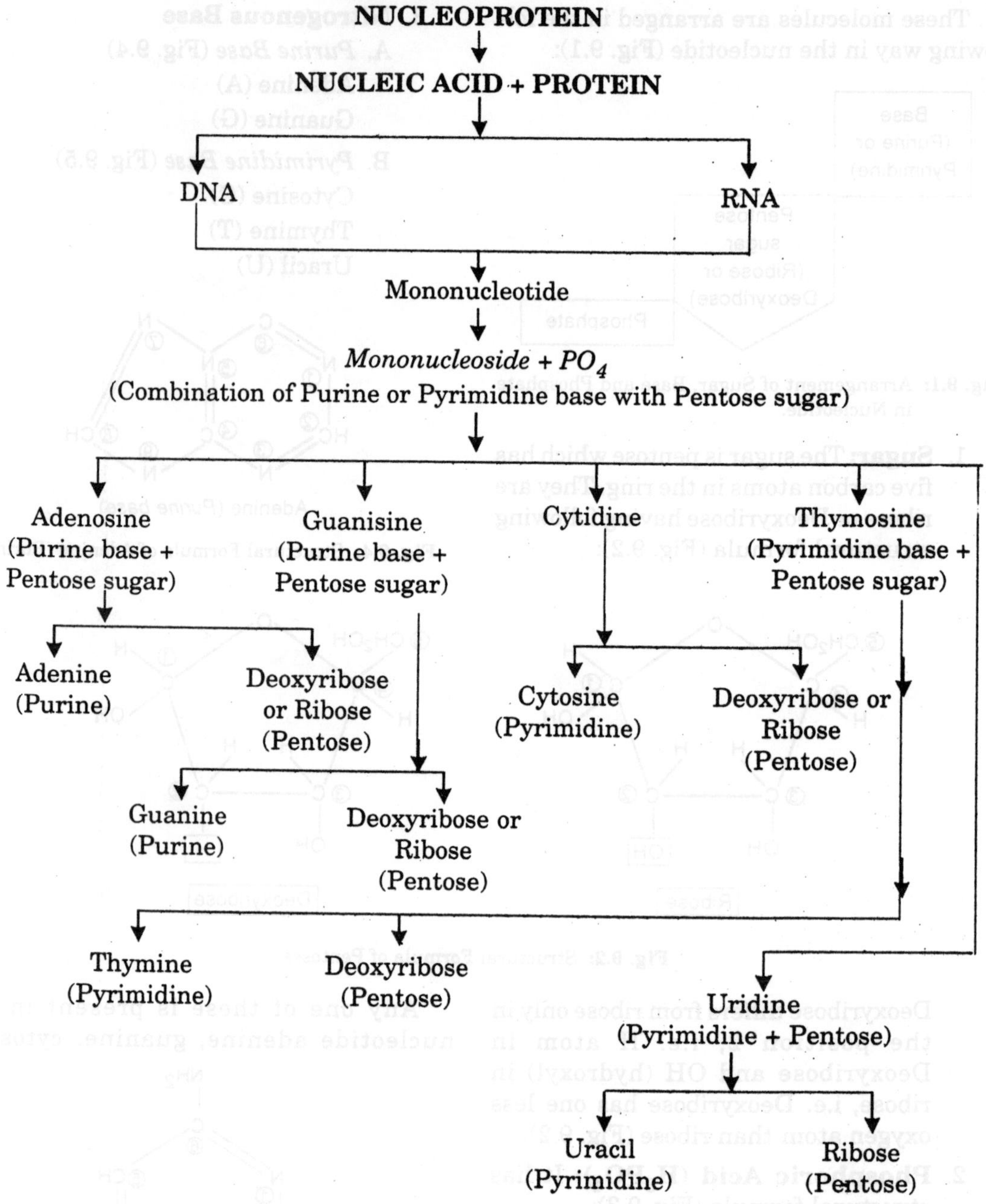


**Fig. II:** Breakdown of nucleoprotein.

## The Chemical Structure of Nucleic Acids

The building unit of a nucleic acid is called *nucleotide*. A nucleic acid consists of several nucleotides in a long chain, i.e. nucleic acid is a polynucleotide which is a polymer and whose monomer is nucleotide. A nucleotide consists of three components (3P) linked together such as:

P : (1) Pentose (5-Carbon sugar)—Ribose or Deoxyribose

P : (2) Phosphoric Acid

P : (3) Purine or Pyrimidine bases

These molecules are arranged in the following way in the nucleotide (Fig. 9.1):

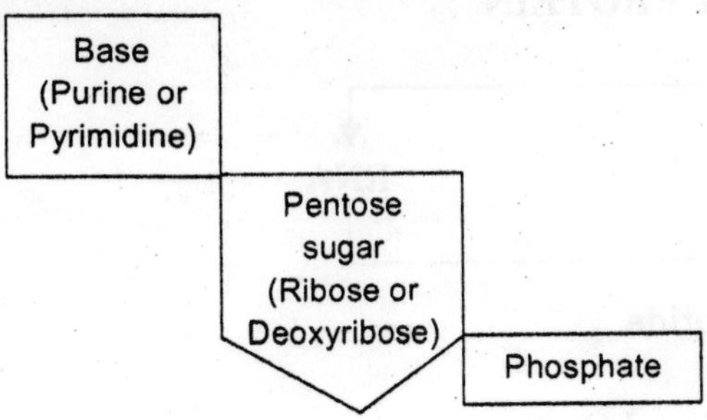


**Fig. 9.1:** Arrangement of Sugar, Base and Phosphate in Nucleotide.

1. **Sugar:** The sugar is pentose which has five carbon atoms in the ring. They are ribose or Deoxyribose having following structural formula (Fig. 9.2):

3. **Nitrogenous Base**
   A. *Purine Base* (Fig. 9.4)
      Adenine (A)
      Guanine (G)
   B. *Pyrimidine Base* (Fig. 9.5)
      Cytosine (C)
      Thymine (T)
      Uracil (U)

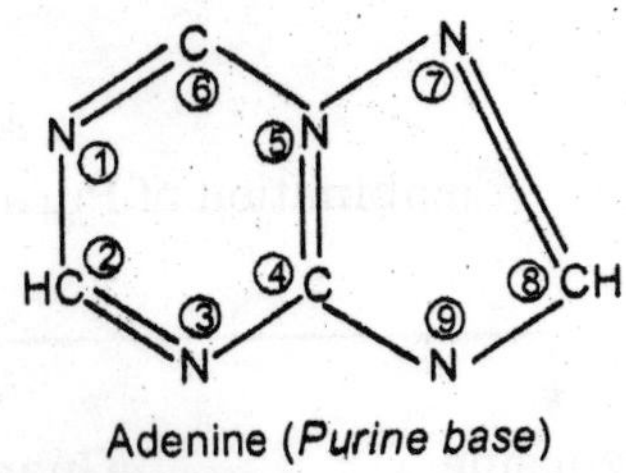


**Fig. 9.4:** Structural Formula of Adenine (Purine).

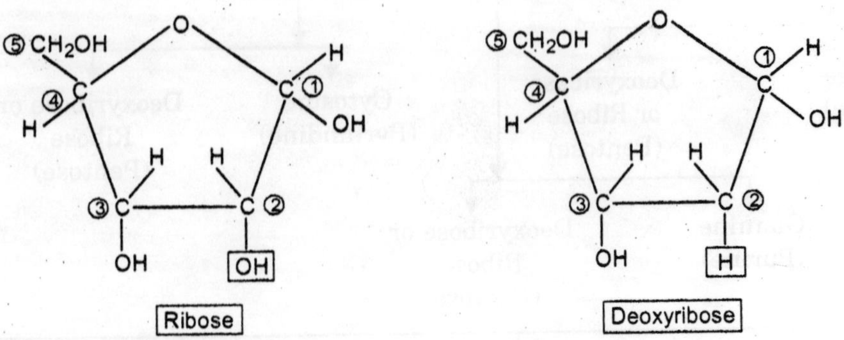


**Fig. 9.2:** Structural Formula of Pentoses.

Deoxyribose differs from ribose only in the position 2, i.e. H atom in Deoxyribose and OH (hydroxyl) in ribose, i.e. Deoxyribose has one less oxygen atom than ribose (Fig. 9.2).

2. **Phosphoric Acid ($H_3PO_4$):** It has structural formula (Fig. 9.3):

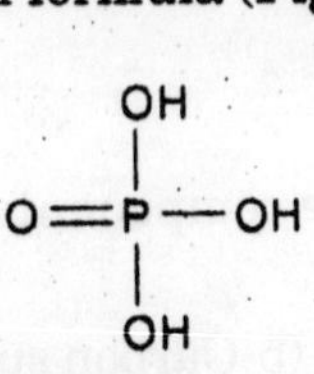

**Fig. 9.3:** Structural Formula of Phosphoric Acid.

Any one of these is present in one nucleotide adenine, guanine, cytosine,

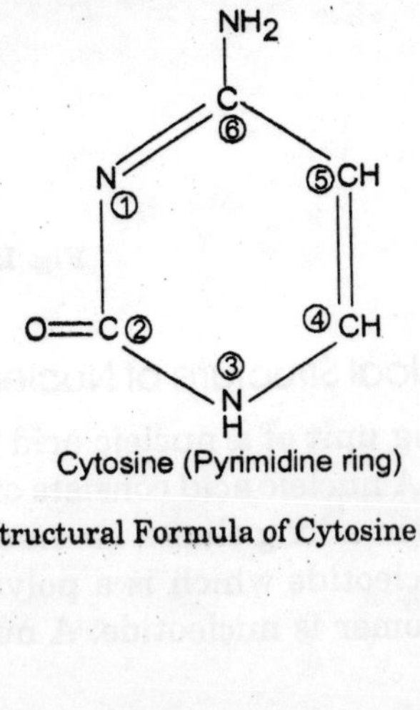


**Fig. 9.5:** Structural Formula of Cytosine (Pyrimidine).

thymine are present in nucleotides of DNA and adenine, guanine, cytosine and uracil (in place of thymine) in RNA. They are abbreviated as A, G, C, T, U, etc.

*Nucleoside:* For formation of nucleoside, sugar attaches position I to the position 9 of purine base and position 3 of pyrimidine base (Fig. 9.6). For formation of nucleoside one mol. of water is removed for attachment of sugar and base (Fig. 9.6).

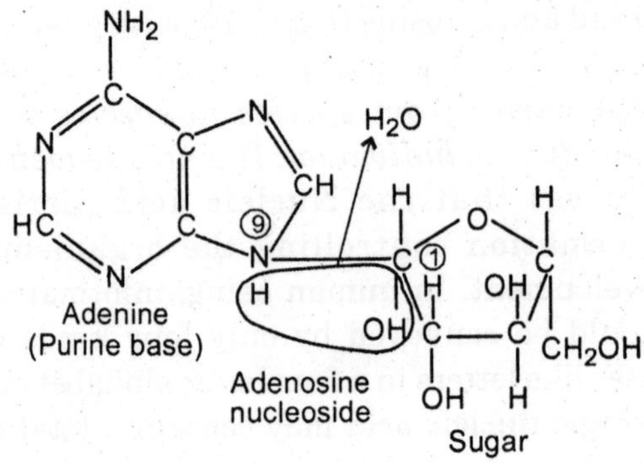


**Fig. 9.6:** Structural Formula of Nucleoside.

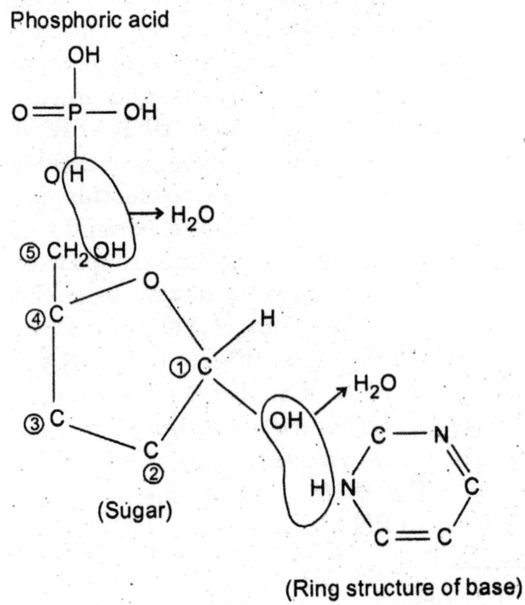


**Fig. 9.7:** Single Nucleotide.

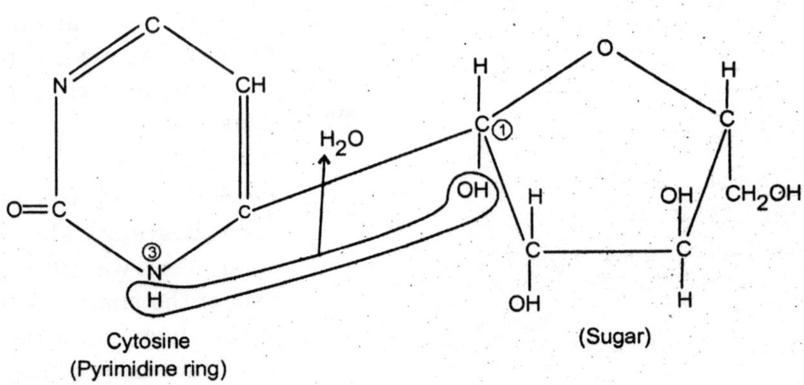


**Fig. 9.8:** Structural Formula of Cytidine Nucleoside.

## Nucleotides (Fig. 9.7)

Now the sugar of nucleoside at position 5 attaches to the phosphoric acid by elimination one mol. of water. So, formation of single nucleotide. 2 mols of water are eliminated and it is a process of condensation. Sugar of one nucleotide attaches with $H_3PO_4$ of another nucleotide by a diester bond and eliminate one mol of water to form dinucleotide. It is also condensation process and catalyzed by the enzyme "DNA or RNA Polymerase". Addition of further nucleotides produces a long chain of polynucleotide, i.e. nucleic acid shose backbone consists of alternating sugar and phosphate groups with the bases projecting out sideway from the sugars. The *sugar* and *$H_3PO_4$ maintain uniformity in the chain* (Fig. 9.9). In DNA it is Deoxyribose and in RNA it is Ribose. However the *bases are not uniform.* In case of *DNA,* it is *double-stranded helical structure* (Fig. 9.10) In DNA, the bases are A, G, C, T. In RNA, there are A, G, C, U. The

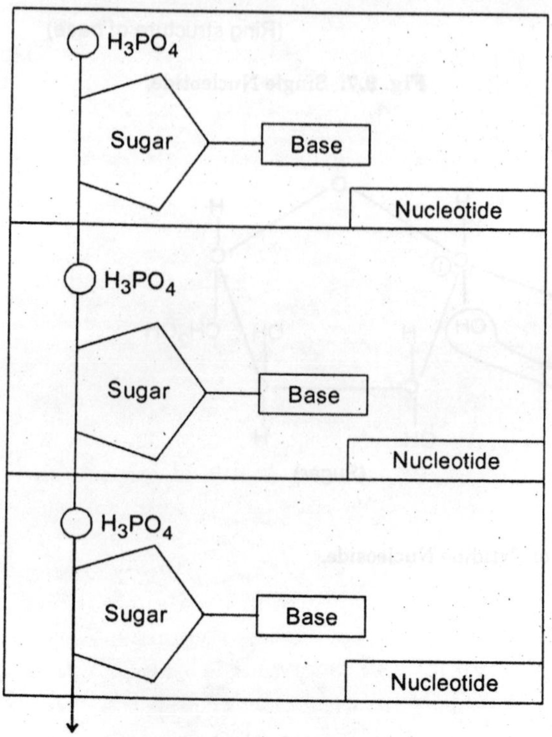


**Fig. 9.9:** Nucleic Acid Chain.

double strand make linkage between the two, through the bases by loose hydrogen bonds and the strands being held together by hydrophobic or van der Waals forces. Purine base of one strand, i.e. Adenine always links with pyrimidine base i.e., thymine of opposite strand and cytosine (Fig. 9.10) with guanine (A = T, C = G). This *sequence* of bases in the *DNA varies from species to species* and *individual to individual.* It is this sequence of bases that the Nucleic acid carries information controlling the organism's development. In human being, information should be conveyed by only four kinds of bases like letters in a four-letter alphabet but a single nucleic acid may contain a total of five million nucleotides, the number of possible combinations is almost infinite. From these four letters, an infinite number of words may be constructed which is theoretically may be meaningless but may convey meaningful information in the cell.

## Molecular Structure of DNA

*The Watson-Crick Hypothesis for DNA Structure (Model of DNA)*

1. *Chargaff* (1950) had noted that in the DNA series the amount of guanine in gram moles always equalled that of cytosine and that of gram molar quantity of adenine always equalled that of thymine.
2. *Pauling and Corey* (1953) by X-ray crystallography showed that nitrogenous bases were at right angles to the axis of the chain and that the planes of the sugar molecules and the bases were mutually at right angles. They postulated a *helical structure* for the DNA molecule which was incorrect.
3. *Mauric Wilkins* had tried to establish DNA structure by X-ray diffraction analysis despite technical difficulties, such DNA's reluctance to crystallize. Though clear photograph was obtained but interpretation was not easy.

4. Above findings led *Watson and Crick* (1953) to put forward a hypothesis for DNA model structure. This was a *turning point* in our *concept of gene.* The work was based on X-ray patterns and came to conclusion the DNA molecule consists of *two parallel chains twisted* to form a *double helix,* each of which were of equal pitch and subtended an angle of 120 degrees at the axis. The two strands of double helical molecule, each of which possesses a polarity, *i.e. "antiparallel* means two helices run in opposite directions differing in the direction of their sugar-phosphate linkages. Thus, if one chain has the linkage from the 3′ position of a sugar to the 5′ position of the next in a given direction, the other will have the linkage 5′ to 3′ in that direction. This is analogous to 2 parallel streets, each running one way but carrying traffic in opposite direction. In the double-stranded DNA molecules, since the *information resides* in the sequence of nucleotides on one strand, which is called *"sense" strand,* the opposite strand might be considered as *"antisense",* i.e. the complement of the *"sense"* strand (Fig. 9.10, 9.11, 9.12).

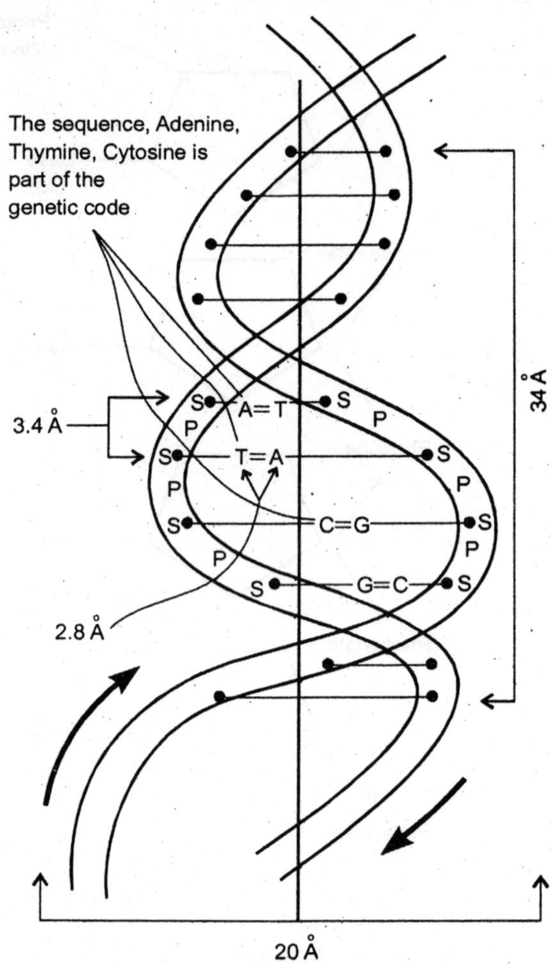


**Fig. 9.10:** Watson-Crick Model for DNA.

## Forms of DNA

There are 3 forms of DNA:

1. **A-DNA:** Double-helical DNA contains about 11 residues per turn. The planes of the base pairs in this right—handed helix are tilted 20 degrees away from the perpendicular to the helix axis. It is formed by the dehydration of B-DNA.
2. **B-DNA:** The Watson-Crick double helix contains about 10 residues per turn. The planes of the base pairs in this right-handed helix are perpendicular to the helix axis.
3. **Z-DNA:** A left-handed double helical form of DNA contains about 12 residues per turn.

- **Physical Properties of DNA Solution**

1. **Acid-Base Properties:** The recurring phosphate groups of DNA making bridges between the mononucleotides are fully ionized at any pH above 4. DNA is thus strongly acidic. They strongly bind divalent cations $Mg^{2+}$ and $Ca^{2+}$ and polyamines "spermine" and "spermidine". The binding of polyamines in the groove of double-halical DNA, stabilizes the DNA molecule and makes it more flexible. The stability of the hydrogen-bonded

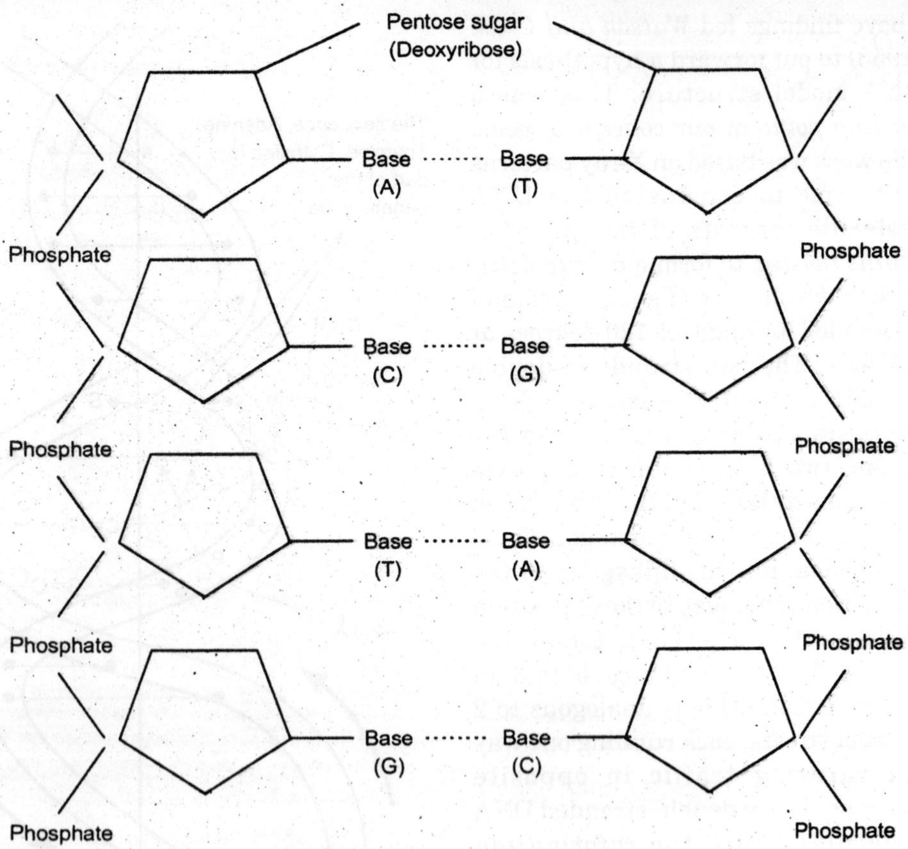


**Fig. 9.11:** Part of DNA Molecule.

base pairs of double-helical DNA is a function of pH and maximally stable between pH4 and 11.0 and outside these, DNA becomes unstable and unwinds.

2. **Viscosity:** Because of the rigidity of the double helix and the immense length of DNA in relation to its small diameter, even very dilute solution of DNA are highly viscous. Viscosity measurement is used to follow the course of unwinding a denaturation of DNA. When they diffuse, they sweep with these large volume of solution and for this DNA shows ideal behaviour as a solute only in extremely dilute solutions.
3. **Sedimentation Behaviour:** Equilibrium sedimentation in Csl (Cesium Chloride) gradients is widely used to determine the "buoyant density". Single-strand DNA is denser than double-strand DNA which in turn denser than protein. RNA can be distinguished from DNA since it is denser than either single or double strand.
4. **Denaturation:** Solution of double-helical DNA shows significant changes in physical properties when subjected to:
   (a) Extreme of pH
   (b) Heat
   (c) Decrease in dielectric constant of the aqueous medium by alcohols, ketones, etc.
   (d) Exposure to urea, amides and similar solutes.

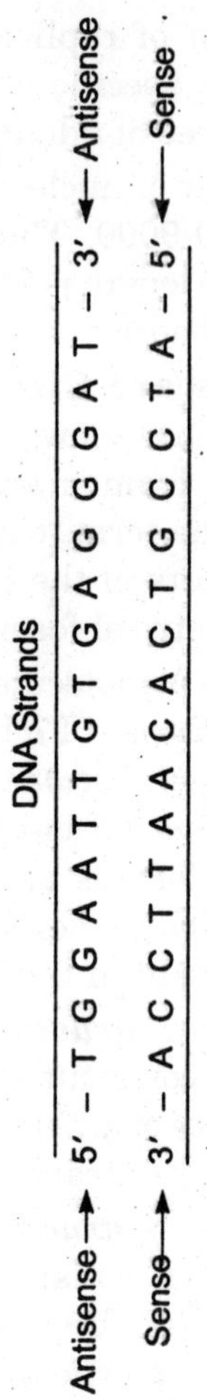


**Fig. 9.12:** DNA Strands show sense and antisense strands.

The changes of physical properties are:

(a) The viscosity of DNA solutions decreases.

(b) The light absorption at 260 nm increases.

(c) The optical rotation becomes more negative.

(d) Buoyant density increases.

These physical changes in DNA is called "*denaturation*".

The double strands of DNA are held together by:

(a) Hydrogen bonding between base pairs.

(b) van der Waals force or stacking interactions between successive bases.

When one or both forces are interrupted, the double-helical structure undergoes transition into a randomly loop form, denoted as "*single-stranded*" or "*denatured DNA*".

**Stages of Denaturation**

There are two stages:

(a) The two strands are partially unwound but remain united by at least a short segment of double-helical structure, with some complementary bases still in register. The unwound segment of each strand assumes a random, changing conformation.

(b) The two strands become completely separated from each other. When quickly cooled at this stage, each free strand doubles back on itself to form short, imperfect intrachain double-helical segment.

5. **Hyperchromic Effect:** Denaturation of DNA is measured by the increase in light absorption at 260 nm, termed as "*hyperchromic effect*".

   The purines, pyrimidines and their corresponding nucleotides absorb ultraviolet light at 260 nm. The native double-strand DNA absorbs much less light at 260 nm and denatured DNA absorbs more light at 260 nm.

6. **Melting Point:** Native DNA is denatured within a very small increment of temperature and this thermal denaturation often designated as melting point. DNA specimens from different cell types have different melting points defined as $T_m$, i.e. the temperature at the midpoint of the melting curve.

7. **Molecular Weight:** It has a molecular weight of 5 to 10 million.

## Important Features of DNA

1. DNA is found in the chromosomes in association with noncovalently bound to basic protein which is either protamines or histones to form nucleoproteins by ionic bonds from which it may be separated by chemical methods.
   The functions of basic protein histone attached to DNA may be the following:
   (A) Serves to *fold* and *compact* the DNA strands into limited sizes.
   (B) Protective or structural function:
      (a) Preventing alteration or
      (b) Denaturation of the DNA,
      (c) Controlling coiling.
   (C) Repressor Function:
      (a) Interfering with the template activity of DNA.
2. It has a molecular weight of 5-10 million.
3. It is a polynucleotide made up of long chains of mononucleotides, i.e. monomeric units of DNA. The four types of monomeric units or mononucleotides are:
   (a) Deoxyadenylate = Deoxyribose + Adenine + $PO_4$
   (b) Deoxyguanylate = Deoxyribose + Cytosine + $PO_4$
   (c) Deoxycytidylate + Deoxyribose + Cytosine + $PO_4$
   (d) Deoxythymidylate + Deoxyribose + Thymine + $PO_4$
4. The monomeric units, i.e. mononucleotides form a *single strand of DNA* which are held in polymeric form by 3′, 5′-phosphodiester bonds.
5. The informational content of DNA is available in the sequence in which these monomers are arranged.
6. The polymer possesses a polarity; one end has a 5′-hydroxyl or phosphate terminal while the other has a 3′-phosphate or hydroxyl moiety.
7. The genetic information is available in the monomeric units and there is mechanism of replicating or reproducing this specific information with a high degree of fidelity.
8. The number of nucleotides in DNA is about 1600-9000. The molecules are long and its length is 250 times greater than its breadth.

8A. **Molecules are Circular:** Electron Microscopy has shown that intact DNA molecules from many sources are circular. The term "circular" refers to the continuity of the DNA chain, not to its geometrical form.
   Not all DNA molecules are circular, some are *linear* (T7 bacteriophage), *interconvert between linear and circular forms* (virus-λ bacteriophage). The linear form is present inside the virus particle, whereas the circular form is present in the host cell.

9. The *primary structure* of the DNA molecule is constituted by the number and sequence of different deoxyribonucleotides in its strands.

10. The *secondary structure* of *B form* of nuclear DNA consists of a *double-stranded helix,* formed by the coiling of two linear polydeoxyribonucleotide strands around an identical central exis (Watson-Crick model). This B. form is believed to be *predominant form in the living cells.*

10A. The diameter of the helix is 20 Å. Adjacent bases are separated by 3.4 Å along the helix axis and related by a rotation of 36 degrees. Hence, the helical structure repeats after ten residues on each chain; that is, at intervals of 34 Å.

11. The two strands of the double helix are held together by *interchain hydrogen bonds* between purine and pyrimidine bases. While each strand is compactly held by *intrachain hydrophobic forces* between the purine and pyrimidine bases of the same strand. Thus, the two chains of DNA in the double helix

of DNA and double-helical portions of RNA is determined by weak interactions—hydrogen bonds, electrostatic and dispersion forces.

11A. The purine and pyrimidine bases are on the inside of the helix, whereas the phosphate and deoxyribose units are on the outside. The planes of the bases are perpendicular to the helix axis. The planes of the sugars are nearly at right angles to those of the bases.

12. The interchain hydrogen bonds bind a purine or pyrimidine base of one strand with a pyrimidine or purine of the other strand—according to the *base-pairing rule*. Each adenine (Purine) residue of one strand is bound to a thymine (Pyrimidine) residue of the other by *two hydrogen bonds* while each guanine (Purine) of one strand is bound to a cytosine (Pyrimidine) or methyl cytosine of the other strand by *three hydrogen bonds* because of steric and hydrogen-bonding factors.

13. The consecutive interconnected base-pairs lie in parallel planes, a distance of 0.34 nm separating each base-pairs from the next. There are ten base-pairs in each turn of a helix and the pitch of the helix is 3.4 nm. The base-pairs are always purine-pyrimidine pairs; but not the purine-purine and pyrimidine-pyrimidine pairs in the double helix as their dimensions cannot fit into the central core.

13A. In DNA, the concentration of adenosine (A) nucleotides equals that of thymidine (T) nucleotides, i.e. A = T. Similarly, concentration of guanosine (G) nucleotides equals that of cytidine (C) nucleotides, i.e. G = C.

14. The two strands of the helix form *right-handed coils* but they are *antiparallel* to each other, i.e. one strand runs in the 5′ to 3′ direction and other in the 3′ to 5′ direction.

15. Since the information resides in the sequence of nucleotides on one strand, the opposite strand is considered as: *"antisense"* i.e. the complements of the *"sense"* strand.

16. The B form has a pitch of 3.4 nm per turn and within the single turn 10 base-pairs exist.

17. The phosphodiester backbone of the strand remains towards the outer side of the coil whose outer surface is formed of the pentose-phosphate residues of the nucleotides. The pentose rings lie parallel to the axis of helix.

18. The diameter of the double helix is 1.9 nm and that of its shallow grooves measures 1.2 nm. The distance of two pentose residues of each base-pair is about 1.1 nm.

19. The sequence of these bases is the important factor in heredity.

19A. The purine and pyrimidine bases of DNA carry genetic information. Whereas the sugar and phosphate groups perform a structural role.

20. The different sequences of bases *down the length* of the DNA molecule seem to act like a code, instructing the cell to make certain proteins the order of bases indicating the sequence of amino acids to be joined up in order to make the protein.

## 21. Z form of DNA

(a) Apart from B form, there is Z form of DNA occurring in short stretches in Drosophila chromosomes and human DNA.

(b) It consists of a double helix of two linear polydeoxyribonucleotide strands each with *alternating* purine and pyrimidine deoxyribonucleotide residues.

(c) Each strand forms a *left-handed coil* around a common central axis.

(d) Each turn of the helix carries *twelve base-pairs* and pentose phosphate backbone follows a zigzag path along the helix instead of the regular smoothly curved spiral path in B form.

(e) Z form seems to regulate the coding activities of much longer stretches of B form.

(f) The helix has a diameter of 1.8 nm and exists only at high ionic concentration.

22. A, C, D and E forms of DNA double helices may also exist, but carry 11, $9\frac{1}{8}$, 8 and $7\frac{1}{2}$ base-pairs perturn respectively. A and C double helices occur at relative humidities of respectively 75% and 66%, and have diameters of 2.3 nm and 1.9 nm respectively.

double-stranded, endless or closed-loop or circular molecule, i.e. monomeric circular DNA which may either:

1. Interlock with another molecule, i.e. dimeric circular DNA or
2. Supercoiled tertiary structure (Fig. 9.13).

## 24. Viral DNA

Some DNA viruses like *E.coli* bacteriophages $\phi \times 174$, fd and MS2 possess *single-stranded circular DNA*. $T_4$ DNA virus, adenovirus and small-pox virus carry DNA double helices.

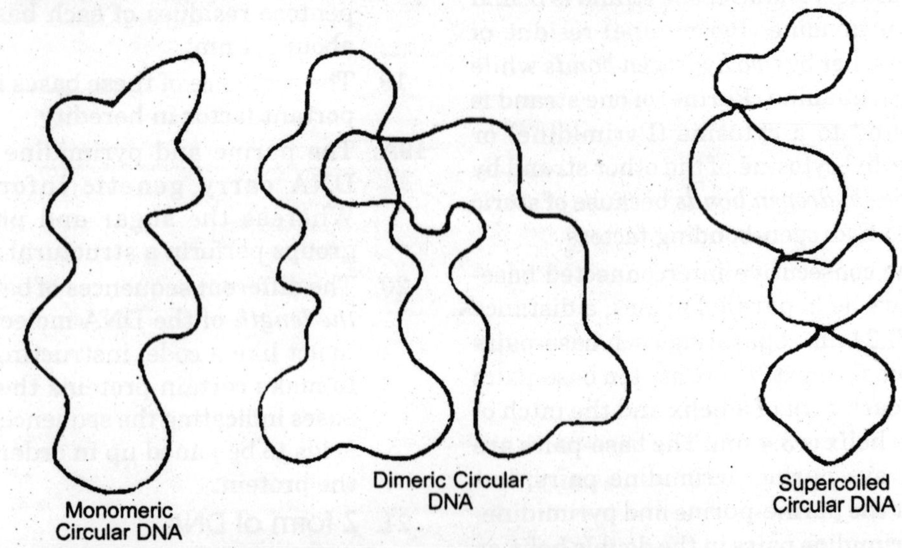


**Fig. 9.13:** Different forms of mitochondrial DNA.

## 23. Mitochondrial DNA

(a) There are fewer and smaller DNA in the mitochondria than the nucleus.

(b) Mitochondrial DNA differs from the nuclear DNA in the following respects:

i. In the primary and secondary structures.

ii. Its individual strands have a primary structure made of only about $10^4$ nucleotides while the secondary structure consists of

## 25. Prokaryotic Bacterial DNA

Prokaryotic bacteria like *E. coli* carry a single chromosome containing a single circular DNA double helix formed of about $4 \times 10^6$ base-pairs.

## Types of DNA Sequences

The kinetics of *DNA reassociation studies* have shown the following types of sequences. 60-70% of human genome consists of single or low copy number of DNA sequences.

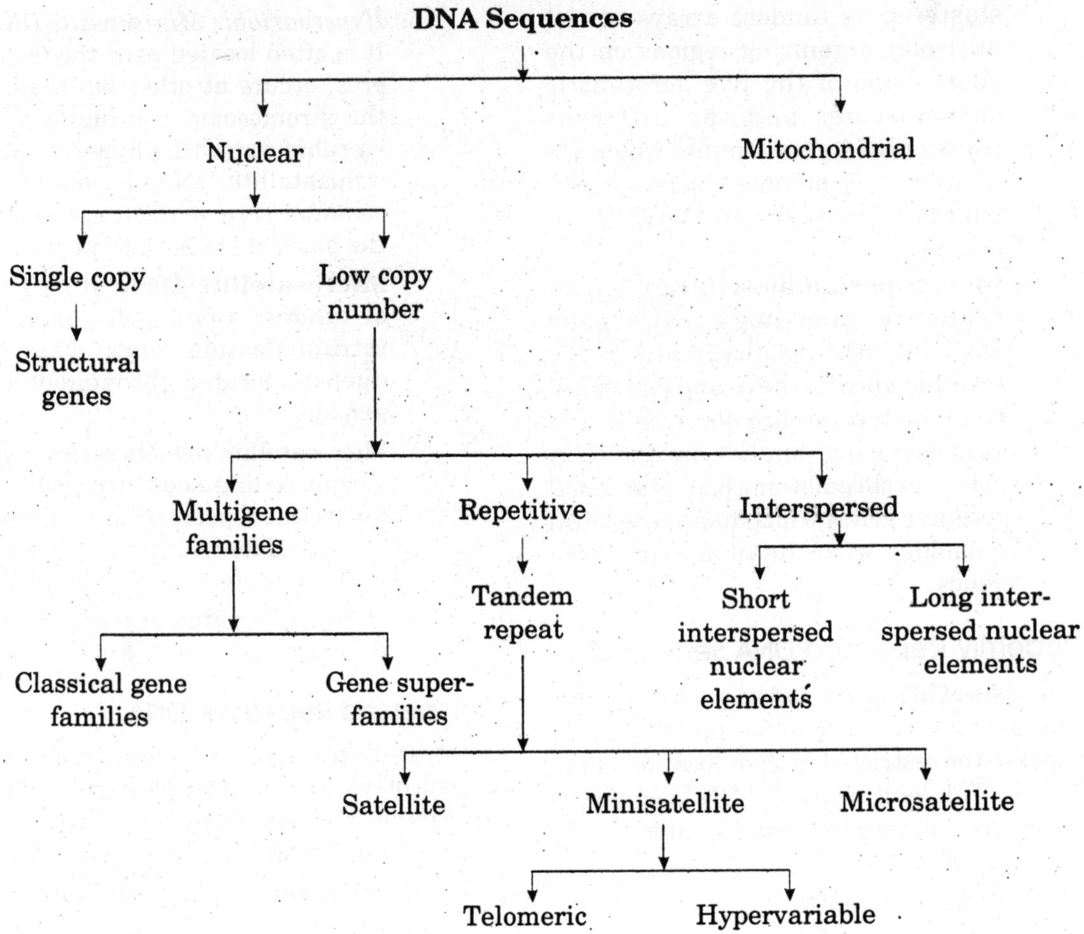


The remainder of the genome, 30–40% consists of either moderately or highly repetitive DNA sequences which are not transcribed. This later portion consists of mainly satellite DNA and interspersed DNA sequences.

## Nuclear Genes

1. There are about 100,000 genes in the nuclear genome code for specific proteins in human. They are distributed between different chromosomes and certain chromosomal regions, with heterochromatic and centromaric regions having few but majority being located in subtelomeric regions.
2. Many human genes are single copy genes coding for polypeptides for cellular functions, e.g. enzymes, hormones, receptors, structural and regulatory proteins.
3. Many genes have similar functions originated through gene duplication events with subsequent evolutionary divergence. Some form clusters and others are dispersed throughout the genome constituting multigene families.

## Multigene Families

Multigene families are of two types:

(a) **Classical gene families:** Having a high degree of sequence homology, classical gene families include the numerous copies of genes coding for the various ribosomal RNAs which are clustered as tandem arrays at the nucleolar organizing regions on the short arms of the five acrocentric chromosomes and the different transfer RNA gene families which are located in numerous clusters interspersed throughout the human genome.

(b) **Gene superfamilies:** Having limited sequence homology multigene families can be physically close together such as the α- and β-globulin gene clusters on chromosomes 16 and 11. Gene superfamilies are the HLA genes on chromosome 6 and the T cell receptor genes which have structural homology with the immunoglobulin genes.

## Tandemly Repeated DNA Sequences

It consists of blocks of tandem repeats of non-coding DNA which can be either highly dispersed or restricted in their location in the genome. Tandemly repeated DNA sequences can be divided into three sub-groups:

(a) Satellite DNA,
(b) Minisatellite DNA,
(c) Microsatellite DNA.

(a) **Satellite DNA:** About 10-15 per cent of the repetitive DNA sequences is satellite DNA in human genome. It consists of very long series of short tandemly repeated DNA sequences which are clustered around the centromeres of certain chromosomes.

(b) **Minisatellite DNA:** It consits of:

i. *Telomeric DNA*: These sequences are necessary for chromosomal integrity in replication and are added to the chromosome by a specialized enzyme known as telomerase. The terminal portion of the telemeres of the chromosomes contain 15 kb of tandem repeats of a 6 bp DNA sequence, known as telomeric DNA.

ii. *Hypervariable Minisatellite DNA*: It is often located near the telomeres, occurs at other locations in the chromosome. It is highly polymorphic consisting hypervariable minisatellite DNA. It contains a common core sequence, which is the basis of DNA fingerprinting.

(c) **Microsatellite DNA:** It consists of tandem single-, di-, tri- and tetranucleotide repeat base sequences located throughout the genome.

Microsatellite repeats rarely occur within coding sequences but trinucleotide repeats in or near genes are associated with certain inherited disorders.

DNA microsatellites are highly polymorphic.

## Interspersed Repetitive DNA

About 10–15 per cent of the human genome consists of two main classes of short and long repetitive DNA sequences which are interspersed throughout the genome.

i. **Short interspersed nuclear elements:** About 5 per cent of human genome consists of 750,000 copies of short interspersed nuclear elements. The most common are DNA sequences of approximately 300 bp.

ii. **Long interspersed nuclear elements:** 5 per cent of the DNA of the human genome consists of long interspersed nuclear elements. The type I consists of 50,000–100,000 copies of a slightly longer DNA sequence of approximately 6000 bp.

## Probe or Nucleic Acid Probe

Nucleic acid probes are *single-stranded DNA or RNA sequences* which have been radioactively or non-radioactively labelled and can be used *to detect DNA or RNA fragments with sequence homology.* Nucleic acid probes are usually DNA probes. They can come from a variety of sources including:

(a) Random genomic DNA sequences;
(b) Specific genes;
(c) cDNA sequences;
(d) Oligonucleotide DNA sequences—produced synthetically based on knowledge of the protein amino acid sequence.

A DNA probe can *be labelled* by a variety of processes, whereby *$^{32}$P-labelled nucleotides* or *non-radioactive labelling systems such as biotinylated nucleotides* are introduced into the DNA molecule. The sites where a radioactively labelled DNA probe hybridizes with complementary DNA sequences on a nitrocellulose filter can be localized by autoradiography while the non-radioactive label biotin can be detected by its high affinity for streptavidin.

## Difference between DNA and RNA

Though there are similarities regarding composition between the DNA and RNA and RNA forms on the DNA strand and sharing many features with each other, RNA possesses several specific differences.

### Difference between RNA and DNA

| RNA | DNA |
|---|---|
| 1. Sugar—Ribose | 1. Sugar—Deoxyribose |
| 2. It does not contain thymine containing nucleotide but uracil is present (thymine may be present in some rare cases). | 2. It contains thymine containing nucleotide. |
| 3. It exists in a single-stranded molecule. But it is capable of folding back on itself like a hairpin and thus acquiring double-stranded character. | 3. It is double-stranded helical molecule. |
| 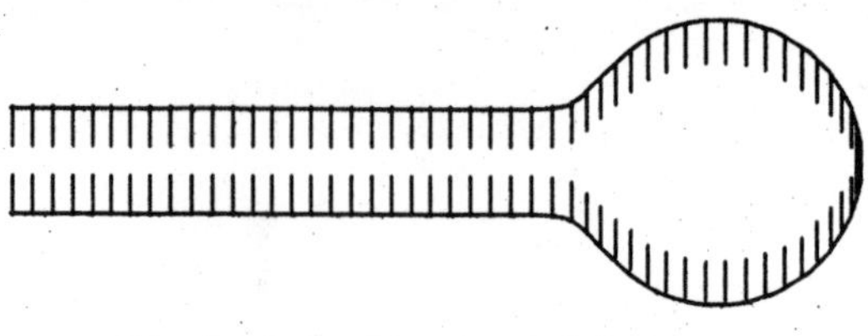 **Fig. 9.14:** Double-stranded character | |
| 4. Guanine content does not equal to its cytidine content nor does its adenine content equal its uracil content. | 4. It has the equality. |
| 5. It can be hydrolyzed by alkali and is useful for diagnostically and analytically. | 5. Not hydrolyzed. |
| 6. It has three forms, i.e. mRNA, tRNA, rRNA. tRNA are of 20 varieties, one for each amino acid. | 6. Single variety but recently P-DNA and Z-DNA have been discovered. |
| **Staining reaction (Histology)** Feulgen—Negative | Positive |
| **Distribution** Mainly in the cytoplasm, but very little amount in the nucleus, nucleolus, Nissl granules, ribosomes, mitochondria. | In the nucleus. |
| **No. of nucleotides** 60-6,000, unbranched structure is less rigid. | **No. of nucleotides** Long chains of up to 200,000 nucleotide pairs but usually 1600-9000. |

| RNA | DNA |
|---|---|
| **Mol. wt.**<br>m RNA— $3 \times 10^4$ - $2 \times 10^6$ daltons<br>t RNA— 2-3 × $10^4$ daltons<br>r RNA— $2.5 \times 10^6$ daltons | **Mol. wt.**<br>50-90,000,00 |
| **Interrelation**<br>1. For synthesis of mRNA in the nucleus, DNA template is required. | 1. Same. |
| 2. RNA primer is required for DNA replication or synthesis as we see in "Okazaki piece". | 2. For DNA replication or synthesis RNA primer is required. |
| 3. For protein synthesis it carries genetic code, i.e. message from nucleus to cytoplasm. | 3. It transcribes the genetic code or message to mRNA for translation. |
| 4. **Function**<br>They are concerned with protein synthesis. | 4. **Function**<br>(a) It is the molecule of heredity. The purines and pyrimidine bases carry genetic information whereas sugar and phosphate groups perform structural role.<br>(b) Protein synthesis. |

# 10

# DNA AND GENE

1. DNA and gene, 2. Functions of DNA in genes.

## DNA and Genes

1. A cell having nucleus contains chromosomes having band-like structure called gene. At the molecular level, a gene is a portion of a DNA molecule arranged in a fixed position in a linear fashion along the DNA chain which is three feet long in 46 chromosomes.
   (a) Okazaki has found that both strands of DNA are replicated to short pieces of up to 2,000 nucleotide residues, which are subsequently joined by DNA ligase.
   (b) On the different chromosomes of mammalian cells there are multiple origin of DNA replication that occur in clusters of up to 100 of these replication units.
   (c) Replication occurs in both directions up and down the chromosomes and on both strands simultaneously. This replication process generates "replication bubbles".
   (d) The two chains separate from each other like undoing zip-fastener, i.e. semiconservative way. The hydrogen bonds linking the base pairs are weak so breakage could easily happen.
2. Genes contain the genetic material which is chemically DNA.
3. DNA of genes has double strands, double helical in nature running parallel in opposite direction and genetic property is limited in one strand.
4. DNA of gene can replicate, i.e. it participates in its own synthesis.
5. DNA of gene regulates not only genetic character but also day-to-day functions of cell, i.e. protein synthesis, enzyme synthesis etc.
6. DNA in genes contains genetic code which is transcribed to mRNA and is sent to cytoplasm, then to ribosomes for translation of protein synthesis.
7. DNA of genes replicates once and once only.
8. DNA of genes is having bases whose

sequence is the important factor in heredity.

(a) The four bases, viz. adenine, guanine, thymine and cytosine are chained in a *wide variety of combination* to form the DNA chain. DNA chains of *no two persons can be identical,* this explains why no two persons look exactly similar in this world except identical twins. Hence, all the cells of individual's body contain the same set of DNA whether it is hair, bits of skin, blood or sperm, and this is the basis of "*DNA finger-printing technique*" and came to be widely recognised.

9. DNA of gene is the molecule of heredity. The purine and pyrimidine bases of DNA carry genetic information, whereas the sugar and phosphate groups perform a structural role.

10. *Storage of genetic information*: Genetic information is stored in DNA, the informational macromolecule of the chromosome. The genetic information flows in the direction of DNA ⟶ RNA ⟶ Protein. Purine and pyrimidine bases carry genetic information, whereas sugar and phosphate groups perform structural role.

11. *Amino Acids sequence of Polypeptide Chain coded by DNA*: The sequence of amino acids in polypeptide chain of each type of protein is specified or coded by the sequence of nucleotide residue in DNA. The segment of DNA molecule specifying and complete polypeptide chain is called *cistron* or *gene.*

12. *One Amino Acid is coded for by three successive Nucleotide Residues*: Each amino acid is coded for by *three successive nucleotide residue.* The different bases down the length of the DNA molecule seem to act like a *code* instructing the cell to make certain proteins, the *order of bases* indicating the *sequence of amino acids* to be joined up in order to make protein, e.g. the sequence *CAA* (cytosine-adenine-adenine) specifies the amino acid *valine,* three thymines in a row, *TTT,* specifies *lysine* while *AAT* (adenine-adenine-thymine) specifies *leucine.* So the sequence of bases CAA-TTT-AAT would direct the cell to link up the amino acids valine-lysine-leucine to make the appropriate peptide and in a similar way more complex proteins.

13. *Relationship between DNA of gene, mRNA and Amino Acids:* There is relationship between DNA of gene, messenger RNA (mRNA) and amino acids. *Each DNA triplet* acts as a *template* for formation of the corresponding *mRNA triplet.* In this way the information in DNA is *transcribed* to mRNA. Each mRNA triplet codes for a particular amino acid. By this process the information incorporated in the mRNA is *translated* into protein structure. RNA triplet and corresponding 20 amino acid are now known and can be seen in the genetic dictionary.

    (a) In some cases as many as four different triplets code for the same amino acid are seen.

    (b) *Methionine* and *tryptophan* are being coded for by only *one triplet* which is exceptional.

14. DNA in genes carry hereditary character from parents to offsprings.

## Functions of DNA in Genes

**A. Genetic Function**

1. Transmission of heredity.
2. Storage of genetic information.
3. Transmission of genetic information.

4. Formation of genetic code.
5. Formation of mRNA triplets for protein synthesis.
6. DNA can replicate.
7. RNA formation is controlled by the DNA of the nucleus by the process called *transcription*.

B. **Cellular Function**
   (a) Cellular reproduction.
   (b) Daily cellular function.
      1. Formation of structural proteins, e.g. organelles—mitochondria, endoplasmic reticulum, lysosomes, Golgi apparatus, etc.
   (c) Formation of enzymes for:
      1. Oxidative reactions for energy
      2. Synthesis of biomolecules, e.g. lipids, glycogen, ATP, etc.

In the formation of RNA, there is separation of two strands of DNA molecule. One of these strands acts as a template on which RNA molecule is assembled which contains the gene and other strand remains genetically inactive. Assembly of the RNA molecule is accomplished by combination of ribose nucleotides with a strand of DNA with the help of RNA polymerase to form molecule of ribonucleic acid (RNA) that carries the DNA code from the gene to the cytoplasm (Fig. 10.1).

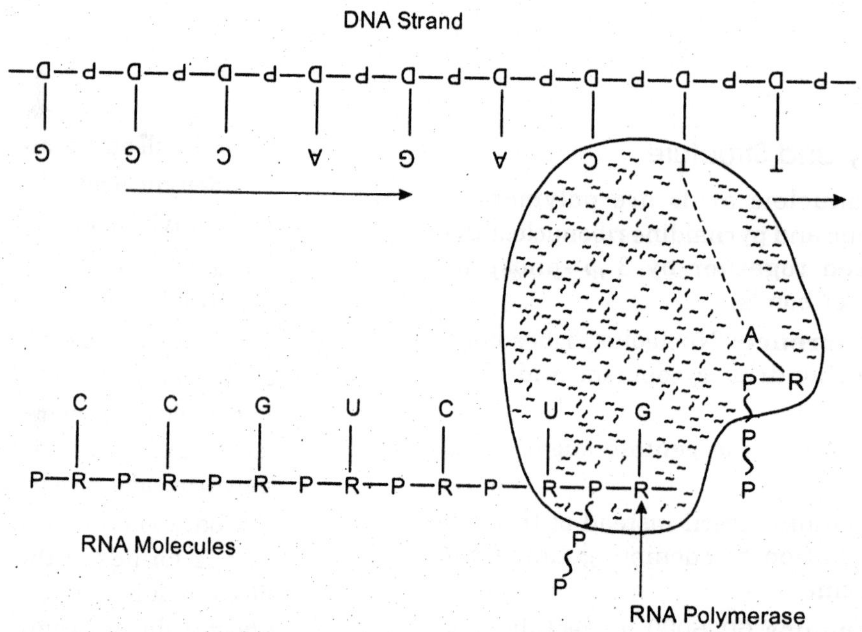


**Fig. 10.1:** Process of transcription. Combination of RNA with one strand of DNA to form RNA which carries DNA code from the gene to the cytoplasm.

# 11

# RNA OR RIBONUCLEIC ACIDS

1. RNA (Ribonucleic acids)—chemistry, structure, structural organisation, 2. Biological importance of nucleic acids.

## Chemistry and Structure

1. Ribonucleic acids are polymers of purine and pyrimidine ribonucleotides linked together by 3′,5′-phosphodiester bonds.
2. **Occurrence:** Nucleolus, ribosomes, mitochondria, cytoplasm and Nissl granules.
3. It contains the pentose sugar called ribose.
4. It contains uracil instead of thymine in addition to adenine, guanine and cytosine.
5. It contains 60–6000 nucleotides and the molecule is unbranched.
6. **Primary structure:**
   (a) It is constituted by the number and sequence of ribonucleotides, i.e. their bases in the strand.
   (b) Each linear strand is composed of many ribonucleotides linked with each other by 3′, 5′-phosphodiester bonds joining the 3′-OH of one nucleotide with the 5′-OH of the next nucleotide.
   (c) Major nucleotides are:
      1. Adenylic acid,
      2. Guanylic acid,
      3. Cytidylic acid,
      4. Uridylic acid.
   (d) Thymine is absent in all ribonucleic acids except tRNA.
   (e) The nucleotides of minor bases like methylpurines, hypoxanthine and dihydrouracil also occur.
   (f) The backbone of the primary structure is formed of the sugar-phosphate residues inter connected by phosphodiester bonds.
   (g) RNA molecule does not hybridize with the "antisense" strand of the DNA of its gene. Therefore, the sequence of RNA molecule (except U being replaced by T) is the same as that of the "antisense" strand of the gene.
7. **Secondary structure:**
   (a) RNA consists of coils of a single polynucleotide strand, stabilized

by hydrophobic interactions between the purine and pyrimidine rings and by hydrogen bonds between complementary bases (e.g. G-C and A-U hydrogen bonding) of the same strand.

(b) The hydrogen bonds involve the same groups of adenine, guanine and cytosine and in the same of DNA, and the $N^3$ as well as the $C^4$-oxo group of uracil (or dihydrouracil) which pairs with adenine in RNA.

(c) Since it is a single-stranded molecule, its guanine content does not necessarily equal its cytosine content and its adenine content does not necessarily equal its uracil content.

8. **Double-stranded helices** RNA have been detected in vertebrate, reovirus, plant, wound, tumor, virus.
9. **Single-stranded RNA Coils** are present in poliomyelitis, influenza and tobacco-mosaic viruses.
10. Hydrophobic and hydrogen bonds between the bases may also produce a compactly coiled globular *tertiary structure* of RNA.

## Structural Organization of RNA

There are 3 main types of RNA molecules exist in all prokaryotic and eukaryotic organisms. These are:

I. messenger RNA (mRNA)
II. transfer RNA (tRNA)
III. ribosomal RNA (rRNA)

### I. The Messenger RNA (mRNA):

1. The mRNA is single-stranded and complementary to the sense strand of DNA.
2. These are large single-stranded fibrous RNA molecules, having molecular weight of $3 \times 10^4$ to $2 \times 10^6$ daltons.
3. It carries $10^3$ to $10^4$ ribonucleotides.
4. It is the most heterogenous in size and stability.
5. It has a large molecular weight of 30,000 to 50,000 to have the coded information corresponding to long polypeptide chain.
6. In mammalian nuclei, the immediate products of gene transcription are another class of RNA molecule which are quite large and heterogenous in size. These heterogenous unclear RNA (HnRNA) may exceed $10^7$ daltons whereas the mRNA molecules are generally smaller than $2 \times 10^4$ daltons. These Hn RNA molecules are processed to generate the mRNA molecules which then enter the cytoplasm to serve as template for protein synthesis.
7. Each mRNA is synthesized as complementary to some part of a DNA template strand so that the genetic message, carried by the base-sequence of that portion of the DNA, is duplicated in the particular mRNA in the form of a complementary base-sequence.
8. The bases are arranged in the mRNA strand according to the *base-pairing rule,* viz. guanine, adenine, cytosine and uracil in the RNA strand correspond respectively to cytosine, thymine, guanine and adenine in the DNA template strand.
9. After modification, mRNA carries adenine, guanine, cytosine and uracil as the major bases along with some minor bases like methylpurines and methylpyrimidines.
10. The 3 end of most mRNA strands consists of a sequence of about 20-200 adenylate residues (*Poly-A tail*).
11. The 5′ end of the mRNA is "capped" by a 7-methylguanosine 5′-triphosphate (*5″-pppG Cap*) which is linked to an adjacent 2-0-methylribonucleoside of

the next ribonucleotide at its 5′ -hydroxyl through the 3 phosphates (Fig. 10.1).

The protein synthesizing machinery begins translating the mRNA into proteins at the 5′ or capped terminus. The 3′-hydroxyl terminus has attached a polymer of adenylate residues 20-250 nucleotides in length.

12. One or more segments of an mRNA strand passes from nucleus to cytoplasm carrying the *genetic code* or *message* to the protein-synthesizing machinery where each serves as a template and is *translated* and specific sequence of amino acids is polymerized to form a primary structure of a protein.
13. Each genetic code consists of many consecutive nucleotide triplets called *codons*, each of which helps to incorporate a particular type of amino acid in the peptide to be synthesized (Table 11.2).
14. mRNA forms about 4 per cent or less of the total cellular RNA.

It remains associated with ribosomes and links several ribosomes to form linear cluster called *polysomes*.

- **Importance of mRNA**

(See biological importance of RNA.)

## II. The transfer RNA or tRNA (sRNA or soluble RNA)

1. It remains large in the cytosol. Cellular concentration of tRNA is $\frac{1}{4}$ to $\frac{1}{6}$th of rRNA and 10-20 per cent of the total cellular RNA molecules.
2. tRNA has relatively small, single-stranded, globular molecules having molecular weight of 25,000.
3. It is a polymer of many ribonucleotides joined by 3′-5′-phosphodiester bonds. It consists of about 80 nucleotides.
4. There at least 20 or more species of tRNA are found in animal cells. One corresponds to each of the 20 amino acids required for protein synthesis. Molecular structure of more than 150 tRNA from various sources have been established.

### A. Primary structure

1. Each tRNA strand consists of sequences of ribonucleotides in it. Different tRNA carry 74-95 nucleotides (or bases).
2. The bases include mainly adenine, guanine, cytosine and uracil; pseudouridine (ψ) is present in some loops and stems while thymine occurs in one loop. Some positions in 90-95 per cent of tRNA molecules are always occupied by specific bases called *invariant or conserved bases*. Some other positions always occupied by any purine base or by any pyrimidine base, but not by specific base; such bases are called *semi-invariant* or *semi-conserved bases*.
3. Still other positions are occupied by *variant bases* whose number and types vary in different species of tRNA and determine the length of tRNA.

### B. Cloverleaf Secondary Structure

1. *Cloverleaf secondary structure* is possessed by each single-stranded tRNA molecule due to folds stabilized by hydrogen bonds between complementary bases in different portion of the same strand; this forms some double-stranded double-helical portions called *stems*.
2. Besides the standard A-U and G-C hydrogen bonding maintaining the base pairing rule, hydrogen bonds occur also between slightly mismatched base-pairs like guanine and uracil in these stems.

3. Unlike DNA, planes of the hydrogen-bonded and paired bases are not perpendicular to the long axis of the double-stranded stem and the double-stranded helical portion has a somewhat deformed structure compared to the B-DNA double helix.
4. Some stems carry at one end a *loop* showing no hydrogen-bonding or base-pairing. Some non-bounded single-stranded segments connect the hydrogen-bonded stems with each other.
5. One of the base-paired stems bears at its farthest end a loop having a sequence of thymine, pseudouridine (ψ) and cytosine; that loop and the stem are called T ψ C or *T loop* and *T stem* respectively. An almost diagonally opposite loop carries the dihydrouracil (D) residue; that loop; and the stem bearing it are called the *D loop* and *D stem* respectively.
6. The 3′ end of each tRNA terminates in an unpaired sequence of cytosine-cytosine-adenine; this unpaired arm and the 3′ terminal at its end are called the *acceptor arm* and the *acceptor end* respectively.
7. The *anticodon loop* is another and non-bonded loop carrying a specific sequence of three-bases constituting an *anticodon.* The bases of an anticodon may be hydrogen-bonded with three complementary bases of a *codon* of an mRNA (Table 11.2). This helps the amino acyl-tRNA to bring the amino acid on a specific codon of mRNA and thereby to incorporate that amino acid at a particular site of a peptide. Counting from the 5′ end, boses 1, 2 and 3 of a tRNA anticodon pair respectively with bases 3, 2 and 1 of an mRNA codon. The base-paired stem leading to the anticodon loop is called the *anticodon stem.*
8. All tRNA molecules have a common CCA at the 3′ terminal. The carboxyl groups of amino acids are attached to the 3′-hydroxyl group of the adenosyl moiety through the ester bond.
9. A *variable loop* or *extra arm* joins the T stem with the anticodon stem and carries only 3–5 bases in the short-chain class-1 tRNA (e.g. alanine + RNA) but 13-21 bases in the long-chain class-2 tRNA (e.g. phenylalanine tRNA).
10. The number of base-pairs or bases in a loop or stem either remains fixed or varies within a specified range in all tRNAs; e.g. T loop–7 bases; T stem–5 base-pairs; D loop-7–11 bases; D stem–4 base-pairs; anticodon loop–7 bases; anticodon stem–5 pairs; acceptor stem–7 pairs; acceptor arm–4 bases; variable loop– 4–21 bases.
11. Some of the loops and stems bear binding sites for ribosomes and amino acyl-tRNA syntheses.
12. The hydogen bonds involved in stabilizing the base-paired stems of cloverleaf secondary structure are called *secondary* H *bonds.*

### C. L-shaped Tertiary Structure

1. An *L-shaped tertiary structure* is also produced in tRNA molecules by further compact folding of cloverleaf form due to *tertiary* H *bonds* between many bases of the unpaired loops, particularly between the T and D loops.
2. The base-paired double-helical stems are consequently arranged into two double-helical columns, continuous with and perpendicular to one another; one column is made of the acceptor T stems, the other

consists of the D and anticodon stems; the latter column holds the anticodon loop at its extreme end, while the bend between the two columns is formed of the T, D and variable loops.

3. Thus, the tertiary structure locates the amino acid acceptor end and the anticodon at the furthest ends of the two columns.

- **Function**

(See biological importance of Nucleic Acids.)

## III. The Ribosomal RNA (rRNA)

1. rRNA forms 60–80 per cent of the total cellular RNA.
2. **Occurrence:** Cytoplasm, granular endoplasmic reticulum, mitochondria, nucleolus and chloroplast in plants.
3. Most of the rRNA remains combined with proteins in macromolecular aggregates called *ribosomes.*
4. Ribosomes are nucleoprotein particles and reticular granules of 100-150 Å in diameter which acts as the machinery for the synthesis of proteins.
5. Ribosomes carry rRNA and protein in the ratio of 1 : 1 in mammals.
6. The mammalian ribosome contain *2 major nucleoprotein subunits*—a large one (60S) and a smaller one (40 S) (S = Sedimentation coefficient).
7. The 607 subunits contain a 5S rRNA, a 5.8S rRNA and a 28S rRNA.
8. The smaller 40S subunit contains a single 18S rRNA and approximately 30 polypeptide chains.
9. The 5S rRNA has its own precursors which is independently transcribed.
10. In the cytoplasm, the ribosomes remain quite stable and capable of many translation.

### A. Primary structure

1. The primary structure consists of the number and sequence of nucleotides i.e. their bases in the rRNA molecule, e.g. mammalian 5S, 5.8S, 18S and 28S rRNAs are made of 120, 160, 1900 and 4700 bases respectively.
2. Besides the major bases like adenine, guanine, cytosine and uracil and a few pseudouridine, methylated nucleotides form about 2 per cent of the nucleotides in mammalian 18S and 28S rRNA.

### B. Secondary structure

1. The secondary structure is formed and stabilized by hydrogen bonds between complementary bases in two different stretches of the same rRNA molecule.
2. This causes formation of many short, hydrogen-bonded, base-paired, double-helical stems, each having less than 8 base-pairs, bearing a non-bonded and unpaired loop at its farthest end and often carrying one or more unpaired bases forming non-bounded bulges. These double-helical stems are connected with each other by non-bonded single-stranded loops to form a number of domains in the rRNA molecule.

- **Function**

The ribosomal RNA (rRNA) always functions in association with both mRNA and tRNA. Transfer RNA (tRNA) transports amino acids to the ribosomes for incorporation into the developing protein molecules, while mRNA provides the information necessary for sequencing the amino acids in proper order for each specific type of protein to be manufactured. The mRNA and tRNA first form a complex with the 40S particle. Then it is believed that the 60S particles provides the enzymes that promote peptide linkages between the successive amino acids. Thus, the ribosome acts as a protein-manufacturing machinery in which protein molecules are formed.

(See biological importance of Nucleic Acid.)

## Biological Importance of Nucleic Acids

**I. Importance of DNA** (Deoxyribonucleic Acid)

(See functions or importance of DNA in genes.)

**DNA in Biotechnology:**

1. DNA is employed for recombinant DNA technology or genetic engineering.
2. DNA is applied for DNA fingerprint technique which is very much useful in forensic medicine.

**II. Importance of RNA** (Ribonucleic Acid)

A. **mRNA (Messenger RNA)**

1. Messenger RNA (mRNA) is produced by DNA which helps to incorporate amino acids in the protein synthesis in the ribosome according to code.
2. RNA functions primarily in the cytoplasm of the cell as a template in connection with the synthesis of proteins as well as in ribosomes. The template formation is directed by nuclear DNA.
3. In mammalian nuclei, the initial products of gene transcription are another class of RNA molecules which are large and heterogenous nuclear RNA (HnRNA) may exceed $10^7$ daltons whereas the mRNA are smaller than $2 \times 10^6$ daltons. These HnRNA molecules are processed to generate the mRNA molecules which then enter the cytoplasm to serve as template for protein synthesis.
4. It passes from nucleus to cytoplasm carrying information of a gene to the ribosomes where each serves as template on which a specific sequence of amino acids is polymerized to form a specific protein molecule.
5. mRNA contains codon that are exactly complementary to the code words of genes.

B. **tRNA (Transfer RNA) or sRNA (Soluble RNA)**

1. They serve as adaptors for the translation of the information in the sequence of nucleotides of the mRNA into specific amino acids.

   The following Fig. 11.1 illustrates a small segment of a molecule of mRNA whose codons are CCG, UCU and GAA for the amino acids proline, serine and glutamic acid.
2. It *transfers* the amino acids molecules to protein molecules during protein synthesis.
3. There are *different types of tRNA*, but each type will *combine specifically with only one* of the 20 amino acids that are incorporated into proteins. Then it acts as a *carrier* to transport its specific type of amino acids to the ribosomes to form protein.

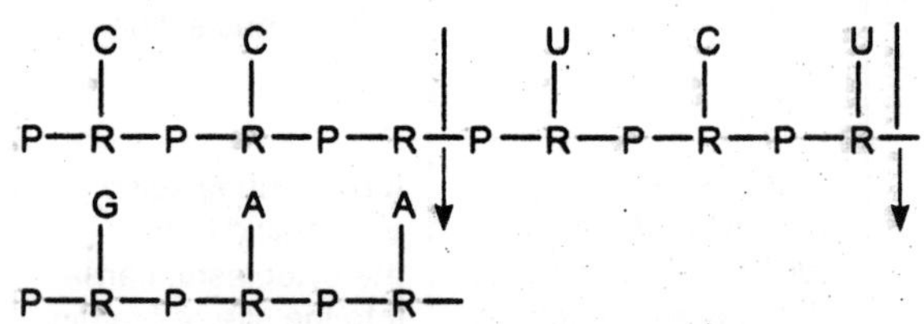


**Fig. 11.1:** Portion of RNA molecule showing three "Code" Words CCG, UCU and GAA representing amino acids proline, serine and glutamic acid.

4. tRNA recognizes a particular codon on the mRNA and thereby delivers the appropriate amino acid to the appropriate place in the chain of the newly forming protein molecule.
5. At one end of tRNA is always an adenylic acid and the transported *amino acid attaches* to a hydroxyl group of the ribose of adenylic acid. A *specific enzyme causes this attachment* for each specific type of tRNA. This *enzyme* also *determines* the type of amino acid that will attach to the respective type of tRNA.
6. Since tRNA causes attachment of a *specific amino acid* to form protein chain, so *each type of tRNA* also have *specificity* for a *particular codon* in the mRNA.
7. The specific prosthetic group in the tRNA that allows tRNA to recognize a specific codon is called *anticodon* which is located in the middle of tRNA molecule, i.e., at the bottom of the cloverleaf configuration.
8. The anticodon bases combine loosely by hydrogen bonding with the codon bases of mRNA. In this way the amino acids are lined up one after another along the mRNA chain maintains sequence of amino acids.

C. **rRNA (Ribosomal RNA):** On the ribosomes, mRNA and tRNA molecules interact to translate into a specific protein molecule, the information transcribed from the gene.

1. The genetic material for some animal and plant viruses is RNA rather than DNA.
2. There are codons in the RNA for 20 common amino acids. Several of the amino acids are represented by more than one codons, e.g., Alanine—4 codons, Arginine—6 codons, Aspartic Acid—2 codons, etc (Table 11.2).

Some codons are represented as *signals* such as "start manufacturing a protein molecule". These codons are designated as CI, i.e. "chain initiating" which

**Table 11.1:** Comparison of mRNA, tRNA, rRNA

| | mRNA | tRNA | rRNA |
|---|---|---|---|
| 1. Amount in percentage | 5% | 15% | 80% |
| 2. Site of synthesis | Nuclear DNA | Nuclear DNA | Nucleolar DNA |
| 3. Molecular weight | $5 \times 10^5$ | $25 \times 10^3$ | $1.1 \times 10^6$ |
| 4. Sedimentation coefficient | 8S | 4S | 28S, 18S, 5-7S |
| 5. Structure | Linear | Hairpin loop and L shaped secondary and tertiary structure | 70% secondary structure |
| 6. Numbers of nucleotide | | 70-85 nucleotide | 28S = 5500 nucleotide<br>18S = 2500 nucleotide<br>5-7S =120 nucleotide |
| 7. Function | It carries genetic information for chromosomal DNA to the cytoplasm where it acts as a template for protein synthesis. | It picks up specific amino acid from the cytoplasm, carries it to the site of protein synthesis. | It is used in the synthesis of ribosomal subunits. |

**Table 11.2:** RNA Codons for different amino acids and for CI and CT. The transcription of these codons from DNA.

| Amino Acid | RNA Codons | | | | | |
|---|---|---|---|---|---|---|
| Alanine | GCU | GCC | GCA | GCG | | |
| Arginine | CGU | CGC | CGA | CGG | AGA | AGG |
| Asparagine | AAU | AAC | | | | |
| Aspartic acid | GAU | GAC | | | | |
| Cysteine | UGU | UGC | | | | |
| Glutamic acid | GAA | GAG | | | | |
| Glutamine | CAA | CAG | | | | |
| Glycine | GGU | GGC | GGA | GGG | | |
| Histidine | CAU | CAC | | | | |
| Isoleucine | AUU | AUC | AUA | | | |
| Leucine | CUU | CUC | CUA | CUG | UUA | UUG |
| Lysine | AAA | AAG | | | | |
| Methionine | AUG | | | | | |
| Phenylalanine | UUU | UUC | | | | |
| Proline | CCU | CCC | CCA | CCG | | |
| Serine | UCU | UCC | UCA | UCG | | |
| Threonine | ACU | ACC | ACA | ACG | | |
| Tryptophan | UGG | | | | | |
| tyrosine | UAU | UAC | | | | |
| Valine | GUU | GUC | GUA | GUG | | |
| CI (Chain initiating) | AUG | GUG | | | | |
| CT (Chain terminating) | UAA | UAG | UGA | | | |

are AUG and GUG, and "stop manufacturing a protein molecule" and designated as CT, i.e. "chain terminating" which are UAA, UAG and UGA.

D. **Other Functions**

1. Adenylic acid after combination with two molecules of phosphates forms ATP which is the biochemical unit of energy exchange in all cells.
2. The hydrogen acceptors are nucleotides, e.g. NAD, FAD. In biological oxidation-reduction, these substances are involved for the transport of hydrogen atoms or electrons through organized system called hydrogen acceptors or electron transport agents.

## DNA Analysis Techniques

- **Introduction**

DNA analysis at present is the standard investigation for an increasing number of Mendelian disorders. It is employed in case of:

1. Genetic state of the family members
2. Pregnancies at risk including:
   (a) Cystic fibrosis
   (b) Haemoglobinopathies
   (c) Duchenne muscular dystrophy
   (d) Huntington's disease

These diseases are usually diagnosed by general procedure like electrophoresis of haemoglobin in thalassaemia, muscle biopsy in Duchenne muscular dystrophy. But DNA analysis clarify the diagnosis specifically in disorders which are associated with specific mutation or gene deletions. The importance created by DNA analysis clinically is to detect carriers of X-linked recessive disorders, presymptomatic diagnosis of autosomal dominant disorders, and prenatal diagnosis of all categories of Mendelian disorders.

- **DNA Analysis**

The standard techniques of DNA analysis which are used in clinical investigation of affected families as a service to families with particular genetic disorders.

- **Techniques**

I. DNA extraction
II. Use of restriction enzymes
III. DNA hybridisation
   A. Electrophoresis
   B. Southern blotting
   C. DNA probe hybridisation
   D. Fluorescence in situ hybridisation.
IV. DNA amplication
   A. Polymerase chain reaction (PCR)
V. Newly developed techniques of DNA analysis
   (a) Use of yeast artificial chromosomes (YACs)
   (b) Pulsed field gel electrophoresis (PFGE)
   (c) Denaturing gradient gel electrophoresis (DGGE)
   (d) Single-strand conformational polymorphism (SSCP) analysis
   (e) Semiautomated gene sequencing
   (f) Making DNA copy from isolated mRNA using reverse transcriptase enzyme.

1. **DNA Extraction:** DNA can be extracted by standard techniques from any tissue having nucleated cell, e.g. whole blood (10–20 ml), buccal cells and chorionic villus material. Extracted DNA is stable and can be stored for long time. And the samples from the subjects having genetic disorders can be collected and preserved for the future investigation of other family members and benefitted many families whose elderly persons or affected relatives are no longer alive when DNA testing is done.
2. **Bacterial Restriction Enzymes:** Bacterial restriction enzymes recognise specific DNA sequences and cleave double-stranded DNA at these sites. Each enzyme has its own recognition sequence and cuts a genomic DNA into a series of fragments that can then be analysed. The size of each produced fragment is constant in an individual subject but commonly varies between different subjects because of differences in non-coding DNA sequences. This variation forms the basis of some predictive tests.
3. **Electrophoresis:** The cutting DNA fragments from genomic DNA by restric enzymes can be ordered according to their size by electrophoresis in an agarose gel. The small fragments migrate faster down the gel than the larger ones, giving a track of DNA fragments progressively diminishing size. The length of a particular fragment can be determined from the distance of its migration in the gel with reference to marker fragments of known size.
4. **Southern Blotting:** The DNA fragments in a gel are denatured into single strands and transferred by the southern blotting technique onto a nitrocellulose filter or nylon membrane. Fluid rising through the gel transfers the DNA onto the membrane, retaining the alignment of the fragments. Although convention-ally performed with a reservoir of buffer solution, the procedure may also be performed without buffer as fluid absorbed directly from the gel by the

paper towels in sufficient to permit transfers of the DNA. The DNA binds to the membrane, providing a stable array of DNA fragments that can be analysed by mixing with a DNA probe in a hybridisation reaction. The basis of this reaction is the ability of complementary DNA strands to bind together.

- **DNA Probe Hybridisation**

A probe is a piece of single-stranded DNA, labelled with radioactive isotope $P^{32}$, which is applied to detect homologous sequences in a sample of genomic DNA. Probes used to study Mendelian disorders display perfect sequences which occur only once in the genomic and may correspond to the gene of interest, to flanking DNA, or to more distant DNA sequences. Gene specific probes derived from genomic DNA contain both coding and non-coding sequences. Complementary DNA (cDNA) probes are synthesized from the messenger RNA of the gene under study with reverse transcriptase and contain only coding sequences. Oligonucleotide probes, containing around 19 nucleotides, can be synthesized for a specific region of a gene whose sequence is known. Randomly generated probes do not correspond to specific genes but can be used to study families they are shown to be located close to a gene of interest.

Probes are prepared by cloning the sequence of interest with recombinant DNA technology. The DNA fragment to be used as a probe is incorporated into vector DNA, usually a bacterial plasmid. The recombinant vector plasmid is then amplified in *E. coli*, allowing large quantities of the probe DNA to be retrieved. The probe is radiolabelled and added to a solution for hybridisation with a membrane blotted with the patient's DNA. The single-stranded DNA probe will bind to any DNA fragment on the membrane that has a matching DNA sequence. These fragments can be identified as visible bands in an X-ray film placed in contact with the membrane, i.e. autoradiography.

- **Fluorescence in situ Hybridisation**

In this technique a DNA probe is labelled with flurochrome and hybridised direct with a metaphase chromosome spread. The presence or absence in an individual of DNA corresponding to the probe DNA can be ascertained by the presence or absence of a fluorescent signal produced by hybridisation to the relevant chromosome, visualised direct using a fluorescence microscope.

- **DNA Amplification**

The *polymerase chain reaction* (PCR) is used to amplify the DNA of a target sequence many times to allow direct analysis. This enables rapid results to be obtained from very small initial samples, and is particularly useful in prenatal diagnosis. Using the PCR technique, it is possible to analyse DNA from single hair bulbs, buccal cells obtained from mouthwash samples, and dried neonatal blood spots. Analysis can also be performed on single cells, making pre-implantation diagnosis of specific genetic disorders a possibility. Flow chart of PCR is given below:

| | | |
|---|---|---|
| First step (Denaturation) | → | Application of heat to denature DNA into two single strands |
| Second step (Annealing primers) | → | Two specific oligonucleotide primers that flank the region of interest are then annealed to their complementary strands |
| Third step (DNA synthesis) | → | In the presence of thermostable polymerase these primers initiate the new DNA strands. |
| ↓ | | |
| Repetition of the cycle | → | The *cycle* of denaturation, annealing and synthesis repeated 30 times will amplify the DNA from the region of interest 100,000 fold |

| | |
|---|---|
| ↓ | while the quantity of the remaining DNA is unchanged |
| Fourth step (Analysis) | → The amplified sequence can be analysed direct by gel electrophoresis and staining with ethidium bromide |

- **Prenatal Diagnosis by DNA Analysis**

1. Detection of known point mutations, splicing mutations or small insertions or deletion by:
   (a) Restriction endonuclease analysis to detect mutations that alter restriction sites
   (b) Allele specific oligonucleotide hybridisation
   (c) PCR amplification and direct automated sequencing.
2. Detection of unknown mutations by PCR amplification followed by:
   (a) Single-strand conformation polymorphism analysis
   (b) Heteroduplex analysis
   (c) Mismatch cleavage
   (d) DNA sequencing.
3. Detection of large insertions, expansions, deletions, or other major structural rearrangements by:
   (a) Southern blot analysis
   (b) PCR amplification.
4. Linkage based analysis using intragenic DNA polymorphisms to diagnose diseases in which there are many pathogenic mutations and the specific mutation in a family is unknown.

- **Elected Single-Gene Diseases Amenable to Prenatal or Presymptomatic Diagnosis by DNA Analysis**

I. **Autosomal dominant**
   (a) Myotonia dystrophy
   (b) Adult polycystic kidney disease
   (c) Huntington disease
   (d) Neurofibromatosis I
   (e) Familial breast cancer

II. **Autosomal recessive**
   (a) Sickle-cell anemia
   (b) β-thalassemia, α-thalassemia
   (c) Cystic fibrosis
   (d) Phenylketonuria
   (e) $\alpha_1$-Antitrypsin deficiency
   (f) Tay-Sachs disease.

III. **X-linked recessive**
   (a) Haemophilia A and B
   (b) Duchenne and Becker muscular dystrophy
   (c) Fragile X-syndrome
   (d) Ornithine transcarbamylase deficiency.

## QUESTION

1. What are nucleic acids? Describe briefly the occurrence, types and constitutents of nucleic acids.
2. Describe the chemical structure of nucleic acids.
3. Describe briefly the molecular structure of DNA.
4. Describe the physical properties of DNA.
5. Describe the chemistry of DNA.
6. State briefly the mitochondrial and viral DNA. How does it differ from nuclear DNA?
7. What are the differences between DNA and RNA?
8. Describe the relationship of DNA in genes.
9. Describe the functions of DNA in genes.
10. What is RNA? Describe the chemistry and structure of RNA.
11. Describe the structural organization of RNA.
12. What is mRNA? Describe the structure of mRNA. What are the functions of mRNA.
13. What is tRNA? Describe the structure and functions of tRNA.
14. What is rRNA? Describe the structure and functions of rRNA.
15. Describe briefly the biological importance of nucleic acids.
16. State briefly about DNA probe hybridisation.

# 12

# INDIVIDUAL GENETIC DIVERSITY AND DETERMINATION

Testing method of genetic differences among individual called *Representational Oligonucleotide Microarray Analysis* (ROMA).

This has been worked out by the scientists at Cold Spring Harbor Laboratory in New York and at the Karoliska Institute in Stockholm. This test method can detect differences in DNA from any two sources. This new DNA test has revealed amazing genetic differences among normal healthy person which was unknown before some person have large chunks of DNA missing from their genome while others have extra copies of DNA stretches. These differences called *Copy Number Polymorphisms* (CNP) shows up in genes linked with cancer risk, with how much people consume drugs and manifestation of reactions to it. Thus, a relationship between CNP's and susceptibility may establish to produce some neurological disorders (e.g. Parkinson's disease), cancer and obesity ROMA can help identify region in the cancer genome that have undergone copy number fluctuation. This method is highly sensitive and quantitative, and can detect loss of a single allele.

Current studies with two types of tumours, i.e. ovarian and pancreatic cancer with the idea of two achievements such as:

1. Analysis of a large set of tumours to identify the genomic regions that are mutated frequently. These regions will be searched for gene candidates and these candidates will be functionally characterised.
2. Changes that occur in a tumour may constitute a genetic fingerprint of that tumour.

   Scientist's aim is to identify similarities in genetic fingerprints using informatics methods. The data is to be compared to information obtained from clinical collaborators to correlate tumour genetic fingerprints and disease status ultimately leading to improve diagnosis and treatment strategies.

Investigation for genetic diversity linked with cancer, the scientists compared DNA of a cancer patient with DNA of a normal subjects revealed encouraging results. The scientists observed that more than 70 large chunks of DNA were altered than the normal subjects. These differences were unknown before and these alternations were much more in DNA than the single nucleotide polymorphism. Single nucleotide polymorphism means changes in the purine and pyrimidine bases, i.e. adenine (A), cytosine (C), thymine (T) and guanine (G) nucleotide code in the DNA.

The scientist Craig Venter who first sequenced the complete human genome opined that these observations will help to explain how, with very little sequence differences, how we human come up with big differences among us.

The scientists have succeeded to locate the genes on chromosomes 5, 16, 19, 21. This investigation has revealed a number of diseases caused by genes associated with these chromosomes. For example, genes identified on chromosome 5 have been linked in colorectal cancer, basal cell carcinoma and a type of dwarfism. Gene in chromosome 16 is related with adult polycystic renal disease.

The results obtained by ROMA have revealed the possibility of mistakes or missing links in the human genome map published by the Human Genome Project. As for instance, scientists at Cold Harbor Laboratory in New York have observed a stretch of DNA on chromosome 16 does not appear there in the published sequence of the human genome, but rather on chromosome 6. According to Sherwood, it is extra copies of a gene that no one knew about. When comparison made between human and chimpanzee genomes revealed similar swaps, when a gene migrates from one chromosome to another. Everybody has two copies of each chromosome except the X and Y chromosomes that separate men from women. The scientists found that each of their healthy volunteers had just one copy of each CNP (Copy Number Polymorphism).

It has been reported by the research workers in the journal *Science* that just as chromosomal rearrangements have played a significant role in primate evolution and human disease, structural polymorphism may play an analogous role in determining genetic diversity within the human population.

# 13

# NUCLEIC ACID AS A GENETIC MATERIAL

1. Nucleic acid as a genetic material—Introduction, location of genetic substance, 2. Evidence that nucleic acid (DNA) as genetic material, 3. Evidence for DNA as a carrier of genetic information—Cytooptical method, Cytochemical method, Microbiological method.

## Introduction

Several attempts were made to find out the probable substance that acts as bearer of heredity and attention was focussed first on the *nucleus,* then on the *chromosomes* and thirdly on *genes.* The genes control the hereditary character from parents to children but it also control the cellular reproduction and day-by-day function, i.e. structural, chemical synthesis, etc.

Each gene, which is a nucleic acid called DNA, controls the formation of RNA which spreads throughout the cell and controls the formation of a specific proteins, some are structural proteins and some are enzymes which controls metabolic reactions producing energy and synthesis of lipids, glycogen, ATP, etc.

**Gene (DNA)**
Transmission of heredity

↓

RNA formation

↓

Specific protein formation

↓

*Structural proteins* for cell, e.g. organelles e.g. Mitochondria, Endoplasmic reticulum, Lysosomes, Golgi apparatus

*Enzymes* in the cells for:
1. *Oxidative reaction* for energy
2. Synthesis—Lipids, Glycogen, ATP, etc.

↓

Cellular function

**Fig. 13.1:** Function of gene.

For formation of cellular proteins there is only one gene pair, i.e. one gene on each chromosome of a chromosomal pair in the cell. If all the DNA of the cell functioned as genes there could be more than 2 million of these. However, since most of the DNA subserves other functions beside as genetic function, it has been estimated that there are 100,000 functional genes in the human cell which can produce 100,000 types of *proteins formed in different cells* but *not in the same cell* for various reasons (Fig. 13.1).

## Location of Genetic Substance

1. **Nuclear Theory:** Transmission of heredity through sperm and ovum led Hacckel (1860) to postulate that sperm consists largely of nuclear materials hence nucleus is responsible for transmission of heredity.
2. **Chromosome Theory:** In subsequent years, process of mitosis, meiosis, behaviour of chromosomes during cell divisions and fertilization focussed attention to the chromosomes in the nucleus. Sutton (1903) in his paper "The Chromosomes in Heredity" stated that the chromosomes group in diploids (2n) consists of two similar sets and at the end of meiosis every gamete receive only one chromosome of each homologous pair. He assumed that genes (Mendel's factor) are parts of the chromosome.
3. **Gene Theory:** After chromosomes the genes were pin-pointed as the probable hereditary material. Then efforts were made regarding chemical structure and mode of action of gene. Though it was known in later stage gene contains protein and nucleic acid but structure of both was unknown. But speculations established that:
   (a) The genes must be self-duplicating, i.e., their structure must be exactly copied every time one chromosomes become two.
   (b) The genes must be stable so that heritable changes (Mutations) may occur very rarely.
   (c) It must have the potentiality to carry all kinds of necessary biological information.

## Chemistry of Gene

1. Miescher (1869) was the first to isolate chemical products of cell nuclei using pus cells. He found that the main chemical extract contained lot of phosphorus and named it "nuclein". Then he used Salmon spermatozoa and found nuclein contains 50% organic material and 9.6% phosphorus and was of acid in nature. He concluded that this material *might be genetically active.*
2. In 1896, nuclein was named as nucleic acid. It had been isolated from yeast called *yeast nucleic acid* containing Phosphate, Pentose (Ribose), four nitrogenous base—adenine, guanine, cytosine, uracil and from thymus known as thymonucleic acid containing phosphate, pentose (Deoxyribose) and four bases—adenine, guanine, cytosine and thymine.

These two nucleic acids were finally named as DNA and RNA.

## Evidence that Nucleic Acid (DNA) is Genetic

Avery and co-workers (1944) produced evidence that nucleic acid rather than protein is the carrier of genetic information.

The bacteria that cause pneumonia (Pneumococcus) exist in two different strains, capsulated (virulent) and non-capsulated (non-virulent). They prepared a media containing chemical extract of disrupted capsulated bacteria without any living bacterial cell. In this media, non-capsulated bacteria were introduced for culture. The remarkable thing was that non-capsulated bacteria developed capsule and become virulent. The conclusion is that a chemical agent in the extract of capsulated bacteria is responsible for permanent and inheritable change in the descendants. The question is: What particular component of the extract is responsible? By systematically isolating, purifying and testing the different chemical components, Avery and his colleagues were able to show the nucleic acid (DNA) is the active agent (Fig. 13.2).

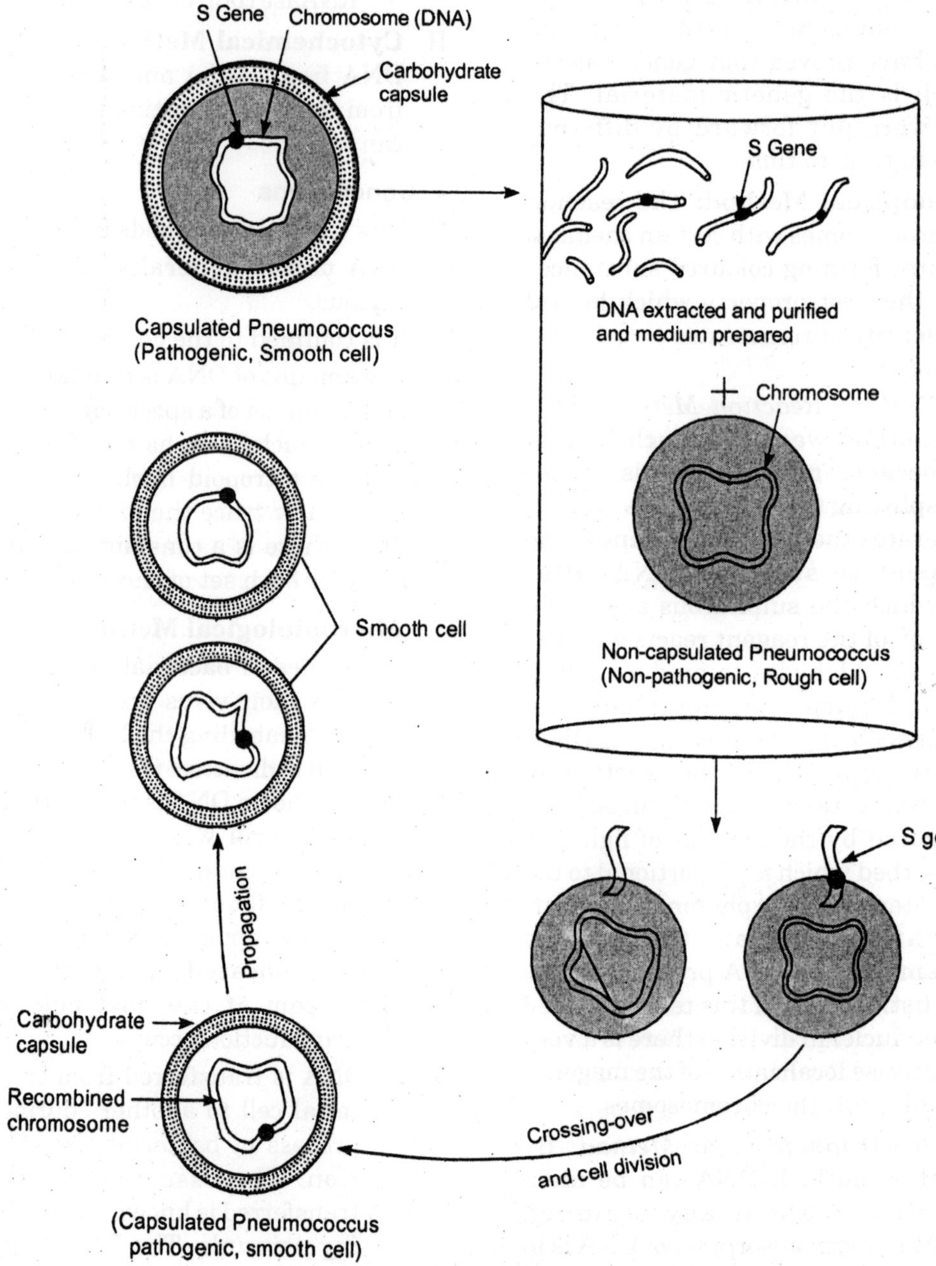


**Fig. 13.2:** Avery-MacLeod-McCarty Experiment.

## Evidence for DNA as Carrier of Genetic Information

- **Introduction**

In early period, the nucleus was thought to be carrier of genetic information. Later on chromosomes then ultimately it was found that gene is responsible for this work. Advance researches using modern techniques have proved that genes contain DNA which is the genetic material. The evidences were put forward by different workers from time to time.

I. **Cytooptical Method:** The reaction of chromosomes with certain chemical groups, forming coloured substances was the first property which helped to identify chromosomes (= coloured bodies).

(a) *Feulgen Reaction Method*: This method was very much helpful because mild hydrolysis causes splitting the purine bases and liberates the aldehyde groups of the pentose sugars of DNA, after which the sulphurous acid fuchsin of the reagent reacts with the free aldehydes to produce a reddish-purple pigment indicates specifically the presence of DNA. The intensity of the reaction on thin sections of nuclei can be measured by the amount of light absorbed which is proportional to the intensity. This photometric observations indicate the relative amounts of DNA present and its distribution. If this test is applied to nuclei in division there is a very precise localization of the magenta colour in the chromosomes.

(b) *Spectrophotometric Method*: By this method, DNA can be identified without any staining. Maximum absorption of DNA is in the vicinity of the wavelength 2600 Angstrom Units.

(c) *Pyronin and Malachite Green for the Differentiation of DNA and RNA:* By this technique, DNA can be differentiated from RNA. DNA gives blue/green colour and RNA gives red colour. But this is not a strict histochemical test unless an RNAase control extraction is done.

II. **Cytochemical Method:** Extracting DNA from nuclei and separating it from its protein, estimation of DNA can be done.

- **Conclusions**

Results from both methods indicate:

1. DNA occurs generally only in the nucleus.
2. It is confined to the chromosomes.
3. The amount of DNA is constant in the diploid nuclei of a species in which the haploid nuclei have half of this amount and the tetropoid nuclei, in the few cases show twice the diploid content. Thus, there is a constant quantity of DNA for each set of chromosomes.

III. **Microbiological Method**

1. Studies of bacterial viruses show that their virus proteins contain DNA, labelling the DNA of a virus with radioactive isotopes. Hershey has shown DNA is transferred into a bacteria when infection with phage occurs, the coat of the virus being left outside at the surface of the bacteria. The viral DNA is incorporated into the genetic system of the host cell when transduction occurs.
2. DNA is transferred from one bacterial cell to another during the process of bacterial transformation. The essential substance transferred is known as *transforming principle.* The DNA particles thus introduced behave like genes that have been incorporated into

the transformed cells. DNA transferred must therefore have genic specificity and be able to replicate when the transformed cell reproduces.

- **Conclusion**

From these observations, it is shown that chromosomes and genes have the ability to reproduce their own specific duplicates for specific forms of DNA possessed by individual genes. For this reason, DNA sometimes referred as determiner or bearer of the genetic information.

3. **Avery and Co-workers (1944):** The produced evidence that nucleic acid DNA rather than protein is the carrier of genetic information.

   The bacteria that cause pneumonia, i.e. Pneumococcus exists in two different strains, capsulated and non-capsulated. They prepared a media containing chemical extract of disrupted capsulated bacteria without any living cell. In this media, non-capsulated bacteria were introduced for culture. The remarkable thing was that non-capsulated bacteria developed capsule. The conclusion is that some chemical agent in the medium has brought this permanent and inheritable change. By systematically isolating, purifying and testing the different chemical components, they were able to show it was nucleic acid, DNA is the active agent for genetic change.

4. **Griffith's Experiment (1928):** He injected heat-killed bacteria of type IIIS (i.e. smooth, capsulated, virulent but dead) into mice carrying type IIIR (i.e. Rough, non-capsulated and not virulent) bacteria and recovered type IIIS (i.e. smooth, capsulated and virulent) bacteria which could be cultured.

   From the result he postulated that it has probably occurred through the transfer of some active substance. Following Griffith's method, a number of workers have shown that such transformation occurs both in vivo and vitro.

5. **Avery, MacLeod, McCarthy (1944):** The steps in their experiments are:
   (a) Extraction of DNA from heat-killed Pneumococcus bacteria of type IIIS.
   (b) Mixing of this extracted DNA directly with in vitro culture of pneumococcus bacteria of type IIR.
   (c) Addition of serum whose antibodies react with the type II R cells and force them to precipitate at the bottom.
   (d) To collect type IIIS cells in cases where transformation has occurred. This becomes possible as type IIIS cells not being precipitated grow diffusely throughout the medium.

- **Conclusions**

DNA extracted from heat-killed type IIIS cannot give rise to live type IIIS nor any change in type IIR cells or mutate to type IIIS. But the presence of type IIIS causing transformation of type IIR to type IIIS occurred due to DNA extracted from heat-killed type IIIS has *induced* the transformation (Fig. 13.3).

6. **Experiment on the Alteration of Inherited Metabolic Properties by DNA:** Experiment have established that DNA from one strain of bacteria can alter the inherited metabolic properties of a

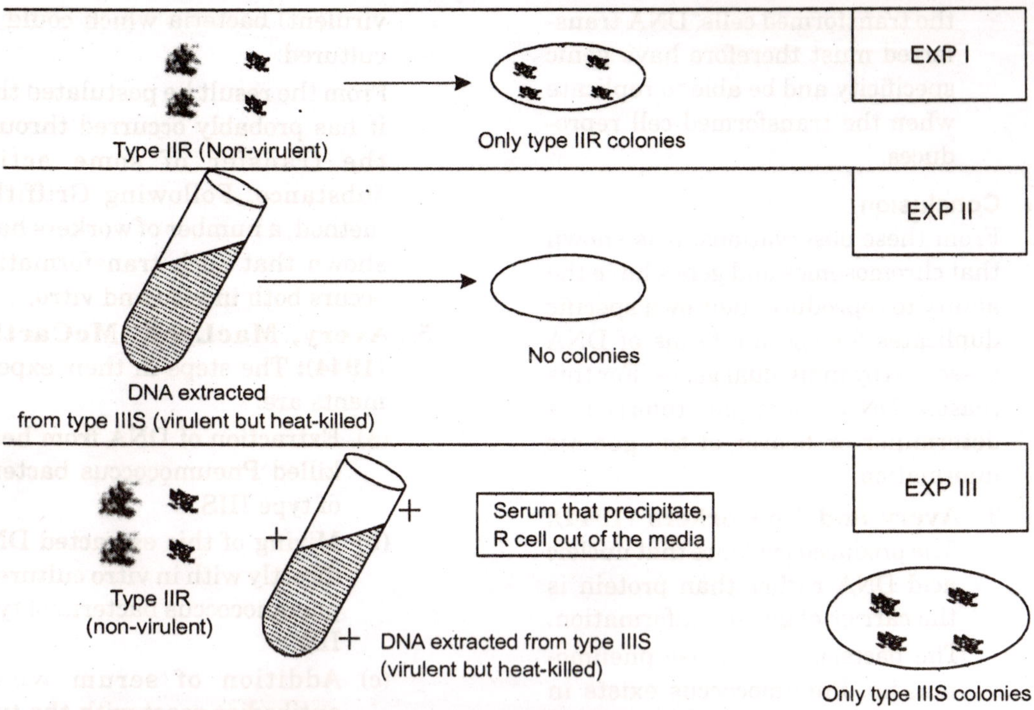


**Fig. 13.3:** Transformation of Pneumococcus Bacteria.

second strain of bacteria. By the application of mutation technique a strain of Pneumococcus can be produced which is incapable of utilising Mannitol. These bacteria lack the enzyme, "Mannitol-phosphate dehydrogenase" and are designated as $M^-$ cells. If DNA is isolated from $M^+$ cells and is introduced in a culture of $M^-$ cells, a few $M^-$ cells become transformed into $M^+$ cells and capable of producing the enzyme for utilization of mannitol.

- **Conclusion**
  - (a) Hereditary capacity of the cells can be permanently altered.
  - (b) A part of the genetic material in DNA is taken up by the chromosomes of the host and in certain percentage of the cases this corporated material is replicated in subsequent generation.
  - (c) Incorporation of DNA into a permanent functional structure is a *"combination"* process.

7. **Production of Antibiotic Resistant Strain:** If purified DNA, extracted from a Penicillin resistant strain of Pneumococcus is added to a culture media of Penicillin-sensitive Pneumococcus, transformation takes place and the whole culture becomes Penicillin resistant.

8. **Transformation in "Bacillus Subtilis":** A strain of this bacterium is unable to synthesize tryptophan and as such cannot thrive in a culture medium lacking the amino acid tryptophan. If such a culture is treated with DNA from a normal strain, a number of 'Transformants' appear in the Tryptophan-free medium. From these, many colonies of the new strain arise since the trait is heritable.

9. **Bacteriophage Infection (1982):**

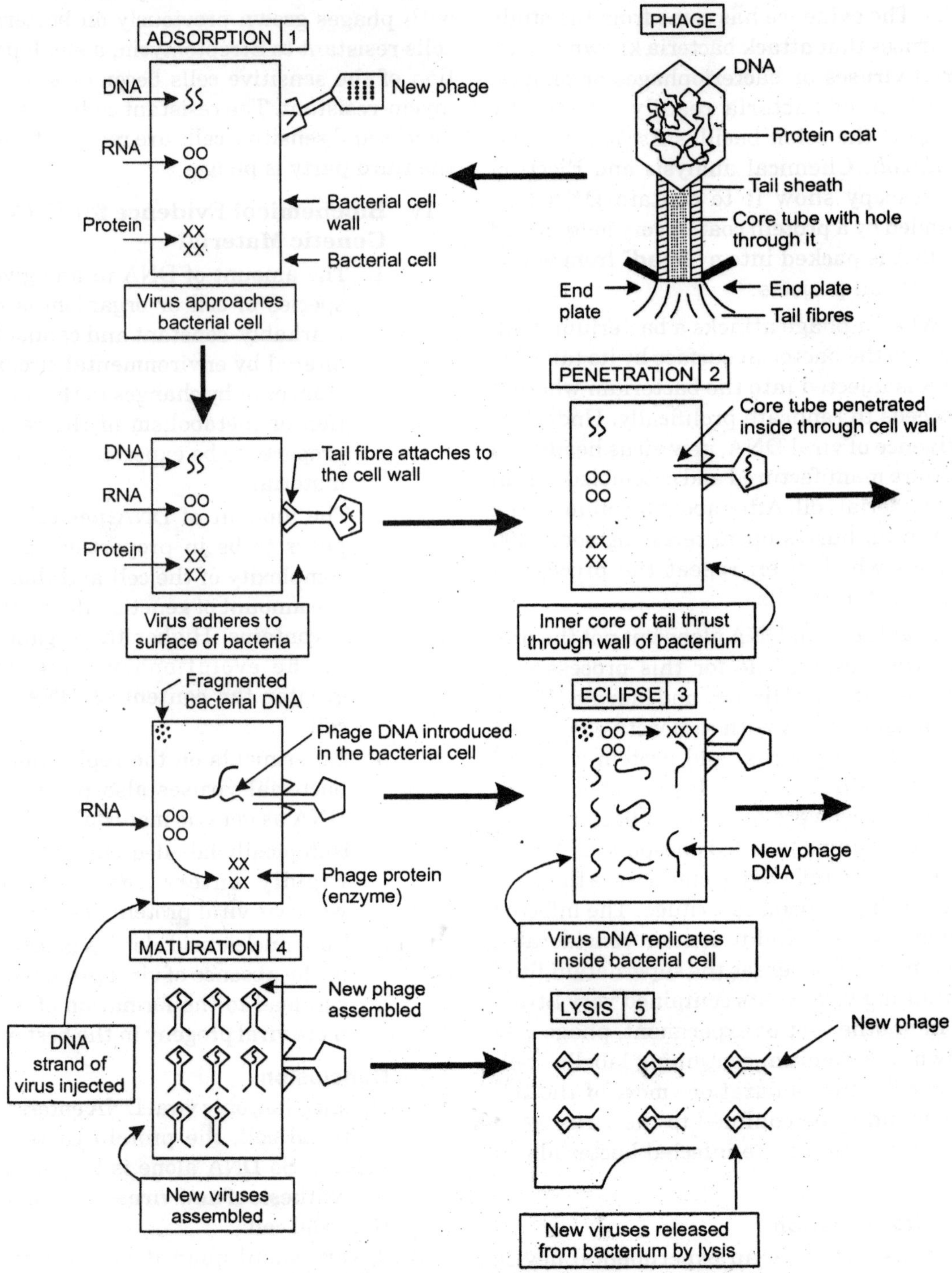


**Fig. 13.4:** Bacteriophage infection.

DNA exerts a potent influence over the biochemical processes which take place in the cells. The evidence has come from the study of viruses that attack bacteria known as bacterial viruses or bacteriophages or phages. One type of bacteria readily infected by phage is the colon bacillus *Escherichia coli* or *E. coli*. Chemical analysis and Electron Microscopy show it to contain DNA surrounded by a protein coat. A very long thread of DNA is packed into a **"head"** from which a short tail projects.

When a phage attacks a bacterium it adheres to the bacterial surface by its tail. The DNA is injected into the bacterium where it proceeds to replicate prolifically. Under the influence of viral DNA, new virus heads and tails are manufactured and assembled within the bacterial cell. After about 30 minutes the bacterium bursts open, releasing some 300 viruses which then repeat the process in other bacterial cells.

That the viral DNA alone but not the protein was responsible for this process was demonstrated by Hirshey and Martha (1952). The labelled DNA with radioactive $P^{32}$ (as it contains phosphorus) and protein with $S^{35}$ (as it contains sulphur). In one set of experiment viruses were grown in a bacterial medium containing labelled phosphorus. The virus coats and infected bacteria were then isolated by high-speed centrifuge. The infected bacteria were found to contain considerable amount of $P^{32}$ whereas the supernatant fluid containing viral coats contained very little.

In another set of experiment, phage was grown in a medium containing labelled sulphur $S^{35}$ after separation, most of the $S^{35}$ was found to be confined to the virus coats and very little in the infected bacterial cell (Fig. 13.4).

- **Transduction**

Transduction is an experimental device by which genetic characteristics of one bacterium is transferred to another bacterium by using a third party. This third party in transduction is a phage. Thus when a bacterium sensitive to Streptomycin is injected with phages grown previously on bacterial cells resistant to Streptomycin, a small portion of the sensitive cells becomes streptomycin-resistant. The resistant cells here are donors and sensitive cells are recipients and the third party is phage.

## IV. Biochemical Evidence for DNA as Genetic Material

1. The amount of DNA in any given species of cell or organism is remarkably constant and cannot be altered by environmental circumstances or by changes in the nutrition or metabolism of the cell, a property to be expected of genetic material.
2. The amount of DNA per cell appears to be in proportion to the complexity of the cell and thus to the amount of genetic information it contains. Higher the organism in the evolutionary scale, the greater the content of DNA per cell.
3. Experiments on the replication of bacterial viruses also pointed to DNA as genetic material.
4. Isotopically labelled DNA of virus rapidly enters the host cell whereas viral protein does not.
5. Viral nucleic acid alone is infectious in the absence of viral protein and can lead to the formation of complete viral progeny in the host cell.

- **Conclusion**

A. On infection, only viral DNA enters the bacterial cell, the protein coats left behind. So DNA alone is responsible for synthesis of new viruses within the bacterial cell.

B. A very small quantity of protein is injected into the bacterium along with DNA and this consists of the enzymes required to start the process.

C. The viral DNA takes over the entire metabolic machinery of the bacterium suppressing its normal control processes and causing it to manufacture new protein coats and nucleic acid identical with those of the invading phage.

D. One theory of cancer is that cells may harbour foreign genetic material possibly from viruses or proviruses which exists in the cell in temperate form and suddenly becomes virulent. In this case genetic material does not manufacture new viruses but instead causes rapid multiplication of cells to become malignant tumour. This suggests that although DNA may exert a powerful influence in a cell, it can be suppressed or inactivated, at least temporarily. This is very important because if all the DNA present in a cell were active all the time chaos might result.

E. This experiment also confirms the Watson-Crick hypothesis and establishing the mode of action of DNA.

F. Although DNA is normally the primary form in which genetic information is stored and from which it is passed to the daughter cell by replication, we know today that RNA can store genetic information in some viruses. Moreover, the genetic information may sometimes be transferred from RNA to DNA by *"reverse transcription"* by *"Reverse Transcriptase"* enzyme.

- **Discussion**

The genetic material for some animal and plant viruses is *RNA* rather than DNA. Although some RNA viruses do not ever have their information transcribed into a DNA molecule, many animal RNA viruses—specifically the retroviruses—are transcribed by an RNA-dependent DNA polymerase to produce a double-stranded DNA copy of their RNA genome. In many cases the resulting double-stranded DNA transcript is integrated into the host genome and subsequently serves as a template for gene expression and from which new viral RNA genomes can be transcribed.

## QUESTION

1. What is your idea about location of genetic substance?
2. State briefly with experimental evidence and suitable diagram that nucleic acid is genetic material.
3. What are the evidences that DNA is the carrier of genetic information?
4. Describe the experiment to prove DNA's control over the cell or give experimental evidences that DNA exerts potent influence over the biochemical processes in the cell.
5. Describe DNA as a carrier of information with experimental evidence of bacteriophage infection. What conclusion can be drawn from the experiment?
6. What are the biochemical evidences that DNA is a carrier of genetic information?

# 14

# REPLICATION OR DUPLICATION OR SYNTHESIS OF GENES OR DNA

1. Replication or duplication or synthesis of DNA or genes—Introduction, Functions of DNA as carrier of genetic information, Steps of replication, Theories of replication, evidence for semiconservative theory, Replication and mitosis, Regulation of replication.

## Introduction

After the study of the molecular structure by the X-ray diffraction analysis of the conformation of fibrous and globular protein molecules yielded the concept that each type of protein molecule has a *specific conformation of precise dimensions, which determines its biological function.* This concept triggered the development of *molecular genetics* after X-ray analysis of the three-dimensional structure of DNA by Watson and Crick.

They postulated:

1. The double-helical structure of DNA, which accounted for:
   (a) The molar equivalence of the bases,
   (b) The characteristic X-ray diffraction pattern of DNA,
   (c) The suggestion of the mechanism by which genetic information can be transferred from parent to daughter cells.

After establishment of the Watson-Crick hypothesis, Crick coined a term called *central dogma* of molecular genetics.

It states that genetic information flows from DNA to RNA to protein. But recently this has been modified to define three major processes of preservation and transmission of genetic information such as:

1. **Replication:** The copying of DNA to form identical daughter molecules.
2. **Transcription:** The process by which the genetic message in DNA is transcribed into the form of messenger RNA to be carried to the ribosomes.
3. **Translation:** The process by which the genetic message is decoded on the ribosomes, where RNA is used as a template in directing the specific amino acid sequence during protein biosynthesis. The central dogma was substantiated by:

1. By the discovery of messenger RNA,
2. By demonstrating the sequence of nucleotide in a gene bears a linear

sequence corresponding to the sequence of amino acids in the protein coded by the gene.

## Functions of DNA as Carrier of Genetic Information

As a carrier of genetic informations, DNA has the following functions:

1. **Heterocatalytic function:** When DNA directs the synthesis of chemical molecules other than itself, e.g. RNA, protein, enzyme.
2. **Autocatalytic function:** When direct the synthesis of DNA itself.

During interphase of mitosis two things happen which are essential for cell division.

1. *Firstly,* the genetic material DNA replicates, i.e. double itself because sufficient DNA is made available for each of the two daughter cells.
2. *Secondly,* the cell builds up sufficiently large store of energy, a kind of "energy reservoir" to carry the process. The DNA begins to reproduce about 5 hours before mitosis takes place and the duration of DNA replication is about 4 hours. The all DNA normally duplicate once only. Following replication of the DNA, the nucleus continues to function normally for about one hour before mitosis begins abruptly.

- **Replication**

Essential property of the genetic material is that it (1) should be replicate accurately, (2) must be complete and (3) carried out with high fidelity to maintain genetic stability within the organism and the species.

The *Watson and Crick double-stranded model of DNA* strongly suggest that replication of the DNA molecule occurs in a *semiconservative manner*, i.e. two chains separating from each other rather like *undoing a zip-fastener.* Thus, when each strand of the double-stranded DNA molecule separates from its complement during replication, each can then serve as a template on which a new complementary strand can be synthesized (Fig. 14.1).

*Watson and Crick* envisaged that the DNA molecule as a kind of twisted ladder, the two uprights consisting of chains of alternating sugar and phosphate groups, the rungs as pairs of bases sticking inwards each other and linked up by weak hydrogen bond in a specific relationship: A with T and C with G (Fig. 14.2). For one complete turn of the helix, there are ten nucleotides having length of 34 Å. The width of the helix is 20 Å and distance between the sugar linkage of two nucleotides is 3.4 Å and between purine and pyrimidine linkage is 2.8 Å. The two strands are held together by hydrophobic or van der Waals forces and hydrogen bonds.

## Sequences of Mechanism of Replication

The sequences in the mechanism are:

1. **Separation of chain:** Two chains are separated from each other like undoing a zip-fastener or semiconservative way. The hydrogen bonds linking the base pairs are weak so breakage could easily happen. Arthur Kornberg observed the existence of an enzyme *DNA Polymerase* based on the study of *E. coli* organism. It is now recognized that the major role of this polymerase is to ensure fidelity and to repair rather than to replicate DNA.
2. **Initiation:** Free nucleotides might then come along and take up positions in relation to each of the two chains. The initiation of DNA synthesis is a complex and is done with the help of a *short link of RNA* about 10 nucleotides which called *RNA Primer* to which at its 3′-hydroxyl end is attached the first Deoxy-ribonucleotide. These fragments of DNA with RNA initiator or primer component is called *OKAZAKI*

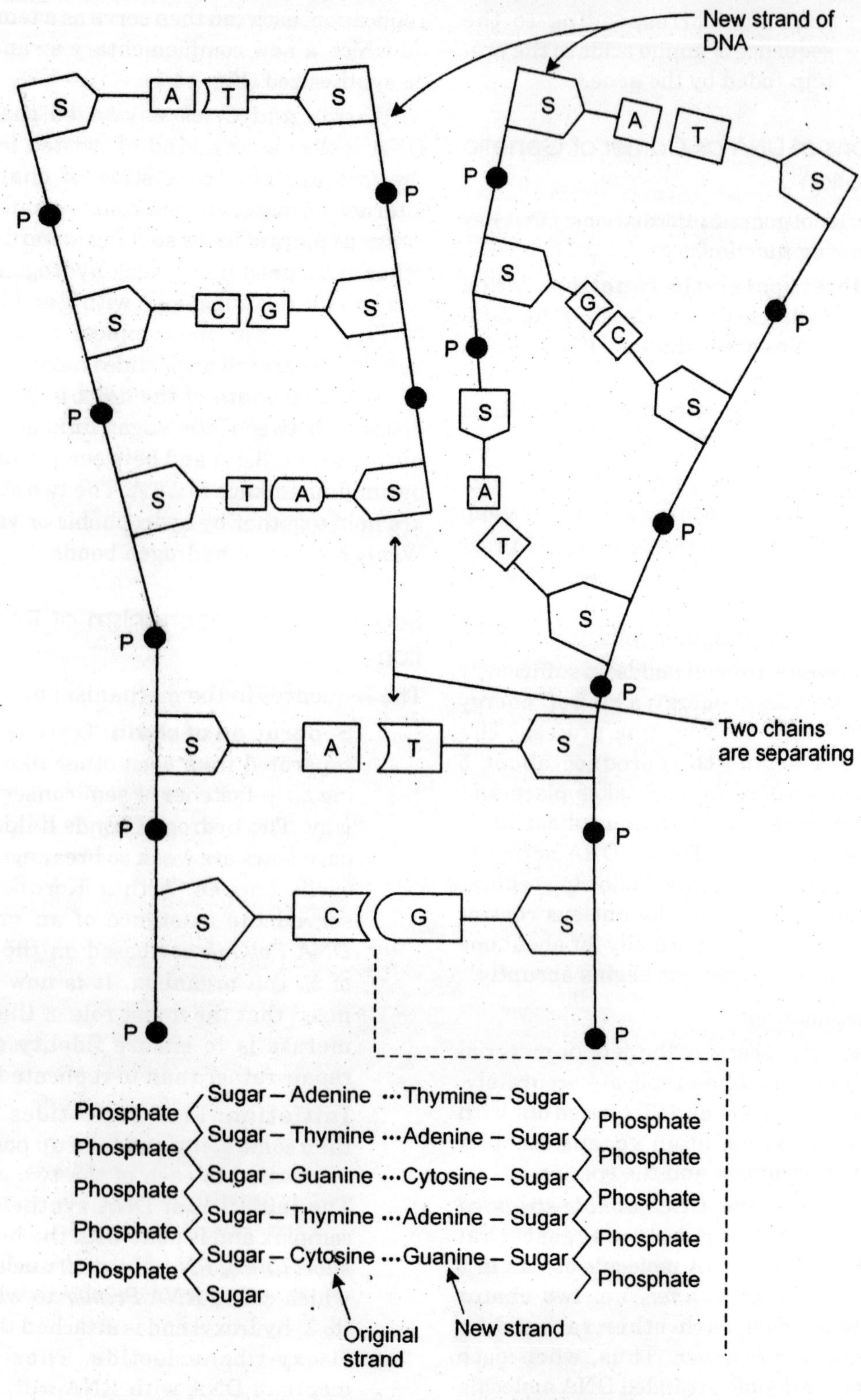


**Fig. 14.1:** Replication of DNA molecule.

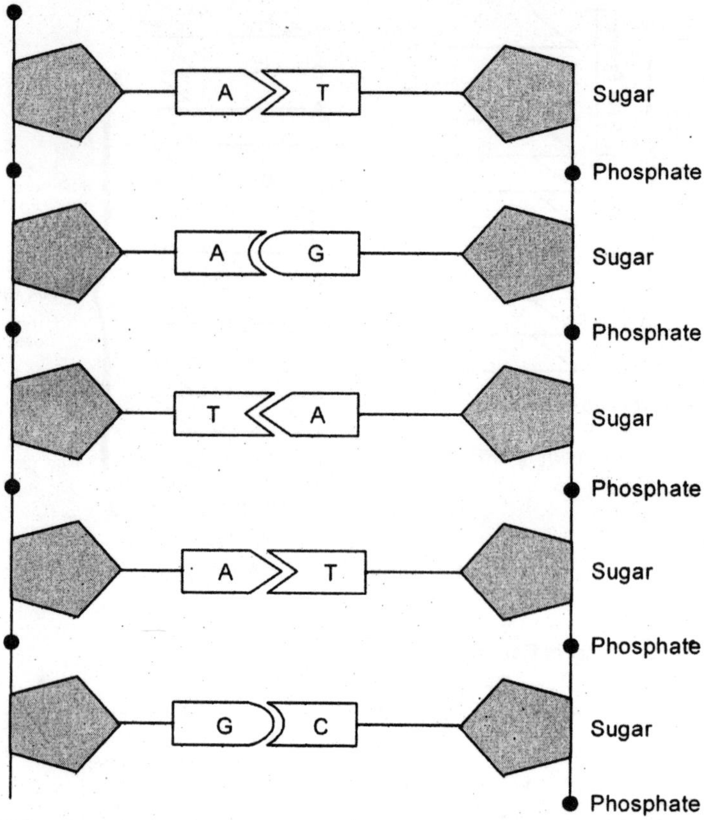


**Fig. 14.2:** A short length of the DNA strand is untwisted to show the position of sugar, phosphate and bases and their pairing.

*PIECES* after the name of discoverer *Okazaki* (Fig. 14.3).

After many "Okazaki pieces" are generated, the replication complex begins to remove RNA primers, the gap is filled up by deoxyribonucleotides and then the fragments of new nucleotides are sealed up by the enzyme *"DNA ligases"*.

3. **Direction:** Though DNA molecules are double-stranded and running in opposite direction, but the replication occurs on both strands simultaneously. However, an enzyme for polymerizing DNA in the 3′ to 5′ direction does not exist in any organism, so that both of the newly replicated DNA strands cannot grow in the same direction simultaneously. But the same enzyme appears to replicate both strands at the same time. The single enzyme replicates one strand called *leading strand* in a continuous manner in the 3′ to 5′ direction, with the same overall forward direction. It replicates the other strand called *"lagging strand"* discontinuously, i.e. the enzyme while polymerizing 150-250 nucleotides in the 5′ to 3′ direction, but at the same time it faces toward the back end of the preceeding RNA primer rather than toward the unreplicated portion (Fig. 14.4).

4. **Replication bubbles:** In mammalian cells there are three types of DNA polymerase enzyme:

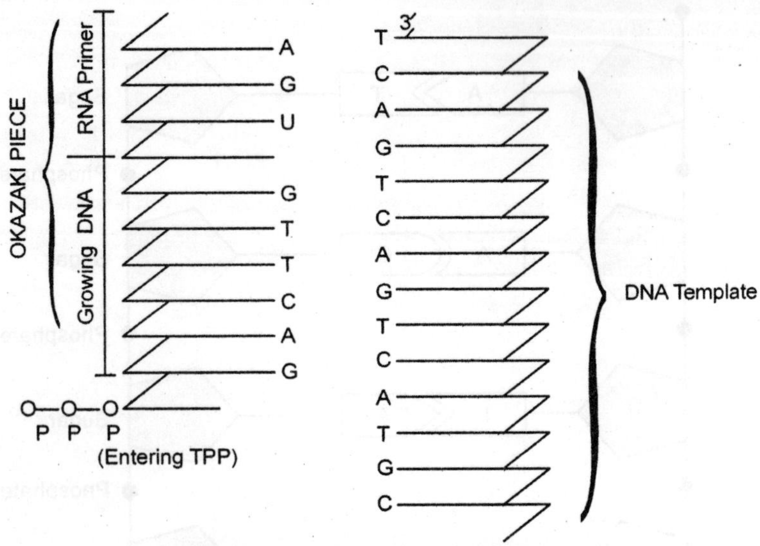


**Fig. 14.3:** Initiation of DNA synthesis.

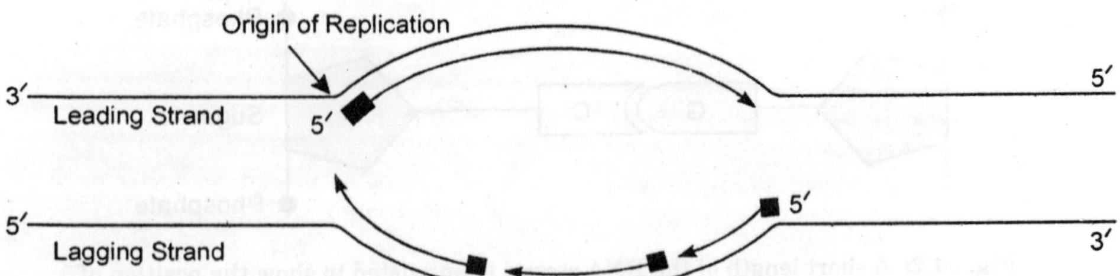


**Fig. 14.4:** Overall direction of replication.

(a) *Polymerase alpha* a maxi or large polymerase, responsible for chromosome replication and one molecule of enzyme can polymerize 100 nucleotides which are 10-fold less than bacterial DNA polymerase and due to interference of nucleosomes that remain attached to DNA during replication.

(b) *Polymerase beta* a mini polymerase having lower mol.wt. present in the nuclei not responsible for DNA replication but for DNA repair.

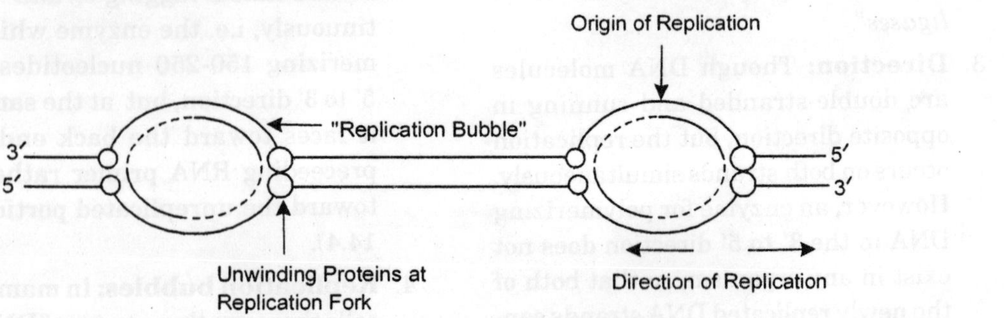


**Fig. 14.5:** Replication bubbles.

(c) *Polymerase gamme* (Mitochondrial DNA Polymerase). It is also known as mitochondrial DNA polymerase, responsible for replication of mitochondrial genome, another DNA molecule in circular form to replicate the entire mammalian genome in 9 hours, there are on the chromosomes "multiple origin" of DNA replication that occur in clusters of up to 100 of these replication units. Portions of DNA between nucleosomes serve as the template sites for Okazaki piece and replication occurs in *both directions up and down* the chromosome and on *both strands simultaneously.* This replication process generates *replication bubbles* (Fig. 14.5).

5. **Stabilization of Single Strand by Protein Molecules:** There must be separation of the two strands to serve as template by hydrogen bindings of its bases with the incoming new nucleotides. The separation is promoted by specific protein molecules that *stabilize the single-stranded structure* as the replication fork progresses (Fig. 14.6).
6. **Unwinding of the Molecule:** In addition to separate the double helix, there must be an "unwinding" of the molecule (once every 10 nucleotide pairs) to allow *the rewinding of the newly formed DNA replicas.* Considering the time factor, i.e. 400,000 turns per second must be unwinded, which is impossible feat. Therefore, there must be multiple "swivels" interspersed in the DNA molecules which is provided by a specific enzyme that introduces *nicks in one strand of the unwinding double helix,* thereby allowing the unwinding process to proceed. The nicks are quickly resealed without requiring energy input due to the formation of a high-energy covalent bond between the nicked phosphodiester backbone and nicking-sealing enzyme (Fig. 14.7). This may be done by two ways by *two sets of enzymes*:
   A. *DNA ligase*: ATP-dependent resealing carried out by it.
   B. *DNA topoisomerase*: It is nicking resealing enzyme and also capable of unwinding supercoiled DNA.
7. **RNA-dependent DNA Polymerase:** Recently a class of enzyme called *reverse transcriptase* has been discovered which can synthesize a single stranded and then double-stranded

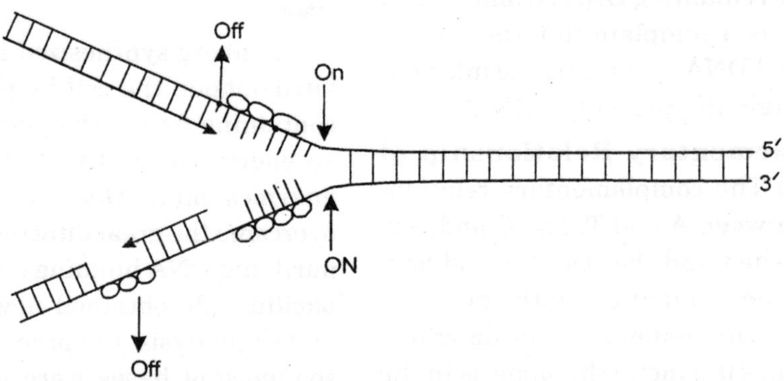


**Fig. 14.6:** Hypothetical scheme for single-strand binding protein action at a replicating fork.

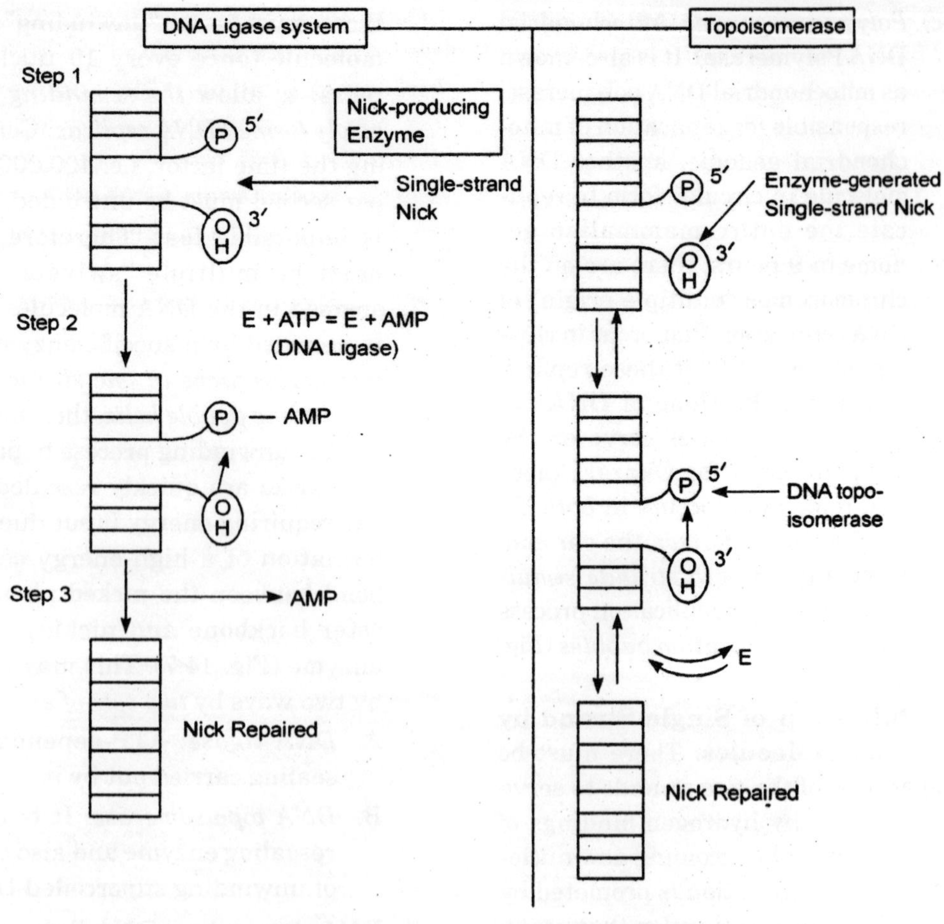


Fig. 14.7: Unwinding of DNA molecule.

DNA molecule using single-stranded RNA template. An enzyme called *RNase H*, degrades the RNA strand and the remaining DNA strand in turn serves as a template to form double-stranded DNA containing the information originally present in RNA.

8. **Complementary Relationship of Bases:** The complementary relationship between A and T, and C and G is established and the result would be a double helix identical with the original one. The sequence in the daughter molecules is exactly the same as in the parent molecule, so the information content of the DNA is unchanged, a feature of genetic replication which is essential.

- **Evidence for Accurate Replication of DNA**

Kornberg synthesized DNA molecules in vitro outside the cell by preparing a media containing free nucleotides of all four types, an energy source (ATP), DNA synthesizing enzymes, intact DNA to act as template. He overcame the great difficulty by isolating and purifying DNA building enzymes from colon bacillus. He obtained new DNA molecules and on analysis, the proportion of bases and sequence of bases were found like parent DNA, a strong indication that accurate replication had occurred.

## Theories of Replication, i.e. Semiconservative Virsus and Conservative Theory

1. **Semiconservative Theory (or Zip-fastener Idea):** The zip-fastener idea, i.e. semiconservative theory is neat and economical way of explaining replication.

   The argument is that the daughter DNA molecule does not contain exactly the same molecule as the original double helix. One of the two strands will be derived from the parent DNA but the other strand will be new. (Fig. 14.8A).

2. **Conservative Theory:** One of the double helix in every generation will have exactly the same molecules as the parental one—in both strands (Fig. 14.8B).

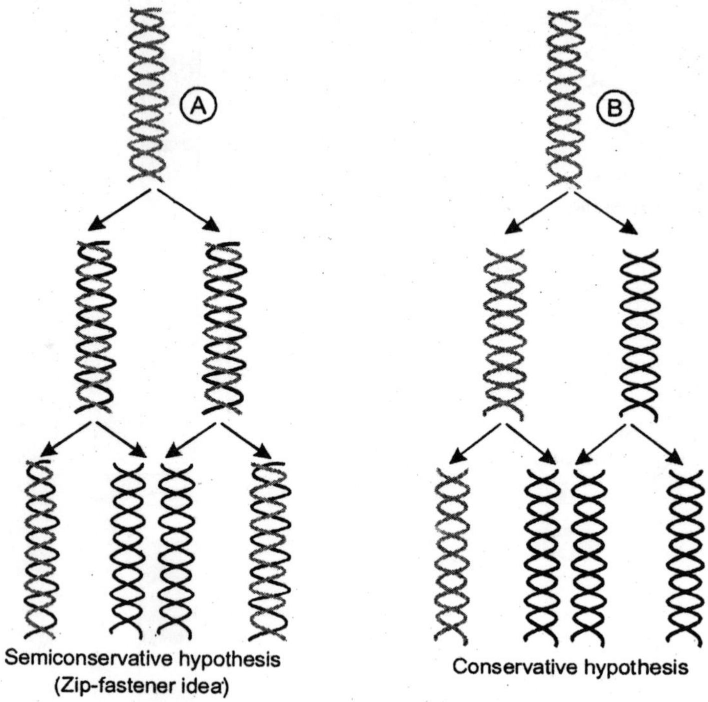


**Fig. 14.8:** Representation of theories of replication.

- **Evidence in Favour of Semiconservative Hypothesis**

The experiment was carried out by Meselson and Stahl (1958).

Cells of the colon bacillus (*E. coli*) were grown on a medium for many generations in which normal nitrogen $N^{14}$ was replaced with the heavy isotope $N^{15}$. This was continued until it was certain that heavy nitrogen $N^{15}$ had been incorporated into all the DNA, after which the heavy nitrogen $N^{15}$ was replaced by normal nitrogen $N^{14}$. Samples of bacteria were taken at intervals and the relative amounts of the two types of nitrogen were estimated after separating the DNA molecules by high-speed centrifuge and the DNA was detected by its absorption of ultraviolet light. In this way 3 fractions were isolated:

(a) *Heavy DNA*: Containing heavy isotope $N^{15}$.

(b) *Light DNA*: Containing $N^{14}$.

(c) *Intermediate DNA:* Containing $N^{14}$ and $N^{15}$.

In the first generation, DNA contained equal amount of $N^{14}$ and $N^{15}$. In the second

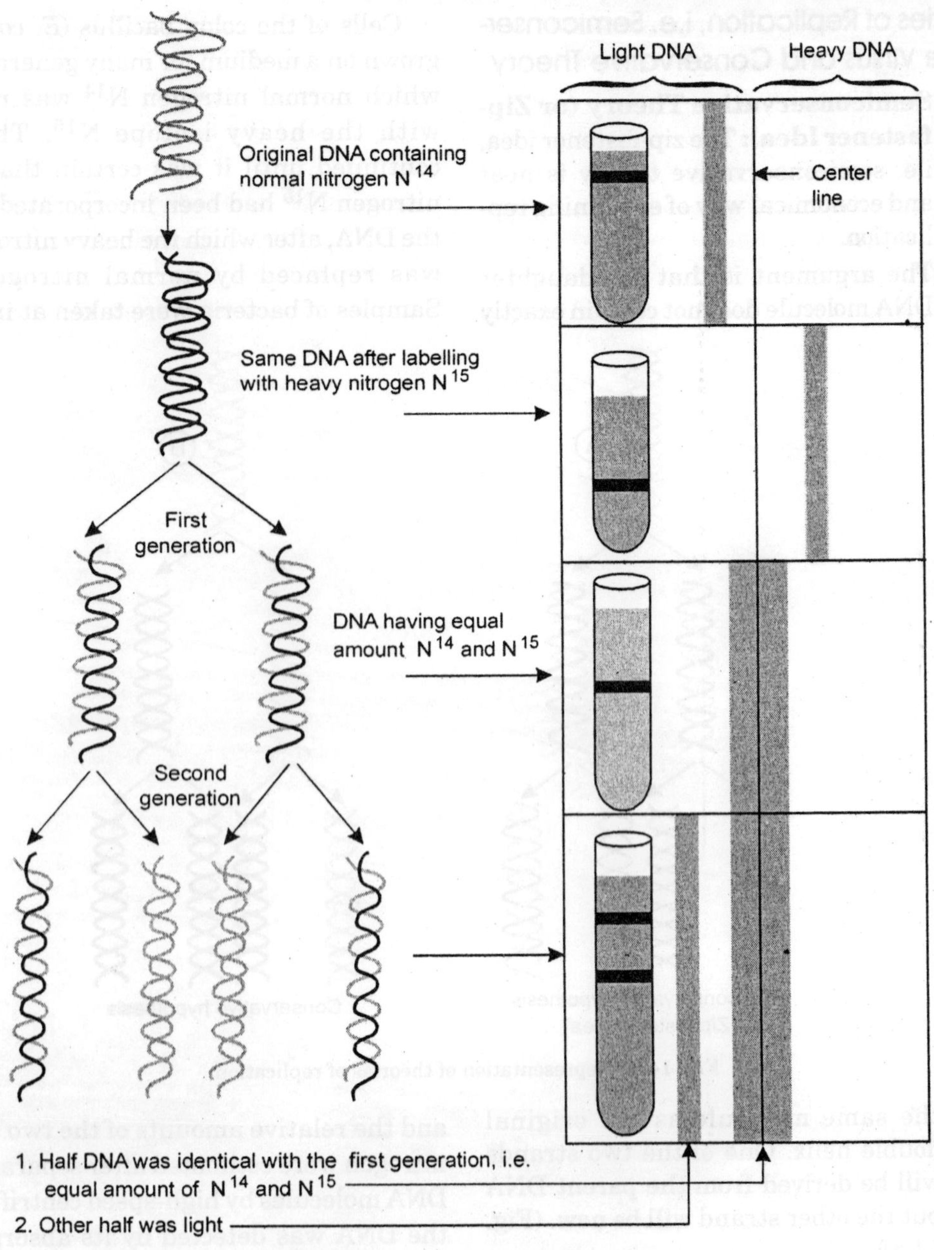


**Fig. 14.9:** Meselson-Stahl experiment.

generation, two sorts of DNA were detected: half was identical with that obtained in the first generation, the other half was light i.e., without heavy Nitrogen $N^{15}$. This experiment favours the semiconservative theory and supports the DNA model of *Watson and Crick* (Fig. 14.9).

## DNA Replication and Mitosis

It is established that by special staining, i.e. Feulgen staining and isotope labelling that in all organisms DNA is confined to the chromosomes. A chromosome can be regarded as an enormously long strand of DNA coded on itself. It is fact that the amount of DNA in the nucleus doubles during interphase and replication takes place before mitosis, i.e. before the chromatids make their appearance. Each chromatid therefore contains a double helix derived from replication of the DNA belonging to the original chromosomes. By separating chromatids, mitosis ensures that the products of cell division contain exactly the same complement of DNA as the parent cell did before replication. It is due to the fact that each of the two double helices formed by replication during interphase is contained in one of the two chromatids that become visible in prophase. These chromatids then become the chromosomes of the daughter cells. The independent behaviour of two separate lengths of DNA belonging to a pair of homologous chromosomes in mitosis, the daughter cell come to contain both lengths of the DNA, just like the parent cells. How would be the distribution of the DNA amongst the daughter cells in meiosis?

## Salient Features of DNA Replication

1. **Multistep Complex Process:** DNA replication is a multistep complex process.
2. **Catalysis by Multi-enzyme Complex:** DNA replication is catalysed by the multi-enzyme complex also known as replication apparatus or replisome and needs several proteins.

   • **Enzymes**

   (a) **Helicases:** To unwind a DNA duplex and separate DNA strands at replication fork.

   (b) **Topoisomerases:** There are two topoisomerases I and II. Some can remove only negative supercoils from DNA, others can remove both negative and positive supercoils.

   (c) **DNA Polymerase:** It makes a new DNA on a template strand. It is of three types, i.e. I, II and III.

   (d) Enzymes that synthesise RNA primers.

   (e) An enzyme to remove RNA primers.

   (f) An enzyme to covalently link successive Okazaki fragments.
3. **Requires proteins:** Proteins that prevent pairing between the complementary strands. It is known as *single-stranded binding proteins* (SSBP) tightly bind to the single-stranded regions produced by the action of DNA gyrase and DNA helicases.
4. **Replication starts at unique and fixed points:** DNA replication always begins at certain unique and fixed points of DNA called *origin*. But in case of eukaryotic chromosomes begins at many origins located at different positions along the chromosomal DNA.
5. **DNA replication is bidirectional:** Replication proceeds from a given point on the chromosome either in one or in both directions. Bidirectional movement involving two growing points or forks appears to be the most common.
6. **Before replication, double-stranded DNA must be separated:** The double-stranded DNA does not replicate. Hence before replication double strands of DNA must be separated from the point of replication origin in the form of replication eye or Y-shaped replication fork.
7. **DNA is replicated by DNA polymerases:** DNA is replicated by enzymes known as DNA polymerase which makes a new DNA on a template strand.
8. **Replication is a semiconservative process:** Each strand of the parental DNA duplex is conserved and copied by base-pairing with matching nucleotides and the result is two duplexes identical to the parental one. Base

**Table 14.1:** Comparison between Unidirectional and Bidirectional Replication

| Unidirectional | Bidirectional |
|---|---|
| 1. DNA replication proceeds in one direction from the point of origin. | 1. Proceeds in two directions from the point of origin. |
| 2. One of the two ends of the replication bubble remains stationary and the other end moves with replication. | 2. Both ends move with replication. |
| 3. Autoradiography reveals labelling is seen only at one fork. | 3. Reveals labelling on both the replication forks. |
| 4. Occurrence—DNA of coliphage $P_2$, mitochondrial DNA. | |

pairing and chain extension create a growing point, or fork, at which the parental duplex is unwound.

9. **Replication of both strands proceeds in the 5′-3′ direction:** In a DNA duplex, two parental antiparallel strands run in opposite directions, i.e. one is $5' \rightarrow 3'$ and other is $3' \rightarrow 5'$. But the two parent strands are replicated in different ways. Synthesis of daughter DNA by addition of nucleotide monomers always proceed in the $5' \rightarrow 3'$ direction on the both parental strands.
10. **Replication occurs in discontinuous pulses: Daughter DNA** is continuously synthesized in its $5' \rightarrow 3'$ direction as the replication fork advances. The replication of the *parental template* strand ($5' \rightarrow 3'$) of the DNA molecule is discontinuous.

## Regulation of DNA Synthesis or Replication

In animal cell including human cells, the replication of the DNA genome occurs only at specified times during the life span of the cell which is called "synthetic or S-phase". This phase is temporarily separated from mitotic phase by non-synthetic periods referred as gap 1 ($G_1$) and gap 2 ($G_2$) occurring before and after the S-phase (Fig. 14.10). The regulation of the entry of a cell into an S-phase involves "cyclic purine nucleotide and probably "dCTP", but the mechanism is not known.

**Table 14.2:** Comparison between continuous and discontinuous replication

| Continuous replication | Discontinuous replication |
|---|---|
| 1. Exhibits in leading strand of a replication fork. | 1. Exhibits in lagging strand of a replication fork. |
| 2. It starts at the beginning of the replication. | 2. It starts later and joins the main body of the growing chain. |
| 3. Unidirectional, progresses towards the replication fork. | 3. Progresses away from the replication fork, i.e. in opposite direction. |
| 4. Synthesis of DNA occurs continuously. | 4. Synthesis of DNA is discontinuous-short fragments, i.e., Okazaki fragments are synthesized and subsequently they are joined. |
| 5. Does not require RNA primer. | 5. Requires RNA primers. |
| 6. Only DNA polymerase III is involved. | 6. Enzymes involved are primase DNA polymerase III, DNA polymerase I and DNA ligase. |
| 7. Leading strand is synthesized continuously. | 7. Lagging strand is synthesized discontinuously is called semi-discontinuous replication. |

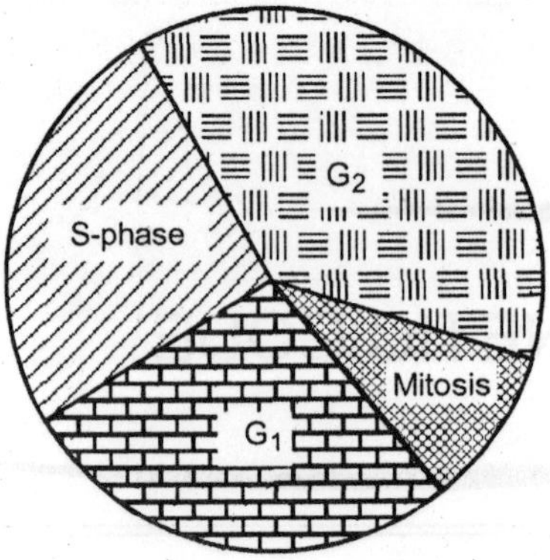


**Fig. 14.10:** Mammalian cell cycle.

During S-phase, the following are seen in the cell.

1. Increase in *polymerase alpha.*
2. Increase in enzyme for substrate for DNA synthesis, i.e. deoxyribonucleoside triphosphate.
3. Nuclear DNA is *completely replicated once and only once.*

   It seems that once chromatin has been replicated, it is marked so as to prevent its further replication until it again passes through mitosis.
4. *A given pair of chromosomes will replicate simultaneously* and within a fixed portion of the S-phase upon very replication.
5. On a chromosome, *clusters of replication units replicate coordinately.*

The nature of signals that regulate DNA synthesis at these levels is unknown, but the regulation does appear to be an intrinsic property of each individual chromosome.

- **Biological Significance of DNA**

1. DNA replication takes place most truly.
2. DNA is sufficiently stable and has minimum chances of mutation and hereditary changes.
3. DNA is able to store all the information related to life.
4. DNA is capable of transmitting this information in the form of reactions in the cell.

## QUESTION

1. What is replication? What functions DNA have as carrier of genetic information?
2. What are the essential properties of genetic material as a carrier of information? How replication of DNA occurs?
3. What is the evidence for accurate replication?
4. What are the theories for replication? Which theory is favourable? What is evidence in favour of semiconservative theory?
5. Describe the relationship of DNA with mitosis.
6. Describe the regulation of replication of DNA.

# 15

# GENETIC CODE

1. Base triplet hypothesis, 2. Codon; Genetic code, Genetic dictionary, 3. Properties of genetic code, 4. Evidences for genetic code, 5. Importance of genetic code, 6. Deciphering the genetic code.

## DNA as a Code for Proteins Synthesis or Mechanism of Instruction of DNA to Cell

The role of DNA is to instruct the cell what proteins to make. How might this be done? A typical protein contains at least twenty different types of amino acid arranged in a specific sequence. There may be as many as 500 amino acids altogether in a single molecule of protein and the way these are arranged gives the particular protein its individuality. Somehow the DNA with its four different bases has got to determine the sequence in which the 20 types of amino acids are put together in the protein. The question is: how can a four-letter code specify a 20-unit protein? It is not possible for a *single base* of RNA (having single alphabet, i.e. A or T or C or G) to specify a single amino acid, for then only four different amino acids could be coded for, and proteins containing only four kinds of amino acid would be formed. Nor is it feasible for *two bases* of RNA to specify a single amino acid since this would that only $4^2$ or → (4 × 4) or 16 doublets are possible and 16 amino acids could be coded for. But *three bases* of RNA, i.e. triplet code would be sufficient to form $4^3$ or → (4 × 4 × 4) or 64 triplets which are sufficient to code for 20 amino acids. This is known as "*Base Triplet Hypothesis*". A triplet of bases specifying an amino acid is known as a "codon" and it forms the basis of genetic code.

20 amino acids are the alphabets of the language of proteins and the 4 bases are the alphabets of the language of nucleic acid. Since the genetic information from DNA is passed through mRNA (messenger RNA) to the proteins, the problem of genetic code was to prepare a "genetic dictionary" for translating the language of RNA into the language of proteins (Fig. 15.1).

The genetic dictionary shows the relationship between the triplets in DNA, mRNA and each amino acid.

## Properties of Genetic Code or General Features of the Code-Word Dictionary

1. **The code is triplet:** The singlet or doublet codes are not enough to code for 20 amino acids, therefore, it was pointed out that triplet code is

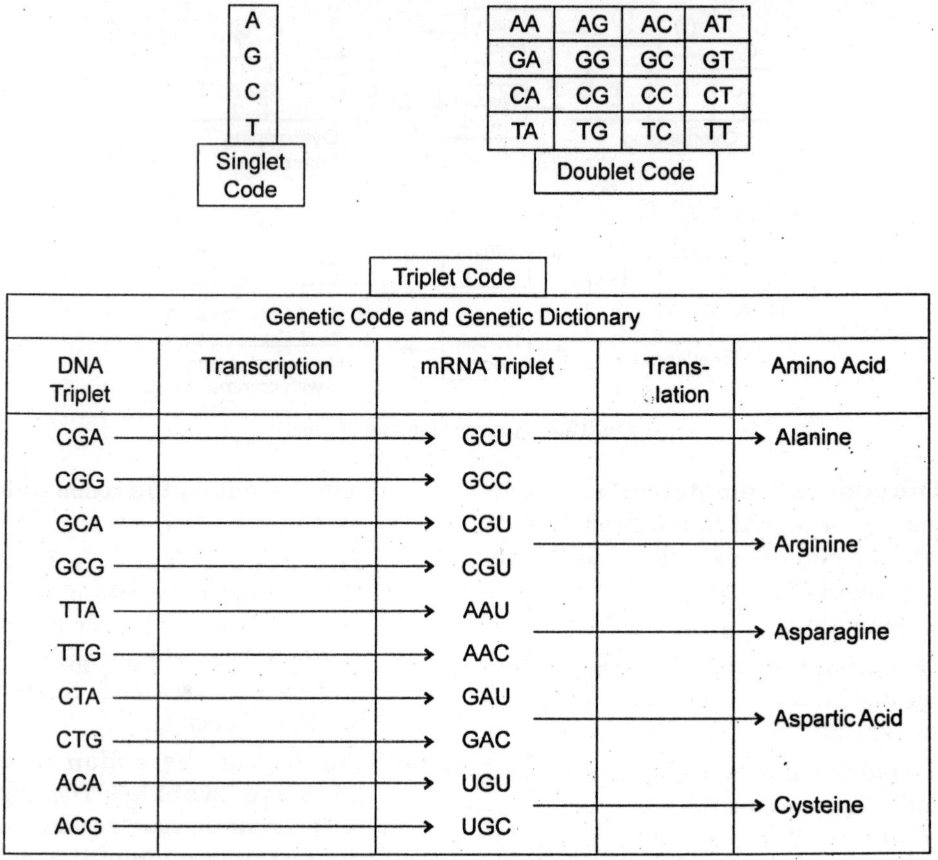


**Fig. 15.1:** Genetic Code and Genetic Dictionary.

minimum required. In a triplet code, there are 64 codons, there is an excess of 44 codons and therefore, more than one codon are present for the same amino acid. This excess will be still greater if a quadruplet is prepared, there will be $4^4 = 4 \times 4 \times 4 \times 4 = 256$ possible words.

2. **The code is degenerate:** There is more than one code word for most of the amino acids. It seems that generally only two of the three nucleotides are required to specify a particular amino acid. So the code contains more potential information than is actually used by the cell: to use the cybernetic term, the code is degenerate.

(a) Degenerate does not mean imperfect, for there is no code word that specifies more than one amino acid.

(b) Degeneracy of the amino acid code has some regularity and provides biological advantages.

(c) Degeneracy is not uniform. Thus, the code for arginine, leucine and serine has six-fold degeneracy, i.e., there are six code words for each of these, whereas the code for other amino acids, such as glutamic acid, tyrosine, histidine and several others, has only two-fold degeneracy of all the amino acids, only tryptophan and methionine have but a single code.

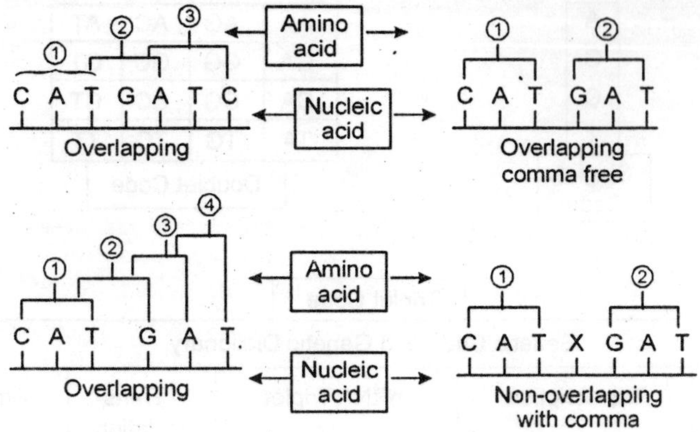


**Fig. 15.2:** Non-overlapping of code.

3. **The code is "commaless":** No punctuation or signal is required to indicate the end of one codon and the beginning of the next, i.e. that the code is commaless.
4. **Third base is less specific in the code:** The third base in the codons is less specific than the first two.
5. **Nonsense codons:** The UAA, UAG and UGA codons do not specify any amino acid so are called "nonsense codons" or termination codons because their inclusion in mRNA results in the abrupt termination of the message though the polypeptide chain has not been completed.
6. **The code is non-ambiguous:** The same may specify more than one amino acid. Such as, UUU codon usually codes for phenylalanine but in the presence of Streptomycin may also code for leucine, isoleucine or serine.
7. **The code is non-overlapping:** That no single base can take part in the formation of more than one codon (Fig. 15.2).
8. **Codon has corresponding anti-codon in tRNA:** The recognition of specific codons in the mRNA by the tRNA molecule is dependent upon their *"anti-codon region"*. Each tRNA molecule contains a specific sequence, complementary to a codon called *"anti-codon"*.
9. **The genetic code is Universal:** Mitochondria require only 22 tRNA molecules to read their genetic code, whereas the cytoplasmic translation system possesses a full complement of 31 tRNA species.
10. **Pairing of the codon and anticodon can "wobble":** The degeneracy of the genetic code resides mostly in the last nucleotide of the codon triplet, suggesting that the base pairing between this last nucleotide and the corresponding nucleotide of the anti-codon is not strict. This phenomenon is referred to as *"wobble"*.

## Evidence for Genetic Code or Method of Cracking the Genetic Code or Evidence for DNA's Control of Protein Synthesis

1. **Experiment of Nirenberg (1961) for cell-free protein synthesis:** **Nirenberg** (1961) made in vitro studies on the incorporation of radioactive amino acids in cell-free protein synthetic system containing artificial, i.e. synthetic mRNA, etc. During this process, the natural mRNA was broken down by mechanical and enzynatic methods. The artificial mRNA was prepared for each nitrogenous base

with the help of enzyme "polynucleotide phosphorylase" obtained from "Azobactor" organism. Thus Nirenberg prepared polyridylic acid (UU UUU ....), polycytidylic acid (CC CCC ...), polyadenylic acid (AA AAA) and polyguanylic acid (GG GGG) from uracil, cytosine, adenine and guanine respectively. He introduced this synthetic mRNA in different cell-free protein synthetic system of *E.coli* containing radioactive amino acids. He found that each artificial mRNA stimulated the synthesis of polypeptides of single kind of amino acids. He concluded that a sequence of UUU code coded for phenylalanine, sequence of CCC code designates proline, AAA for lysine and GGG for glycine.

2. **H. G. Khorana's experiment**

   **Khorana** succeeded in synthesizing mRNA of known sequences and he synthesized an alternating polymer such as UGU GUG UGU. Use of this particular alternating polynucleotide as mRNA in vitro protein synthesizing system gave rise to the formation of the alternating polypeptide (e.g. valine-cystin-valine) evidently directed by the codon sequence UGU-GUG-UGU. Thus Khorana determined the exact *sequence of nucleotides* in the various *genetic codons.*

## Importance of Genetic Code or Significance of Genetic Code

1. In future it would be possible to remove the genes for hereditary diseases and other harmful genes.
2. It would be possible to introduce the desired genes for future generations to obtain the type of persons required.

## Deciphering the Genetic Code

Relationship between sixty-four kinds of code words and twenty kinds of amino acids can be explained by comparing the amino acid sequence of its gene or mRNA. But it was not experimentally proved because the base sequences of genes and mRNA molecules were unknown.

**Experiments**

I. Nirenberg's Experiment (1961)

II. H. G. Khorana's Experiment

Nirenberg discovered that the addition of polyuridylate (Poly U) to a cell-free protein-synthesizing system led to the synthesis of polyphenylalanine. Poly U evidently served as a messenger RNA. The first code word was deciphered: UUU codes for phenylalanine. This study helped to elucidate the genetic code completely.

Nirenberg then found that protein synthesis resumed on addition of a crude fraction of mRNA. Thus, there was a cell-free protein-synthesizing system that was responsive to the addition of mRNA. A variety of synthetic ribonucleotides were added to the cell-free protein-synthesizing system and the incorporation of $C^{14}$-label L-phenylalanine was measured and the results were successful. The same experiment was carried out with a different $C^{14}$-labelled amino acid in each incubation mixture. It was found that poly A led to the synthesis of polylysine and that Poly C led to the synthesis of polyproline. Thus, three code words were deciphered. The code words GGG could not be deciphered in the same way because poly G does not act as a template, probably because it forms a triple-stranded helical structure. Polyribonucleotides with extensive regions of ordered structure are ineffective templates for protein synthesis.

- **Copolymers as Template**

Codon compositions for many amino acids were determined by using copolymer as templates. Polyribonucleotides consisting of two kinds of bases were used as template. And it was concluded that valine, leucine and cysteine are specified by codons that contain 2U and 1G, whereas tryptophan and glycine are specified by codons that contain 1U and 2G.

The same type of experiment was carried out with other random copolymers and the compositions of codons corresponding to each of the twenty amino acids were deduced.

Sixty-four codons have been deciphered. Sixty-one triplets corresponds to particular amino acids, whereas three code for chain termination. Because there are twenty amino acids and sixty-one triplets that code for them, it is evident that the code is highly *degenerate*, i.e. many amino acids are designated by more than one triplet.

Significance of degeneracy—

1. Minimizes the deleterious effects of mutations.
2. Permitting DNA base composition to vary over a wide range without altering the amino acid sequence of the proteins encoded by the DNA.
3. Other eighteen amino acids are coded by two or more.
4. Leucine, arginine and serine are specified by six codons each.
5. In normal conditions, the code is not ambiguous, a codon designates only one amino acid.
6. Codons that specify the same amino acid are called *synonyms* e.g., CAU and CAC are synonyms for histidine.

The synonyms are not distributed haphazardly throghout the table of genetic code. An amino acid specified by two or more synonyms occupies a single box. Most synonyms differ only in the last base of the triplet.

**Table**—showing the position of the bases in a codon, to find out the corresponding amino acid. UAA, UAG and UGA are termination signals. AUG is part of the initiation signal, in addition to coding for internal methionines.

**Table 15.1: The Genetic Code**

| First position (5′ end) | Second position | | | | Third position (3′ end) |
|---|---|---|---|---|---|
| | U | C | A | G | |
| U | Phe | Ser | Tyr | Cys | U |
| | Phe | Ser | Tyr | Cys | C |
| | Leu | Ser | Stop | Stop | A |
| | Leu | Ser | Stop | Trp | G |
| C | Leu | Pro | His | Arg | U |
| | Leu | Pro | His | Arg | C |
| | Leu | Pro | Gin | Arg | A |
| | Leu | Pro | Gin | Arg | G |
| A | He | Thr | Asn | Ser | U |
| | He | Thr | Asn | Ser | C |
| | He | Thr | Lys | Arg | A |
| | Met | Thr | Lys | Arg | G |
| G | Val | Ala | Asp | Gly | U |
| | Val | Ala | Asp | Gly | C |
| | Val | Ala | Glu | Gly | A |
| | Val | Ala | Glu | Gly | G |

## QUESTION

1. What is Base Triplet Hypothesis?
2. What do you mean by codon, genetic code and genetic dictionary?
3. Describe the properties of genetic code.
4. What are the evidences for genetic code?
5. Describe briefly the experiments done by Nirenberg (1961) and Khorana. What they conclude from these experiments? Mention the importance of genetic code.
6. State briefly the DNA as a code for protein synthesis.
7. Describe the mechanism of instruction of DNA to cell.

# 16

# REGULATION OF GENE EXPRESSION

1. Types of gene regulation, 2. Types of temporal response to a regulatory signal, 3. Jacob-Monod hypothesis, 4. The operon and its control, 5. Events in Jacob-Monod experiment.

## Introduction

The genetic information present in somatic cell is practically identical in metazoa except few cells which have amplified and rearranged genes in order to carry out specialized cellular function. The expression of the genetic information must be regulated for:

1. Ontogeny and differentiation of the organism and its cellular components.
2. Adaptation to its environment, to conserve energy and nutrients, the expression of genetic information must be responsive to extrinsic signal.

## Types of Gene Regulation

There are two types of gene regulation:

1. **Positive Regulation:** When the expression of genetic information is quantitatively increased by the presence of a specific regulatory molecule (negative regulator).
2. **Negative Regulation:** When the expression of genetic information is diminished by the presence of a specific regulatory molecule (negative regulator).

   However a double negative has the effect of acting as positive.

## Types of temporal Responses to a regulatory signal

There are 3 types of temporal responses to an induced signal.

1. **Type A:** There is increased rate of gene expression which is *dependent* upon the continued presence of the induced signal.

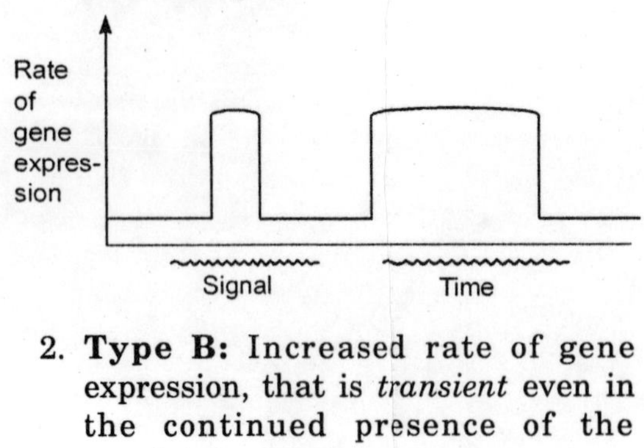


2. **Type B:** Increased rate of gene expression, that is *transient* even in the continued presence of the regulatory signal.

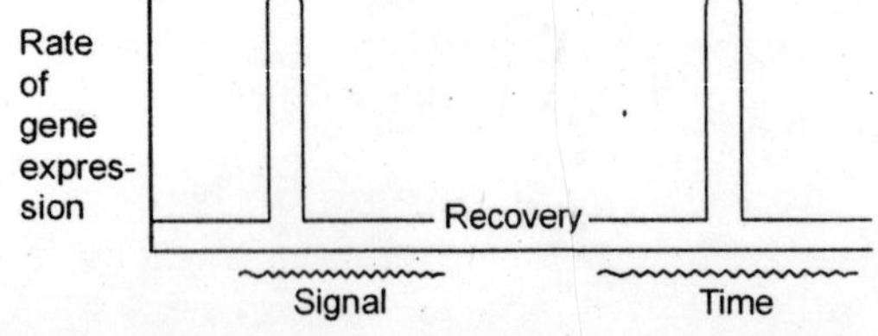

3. **Type C:** An increased rate of gene expression that persists *indefinitely* even after the termination of signal (Fig. 16.1).

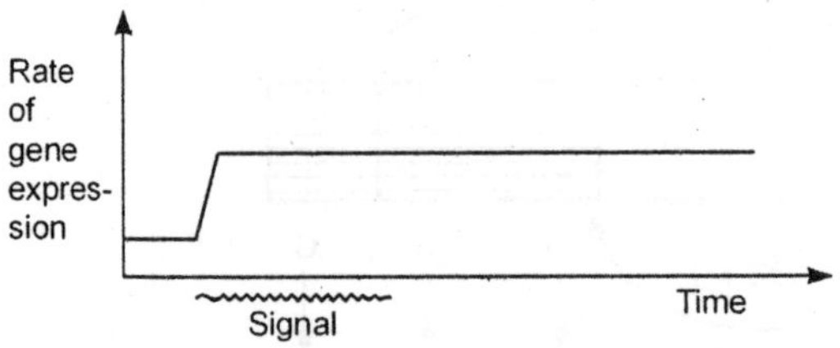


**Fig. 16.1:** Types of temporal responses.

- **Example**

It has been postulated that different parts of the genetic code come into operation at different time as differentiation takes place, i.e., there must be a mechanism which ensures that the correct parts of the code operate at correct time. In other words, a mechanism switches the appropriate genes on or off, as and when they are required.

During human development, two types of haemoglobin are formed. Adult Hb in adult stage contains *two α and two β polypeptide chains* and foetal Hb in foetal stage contains two α and two γ polypeptide chains which causes difference in $O_2$-carrying capacity. At birth production of foetal Hb ceases and is replaced by adult Hb with the β chain. There is genetical evidence mainly studying from sickle-cell anemia that polypeptide chain is controlled by a single gene and it has been traced to a single incorrect amino acid in the β chain. The idea is that the gene specifying chain works during embryonic and adult life. The gene specifying the γ chain works only during embryonic life and is then switched off. The gene specifying the β chain is inactive during embryonic life and is switched on at birth. So it can be predicted there is existence of two other genes one for switching off the γ chain and one for switching on the β gene at birth.

## The Jacob-Monod Hypothesis of Gene Action

Bacteria *E.coli* will only synthesize enzymes if and when it is appropriate to do so. For example, it produces enzymes for breaking down sugar for the production of energy. One such sugar is lactose. If lactose is present in the nutrient medium in which the bacteria are growing, the bacteria produces enzymes β-galactosidase, galactosidase permease, galactosidase transferase to break it down. But if lactose is absent from the medium, these enzymes are not produced. If lactose is added to a medium which previously lacked it, then, and only then, will the bacteria start synthesizing the enzymes. He put forward a hypothesis to explain the mechanism of regulatory function by an *"operon model"* (Fig. 16.2).

## The Operon and Its Control

Formation of all the enzymes for series of biochemical reactions are controlled by series of genes located on the same chromosomal DNA strand. This area of DNA strand is called operon. Operon is named according to the substance used for studying lac operon (lactose), his operon (histidine), ara operon (arabinose) etc.

The operon has three units:

1. **Structural Genes:** Through the intermediacy of mRNA it brings about synthesis of enzyme.
2. **Operator Gene:** Situated close to the structural gene is another section of DNA strand called operator gene. At times when the enzyme is needed the operator gene activates the structural gene, causing it to promote synthesis of enzyme. It is the control gate that can be opened or closed. The adjacent gene called *regulator gene (or promoter)* first combines with *"RNA Polymerase"* and begins moving down along the operon to cause transcription of mRNA. And if the gate is open, it initiates structural gene for enzyme synthesis and if gate is closed, RNA polymerase becomes blocked and operon remains dormant.

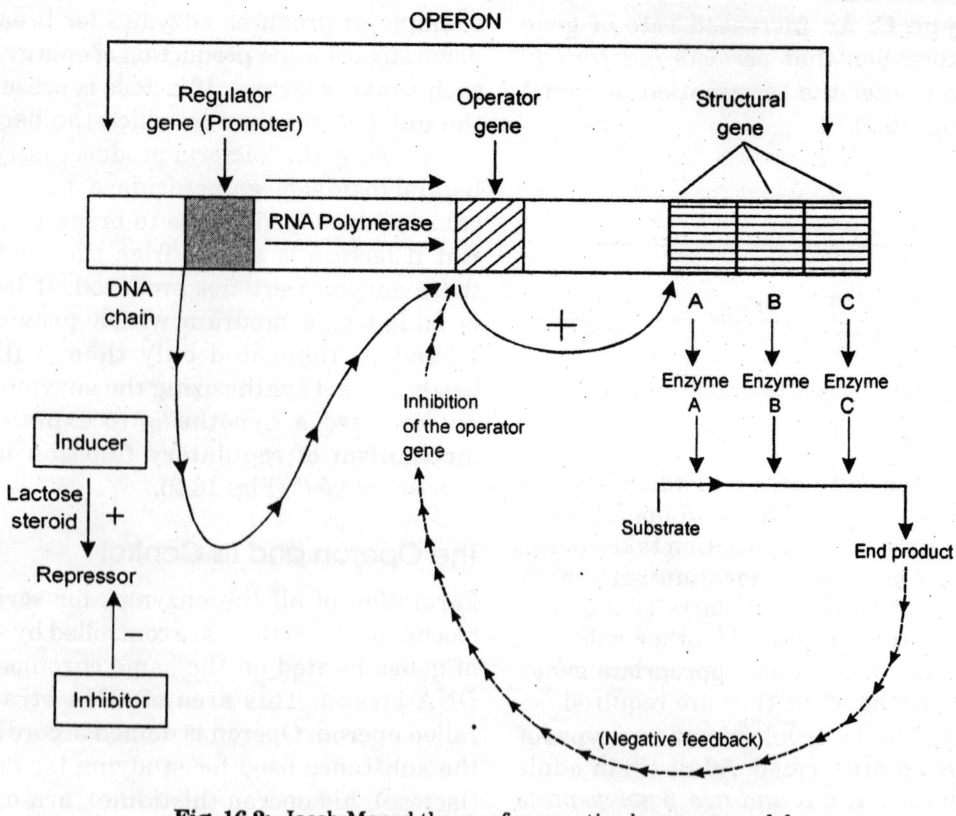


**Fig. 16.2:** Jacob-Monod theory of gene action by operon model.

This gene is controlled by two ways:

(a) *Negative Feedback Inhibition*: If the concentration of the end product is high, the operator gene is inhibited and when concentration falls, the gene becomes active.

(b) *Regulator Gene Control*: It inhibits the operator gene through *"repressor substance"*.

3. **Regulator Gene:** Another gene situated further along the DNA chain, i.e. before the operator gene. At times when the enzyme is not required, this gene causes the synthesis of a protein called *"repressor substance"* which inhibits the operator gene thereby preventing the structural gene from synthesis of enzyme. When the enzyme is required the repressor substance is inhibited by "inducer" (e.g. steroid, lactose), the operator gene becomes functional, the structural gene is activated, and the enzyme is produced (repressor-inducer complex formation).

Another function of regulator gene is to bind first with RNA polymerase and travel towards the operator gene and then to the structural gene. RNA polymerase acts with mediator *cyclic AMP, CRP* (Cyclic AMP receptor Protein).

- **Other Control Mechanism by the operon**

1. Occasionally, many different operons are controlled all at the same time by the same inducer or by the same inhibitor. When this occurs, all of the operons that function together are called *"regulon"*.
2. Other operons are controlled by increasing or decreasing the affinity of

the regulator gene (promoter) for RNA polymerase.

3. Some synthetic processes are controlled not at the DNA level but at the RNA level to control translation process by mRNA such as excess quantities of the amino acid *tryptophan in the cytoplasm* to prevent translation of one or more of the protein enzymes required for tryptophan synthesis.
4. Enzyme induction and positive regulation take place in the presence of specific substrates and hormones, e.g. sex hormones (androgens, oestrogen, progesterone), the glucocorticoids, thyroxine. Some hormones bind two receptor proteins; the resulting hormone-receptor complex binds to specific sites in the chromatin and thus promotes transcription of certain genes.
5. Specific enzyme activities are also turned on and off during the cell cycle, the mediators are *"cyclic AMP" and "cyclic GMP"*.
6. **Cytoplasmic control:** The Jacob-Monod hypothesis proposes gene control occurs at the—
   (a) *Level of Transcription*—in the nucleus, i.e. unwanted parts of the DNA are prevented from being transcribed into mRNA.
   (b) *Level of Translation*—in the cytoplasm where unwanted mRNA is prevented in the cytoplasm from being translated into proteins. This idea was supported by two evidences:
   1. Hammerling's study with "*Acetabularis alga,*
   2. Treating cell with the antibiotic Actinomycin D.
7. **Role of Internal Environment:** Internal environment, i.e. immediate chemical surroundings of the individual cell can regulate gene action. In other words, a chemical molecule which is not itself part of the nuclear material can influence gene action, e.g. lactose.

- **Some Events in Jacob-Monod Experiments**

1. **Constitutive Enzymes and Inducible Enzymes:** constitutive enzymes are those formed at constant rates and in constant amounts regardless of their metabolic state of the organism and inducible enzymes are made only when needed, in response to the presence of their substrate, particularly when the latter are the sole carbon source. According to the hypothesis there are two classes of enzymes, differing in their occurrence and concentration under different metabolic condition. Constitutive enzymes are the basic machinery of the cell, e.g. enzymes of glycolysis.
2. **Coordinate Induction:** Induction of a group of related enzymes or proteins by a single inducing agent is known as *"coordinate induction"*, e.g. lactose induces synthesis of β-galactosidase, β-galactosidase permease and β-galactosidase transferase.
3. **Enzyme Repression and Coordinate Repression:** Some enzymes are no longer made, i.e. they are repressed, when their products are no longer needed by the cell and called *"enzyme repression"*. Sometimes a whole series of enzymes is repressed as a group known as *"coordinate repression"* and it is ordinarily evoked by the end product hence also known as *"end product repression"*.
3. **Catabolite Repression:** Intracellular elevation of some catabolite derived from the substrate or substrate itself causes repression of enzyme called *"catabolite repression"*.

4. **Gratuitous Induction:** The induction of enzyme by substrate analogue but not by substrate, i.e. substance other than substrate called "*gratuitous induction*", e.g. thiogalactoside, a sulfur analogue of methyl galactoside causes induction more effectively that β-D galactoside (lactose) or methyl galactoside.

## QUESTION

1. Describe briefly the regulation of gene expression.
2. Why regulation of gene expression is necessary? What are the types of gene regulation? Explain with a suitable model.
3. What is operon? Describe the mechanism by which it is operated.
4. How gene activity is controlled? Which experiment established the idea of genetic regulation?
5. Discuss Jacob-Monod hypothesis of gene action. What are the events that are seen in Jacob-Monod experiment?
6. Describe the gene switching.
7. Discuss operation of genetic code at different time during differentiation.

# 17

# BIOSYNTHESIS OF PROTEINS AND ENZYMES

1. Introduction, 2. Beadle-Tatum experiment, 3. Zamecnick experiment, 4. Base-triplet hypothesis, 5. Communication of DNA with cytoplasm, 6. Steps of protein synthesis—Replication, Transcription, Translation, 7. mRNA, tRNA, rRNA, 8. Direction and rate of chain growth, 9. Processing and route of synthesized protein, 10. Errors in protein synthesis in aging, 11. Synthesis in mitochondria, 12. Sigma factor.

## Introduction

The genes control the hereditory character from parents to offspring but it also control the cellular reproduction and day-by-day function i.e. structural, chemical, and synthesis, etc. Each gene, which is a nucleic acid called DNA, controls the formation of RNA which spreads throughout the cell and controls the formation of a specific proteins, some are structural proteins and some are enzymes which controls metabolic reactions producing energy and synthesis of lipids, glycogen, ATP, etc.

- **Evidence**

## Beadle and Tatum's Experiment

Beadle and Tatum using bread mold

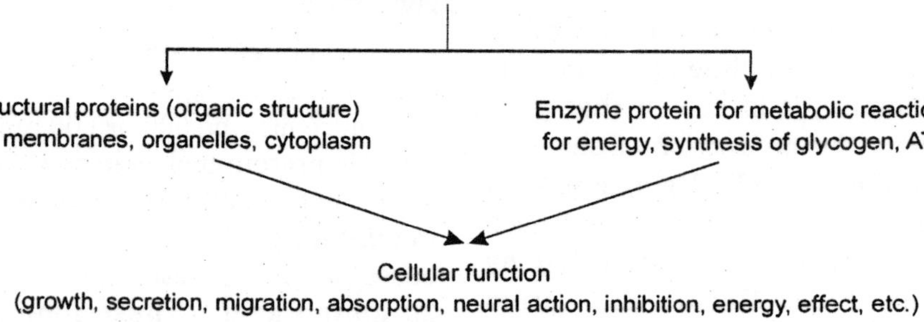


**Fig. 17.1** Flow chart showing gene control.

*Neurospora crassa,* found that this mold would thrive on a minimal medium containing nothing but salts, sucrose and the vitamin biotin, it appeared that it could synthesize all its own amino acids. They then subjected samples of the mold to X-ray radiation treatment and found as a consequence that some of their progeny failed to live successfully on the *minimal medium* which consists of sugar, a nitrogen source such as ammonium nitrate and tartarate, some inorganic salts and acids and a single vitamin biotin but showed growth in standard or *complete medium* containing many more organic compounds, vitamins, amino acids and other substances (Fig. 17.2). By systematically testing the growth of these mutants on different media, they discovered that they had developed an inability to synthesize one specific amino acid. Moreover, crossing these mutants with the normal *wild type* revealed this inability to be transmitted in a normal Mendelian manner. On the assumption that the inability to synthesize the amino acid was caused by the absence of a specific enzyme, Beadle and Tatum put forward the hypothesis that a single gene controls the production of a single enzyme, the *one gene-one enzyme hypothesis*.

Nor is this concept confined to lower organisms, various human disorders known to be due single gene (e.g. phenylketonuria which is associated with a type of mental deficiency) have been shown to be caused by absence of one specific enzyme.

From studies on the genetic control of complex molecules like haemoglobin (e.g. sickle-cell anemia), it is known that gene control not only the production of enzymes but other proteins as well. A single protein consists of several polypeptides and each polypeptide synthesis is controlled by different genes hence Beadle and Tatum hypothesis more accurately can be expressed as *one gene-one polypeptide hypothesis.*

## Zamecnick Experiment to Show Ribosome as the Site of Protein Synthesis

Zamecnick and others injected radioactive amino acids into rats. At successive time intervals the livers were removed and homogenized, and fractionated by differential centrifugation. The intracellular fractions were then examined for proteins with isotopic amino acids. After long duration all the fractions contain protein with isotopic amino acid, but very shortly after injection only the microsome fraction contained labelled protein. They concluded that proteins are synthesized first in the microsome fraction and then transferred to other cell components. They found incubation of labelled amino acids with freshly prepared cell-free homogenates of rat liver also resulted in the incorporation of isotopic amino acids into the microsomal proteins—which requires ATP.

Liver homogenates were then fractionated into nuclear, mitochondrial, microsomal, and soluble fractions (cytosol fraction) and various combinations were tested for incorporation of amino acids into proteins. It was found that two fractions, i.e. microsome and soluble or cytosol fractions were necessary for synthesis of proteins. When labelled microsome fraction was further sub-fractionated, the incorporated amino acids were recovered from ribonucleoprotein particles named as *ribosomes*.

## DNA as a Code for Proteins (Base Triplet Hypothesis)

A typical protein contains at least 20 different amino acids arranged in a specific sequence and there may be 500 or 600 amino acids in a single protein molecule and the way these are arranged gives the particular protein its individuality. DNA with its four different bases has got to determine the sequence of 20 amino acids. These four bases (four letters) code can specify 20 amino acids if three bases code specify single amino acid,

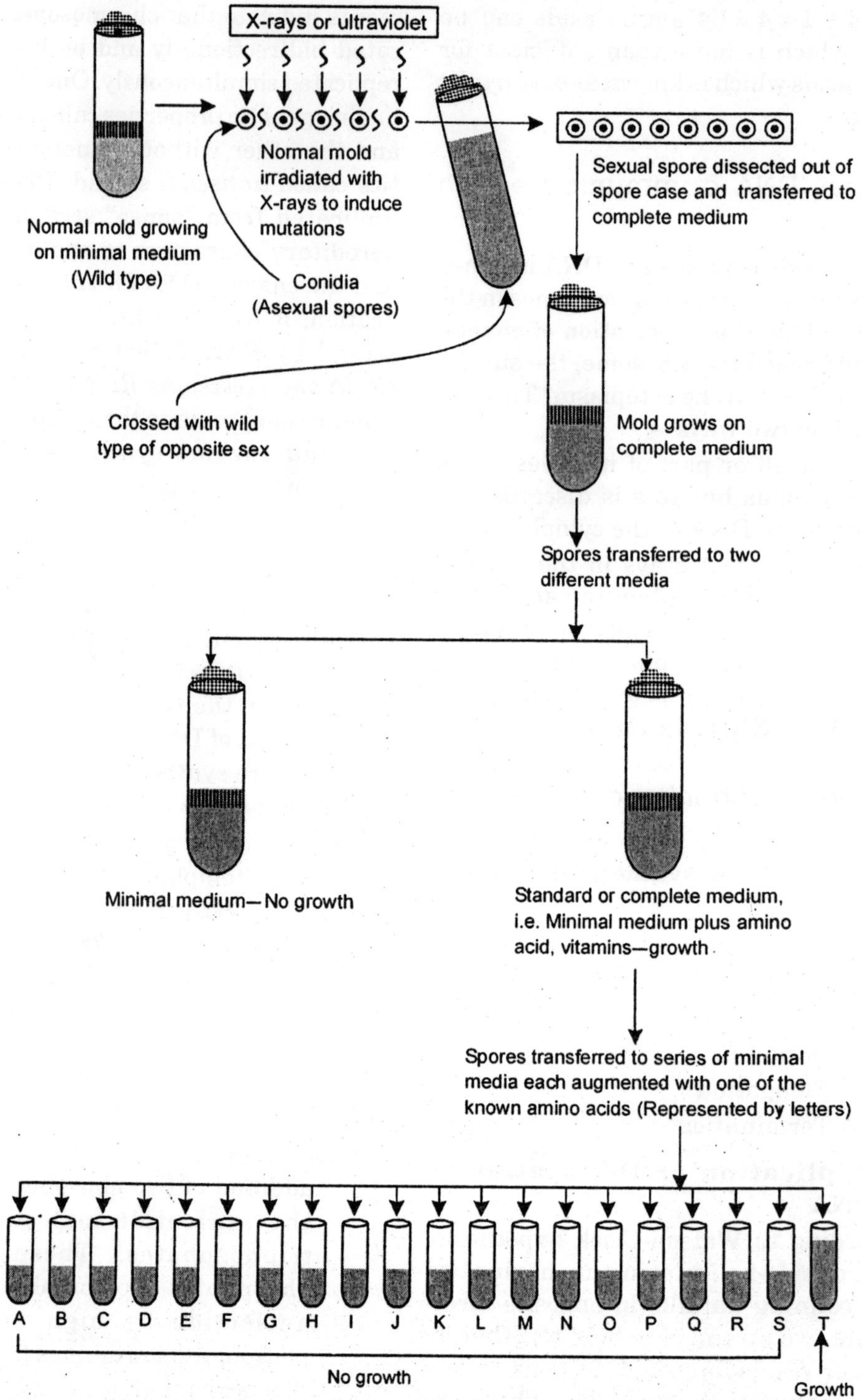


**Fig. 17.2:** Beadle and Tatum's experiment to detect biochemical mutants in Neurospora on which the one gene-one enzyme hypothesis is based.

i.e. $4^3 = 4 \times 4 \times 4 = 64$ amino acids can be specified which is more than sufficient for 20 amino acids which is known as *base triplet hypothesis*.

## How does DNA communicate with cytoplasm?

This triplet code is present in DNA in genes which are present in the chromosomes in the nucleus but how the information of genetic code should reach the ribosome, the site for protein synthesis in the cytoplasm? This can be effected by two ways—

1. DNA itself or part of it moves out of the nucleus but this is discarded as there is no DNA in the cytoplasm.
2. Secondly, DNA stays in the nucleus and another molecule, acting as "messenger" carries instructions from DNA to the cytoplasm and several evidences have proved messenger RNA (mRNA) is the carrier of message.

## Steps in the Mechanism of Protein Synthesis

The stages of protein synthesis are:

I. Replication
II. Transcription
III. Translation
 (a) Activation
 (b) Initiation
 (c) Elongation
 (d) Termination

### I. Replication or Duplication of DNA

According to Watson-Crick hypothesis, DNA is a double-helical structure having two strands running parallel in opposite directions. The two strands are held together by (1) van der Waals force and (2) weak hydrogen bond between the base pair which are broken during replication. Double-strand DNA undergoes semiconservative replication shown by Meselson and Stahl. Autoradiographic, genetic and biochemical experiments indicate that chromosomes are replicated bidirectionally and both strands are replicated simultaneously. One of the strands having genetic properties called *sense strand* and the other without genetic characteristics called *antisense* strand. DNA molecule replicated from "sense" strand transmits hereditary characters to the daughter cell and the enzyme *DNA Polymerase* helps replication. RNA is synthesized in the "sense" strand by *RNA Polymerase* called *tRNA, rRNA and messenger RNA (mRNA)* (carries genetic code) come to the cytoplasm for protein and enzyme synthesis. Many RNA chains may be transcribed simultaneously from a single gene.

### II. Transcription

A given mRNA molecule is required to carry instructions from only a very short section of a DNA molecule. So mRNA should copy exactly the sequence of bases in this short section of DNA. The process by which all RNAs are synthesised for DNA is called *transcription*.

Appropriate part of DNA molecules serves as a template for RNA. The double helix unzips in the relevant region and RNA molecule is assembled. These separate types of RNA are important to protein synthesis: messenger RNA (mRNA), transfer RNA (tRNA) and ribosomal RNA (rRNA) or soluble RNA (sRNA). Assembly of RNA molecule is accomplished in the following stages:

1. **Activation of Nucleotide:** By addition of two phosphate radicals to the nucleotide molecule to form triphosphates. These last two phosphates are combined with the nucleotide by high energy rich phosphate bond derived from ATP. For this derivation process, large amount of energy is available to each nucleotide, for promoting subsequent chemical reactions for formation of RNA chain.

2. **Assembly of the RNA using DNA Strand:** Separation of two strands of DNA and of these strands is used as a template on which the RNA molecule is assembled. This strand contains genetic code while the other remains genetically inactive. Assembly of the RNA molecule is accomplished by the enzyme *RNA polymerase.* The procedure is effected through the following stages:
   a. An activated RNA nucleotide is temporarily bonded with starting base in the DNA strand, then another nucleotide to the next base and so forth.
   b. Each successive nucleotide is pulled into place so that it lies adjacent to the previous nucleotide.
   c. The pyrophosphate radical of each new nucleotide breaks away and in doing so liberate enough energy to cause an ester linkage between the remaining phosphate on the nucleotide and the ribose on the end of the growing RNA molecule.
   d. Immediately after bonding of the ribose and phosphate, the lease of the previous RNA nucleotide breaks its bond with the DNA strand.
   e. Then the next RNA activated nucleotide is added to the RNA molecule in the same manner.

The four different bases of DNA and four bases of RNA, always combine with each other in specific combinations. Therefore, the code present in DNA strand is transmitted in "complementary" form to the RNA molecule. Once the RNA molecules are formed, they diffuse out or probably come out through the pores of nuclear membrane to all parts of the cytoplasm. The *mRNA* carries the genetic code, *rRNA* for formation of ribosome which is the physical and chemical structure on which protein molecules are synthesized and *tRNA* carries activated amino acids to the ribosome.

- **Messenger RNA (mRNA)**

It is long straight strands, suspended in the cytoplasm. It is composed of several hundred to several thousand nucleotides in unpaired strands having molecular wt about 500,000. It contains *codons* that are exactly complementary to the code words of genes specified for particular amino acids. Several of the amino acids are represented by more than one codon. Some codons (AUG, GUG) represent such signals as "start manufacturing a protein molecule" designated as "CI" for *chain initiating* and other codons (UAA, UAG, UGA) for "stop manufacturing protein molecule" designated as "CT" for *chain terminating*.

- **Transfer RNA (tRNA)**

This is another type of RNA responsible for protein synthesis and known as "Transfer RNA" because it transfers amino acid to the site of protein synthesis. There are different types of tRNA and each type combines with one amino acid of the 20 amino acids and acts as carrier to transport amino acids to ribosome. Each specific type of tRNA with *amino acid arm*. One end and *anticodon* arm

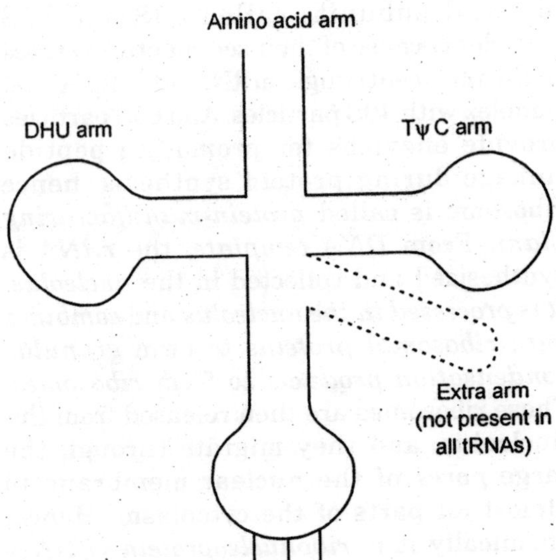


**Fig. 17.3:** Cloverleaf structure of tRNA.

at the other end recognizes particular codon on the mRNA and delivers the appropriate amino acid to the chain of protein molecule. Beside these, there are two arms called Dihydrouridine arm (DHU arm) and *Pseudouridine residue arm* (TUC arm) giving a cloverleaf appearance (Fig. 17.3).

The anticodon arm contains triplet bases. It contains 73 to 93 nucleotides having mol. wt of 25000. It is short molecule, single-stranded like mRNA but 60–70% structure double-backed on itself and twisted into a double-helix form. Unfolded tRNA is not capable of acting as an amino acid acceptor. tRNA must contain at least two other specific sites, i.e. *enzyme recognition site* at which the tRNA is bound to the corresponding activating enzyme, and the "ribosome attachment site".

- **Ribosomal RNA (rRNA)**

Ribosomal RNA (rRNA) constitutes about 60 per cent of the ribosome, the remainder of 40 per cent contains 50 different types of proteins and enzymes needed for protein synthesis. The ribosome is the physical and chemical structure on which protein synthesis occurs with the help of tRNA and mRNA. The ribosomes are composed of two physical subunits called 40S and 60S particles because of their sedimentation rates in the ultracentrifuge. mRNA and tRNA first complex with 40S particles. And 60S particles provide enzymes for promoting peptide linkage during protein synthesis, hence ribosome is called *protein-manufacturing plant.* From *DNA template,* the *rRNA* is synthesized and collected in the *nucleolus. It is processed in the nucleolus* and *combines with ribosomal proteins* to form *granular condensation products* to form *ribosomes.* These ribosomes are then released from the nucleolus, and they migrate through the large *pores* of the nuclear membrane to almost all parts of the cytoplasm. Hence, chemically it is *ribonucleoprotein* (RNA + Protein). Ribosomes are found in *free form* or *bound to the endoplasmic reticulum.* Many of the ribosomes are associated in beadlike strings called *polyribosomes* or *polysomes* which are formed by the attachment of a number of ribosomes to a single molecule of mRNA during protein synthesis.

### III. Translation

When a molecule of mRNA comes in contact with a ribosome, it travels through the ribosome, beginning at a predetermined end specified by an appropriate sequence of RNA bases. However, the protein molecule does not begin to form until a *start* or *chain-initiating codon* enters the ribosome. Then, the mRNA travels through the ribosome, a protein molecule is formed—a process is called *translation.*

Thus, the ribosome reads the code of the mRNA in much the same way that a tape is "read" as it passes through the playback head of a tape recorder. Then, when a *stop* or *chain-terminating codon* slips past the ribosome, the end of a protein molecule is signalled, and the entire molecule is freed into the cytoplasm. The protein synthesis occurs through the following steps:

(a) **Activation**

1. Each amino acid is activated by a chemical process in which ATP combines with the amino acid to form an *adenosine monophosphate complex with amino acid.*
2. The activated amino acid, having and excess of energy, then combines with its specific tRNA to form an amino acid—tRNA complex and at the same time, releases the AMP with the help of *Aminoacyl-tRNA synthetase.*

(b) **Initiation:** The tRNA carrying the amino acid complex then comes in contact with the mRNA molecule in the ribosome where the anticodon of the tRNA attaches temporarily to its specific codon of the mRNA, thus lining up the amino acids in appropriate sequence to form a protein molecule. This process is called *initiation* which requires 3 specific proteins called *initiation factors* (IF-1, IF-2, IF-3), GTP (Guanosine triphosphate) and $Mg^+$.

The complete ribosome contains 2 sites, the *P site* or *peptidyl site* contains peptidyl-tRNA and *A site* or *aminoacyl-site* contains Aminoacyl-tRNA. With the formation of initiation complex for the first codon the aminoacyl-tRNA molecule leaves "A site" and enters the "P site". translation of the codons of mRNA takes place in the $5' \rightarrow 3'$ terminal direction.

(c) **Elongation** $\rightarrow$ 3 steps:

1. Binding of aminoacyl-tRNA
2. Peptide bond formation
3. Translocation

In the elongation process, the polypeptide chain is lengthened by sequential addition of new aminoacyl residues enzymatically transferred from aminoacyl-tRNA esters, each of which is bound to the ribosome in response to a specific codon or base triplet in the mRNA. Two *elongation factors,* EF-T and EF-G, also *proteins* are required in the lengthening of the chain. After formation of each new peptide bond, the ribosome moves along the mRNA to bring the next codon in position for alignment of the next aminoacyl-tRNA, a process which requires energy provided in the form of GTP. This process of elongation, i.e. assembly of amino acids in the chain starts at one end with a free amino group and to proceed amino acid by amino acid, to the other end with a free carboxyl group to form peptide bond (CO — NH) with the next amino acid with the help of enzyme *peptidyl transferase.* After peptide linkage, the peptidyl-tRNA is still bound to "A site" but the ribosome moves to the next codon of the mRNA and simultaneously shifts peptidyl-tRNA to the "P site" is called *translocation.*

**Table 17.1:** Components required in the four major steps of translation

| Stage | Necessary components |
|---|---|
| 1. Activation of the amino acids | Amino acids<br>tRNA<br>Aminoacyl-tRNA synthetase<br>ATP<br>$Mg^{2+}$ |
| 2. Initiation | f Met-tRNA (In Prokaryotes)<br>Met-tRNA (In Eukaryotes)<br>Initiation codon in mRNA (AUG)<br>mRNA<br>CTP<br>$Mg^{2+}$<br>Initiation Factors (IF-1, IF-2, IF-3)<br>30S Ribosomal subunit<br>50S Subunit |
| 3. Elongation | Functional 70S ribosome<br>Aminoacyl-tRNAs specified by codons, $Mg^{2+}$, Elongation Factors (EF-T, EF-G), GTP |
| 4. Termination | Functional 70S ribosome<br>Termination codon in mRNA Polypeptide release factors, $R_1$, $R_2$, $R_3$. |

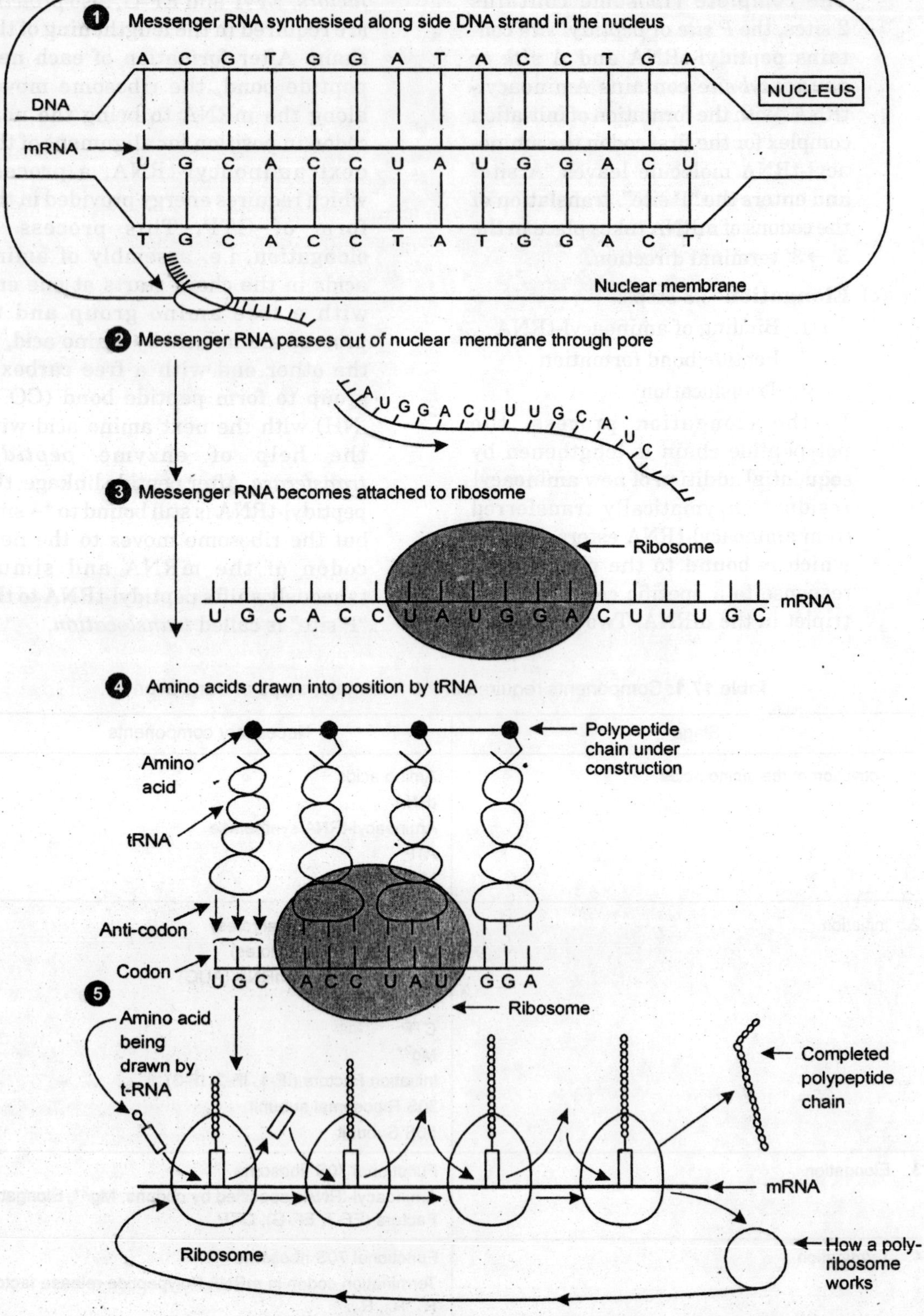


Fig. 17.4: Biosynthesis of proteins and enzymes.

(d) **Termination:** After completion of polypeptide chain, it peals off from ribosome as soon as the "terminal signal" of mRNA reaches the ribosome (Fig. 17.4). The details of completion of polypeptide chain and its detachment is not known. The release of polypeptidyl-tRNA from the ribosome is promoted by three specific proteins called "release factors" $R_1$, $R_2$, $R_3$. Whether the polypeptide chain leaves ribosome in coiled chain or as a folded chain is not known.

## Direction and Rate of Chain Growth during Polypeptide Synthesis

Translation of the codon of mRNA takes place in the 5′→ 3′ terminal direction. Assembly of amino acids in the polypeptide chain starts at one end with a free amino group and to proceed, amino acid by amino acid, to the other end with a free carboxyl group. At 37°C a single ribosome in the intact rabbit reticulocyte can complete an entire α chain of haemoglobin (146 residues) in about 3 min. In *E.coli* and other bacteria growing in an optimal medium, each functional ribosome can insert about 20 amino acids per second.

## Processing and Route of Synthesized Proteins Route

Ribosomes may occur either free or bound to the endoplasmic reticulum to form rough-surfaced endoplasmic reticulum. Free ribosomes are employed for synthesis of proteins for internal use, e.g. haemogobin reticulocyte and ribosomes attached to the endoplasmic reticulum make proteins to be secreted outside the cell. The route of synthesized proteins has been studied by electron microscopy and biochemical methods in the pancreas exocrine cells. The newly formed proteins, e.g. zymogens, trypsinogen and chymotrypsinogen → leave the bound ribosomes → pass through the endoplasmic reticulum membrane → enter the intracisternal space, which serves as a system of collecting channels → reaches Golgi apparatus for concentration and packaging into zymogen grannules, large, dense→ aggregates to zymogen molecules→ which are surrounded by a single membrane→ stored until they are scereted→ membrane of the zymogen granules fuses with the plasma membrane→ opens the lumen of the zymogen granules to exterior.

## Processing

Once a polypeptide has been made by ribosome it may need to undergo covalent modification to yield its biologically active form. The different processes are:

1. N-formyl group of "f Met" residue is removed by "deformylase".
2. Initiating "Methionyl" residue is removed by "methionine amino-peptidase".
3. Sulfydryl group "— SH" is oxidised to disulfide "S — S" linkages.
4. Other modifications are phosphorylation, methylation, hydroxylation, attachment of coenzymes or prosthetic groups, etc.

## Clinical Implication

### Errors in Protein Synthesis in Senescence or Aging

Much interest attaches to the question of the accuracy of protein synthesis in animal tissues and to the consequences of errors. There is some evidence that structurally similar amino acids may replace a given amino acid no more than once in 3000 residues; presumably the chance of errors in incorporation of dissimilar amino acids is much less. It has been postulated that the *error frequency in protein biosynthesis rises with the age*, i.e. in senescence in which the smooth functioning of the organism gradually declines, culminating in death.

The manifestations of senescence in man include greying of hair and balding, wrinkling of the skin, cardio-vascular difficulties resulting in panting, reduction in sexual libido leading to reproductive failure, digestive difficulties due to inefficient action of digestive enzymes, and a reduction in the glomerular filtration rate resulting in water retention. The later stages of senescence may be accompanied by hardening of the arteries muscular weakness due to replacement of muscle fibres by connective tissue, reduction in body size due to progressive atrophy of the bones and other tissue, slower reaction times, mental senility, and ultimately incontinence.

**Reason:** *Cause of these changes* is not clear but it has been postulated that senescence results from gradual accumulation of genetic and biochemical defects resulting the protein molecule may be rendered less active or even inactive. This has been confirmed by estimating catalytic activity of aldolase of skeletal muscle of aged animals and young animals. The total amount of aldolase remained unchanged in both animals but the catalytic activity of aldolase was one-third in aged animals. The implication is that the old animals have shown many errors in biosynthesis of aldolase rendering large fraction of aldolase inactive for deterioration in the accuracy of translation with age. There are three main causes of mistakes in protein synthesis:

1. *Firstly,* there is gradual decline in the efficiency of mitosis with the result that the daughter cell possesses abnormal chromosome constitutions. Since these abnormalities arise as a result of aberrations in the division of the body (somatic) cells as opposed to the reproductive (germ) cells, they are called *somatic mutations* and they are prevalent in old age.
2. *Secondly,* may be due to errors in the functioning of genetic code. There are several possibilities:
   (a) Abnormal replication in dividing cells, e.g. a wrong nucleotide insertion resulting in an alteration in the structure of DNA in the daughter cells.
   (b) Errors may occur in transcription.
   (c) Errors may occur in translation of the code.

   Evidence suggests that the malfunctioning of cell is due to errors in transcription and translation of the code, rather than to changes in the DNA itself.
3. *Thirdly,* is due to genes which have been switched off in a fully differentiated cell may suddenly start operating again. Such derepression of genes after differentiation will result in the wrong proteins being produced.

## Protein Synthesis in Mitochondria

Mitochondria synthesises protein is evident by the following:

1. Isolated mitochondria from many types of eukaryotic cells incorporate labelled amino acids into mitochondrial protein.
2. Mitochondria contain all the essential components required for protein synthesis, including mRNA, ribosomes, tRNAs and aminoacyl-tRNA synthetases, all of which are distinctive and different from those required in extra-mitochondrial or cytoplasmic protein synthesis.
3. The ribosomes of mitochondria vary in size and are not always easily visible in electron microscope but often are located on the portion of the inner membrane between the cristae.
4. Protein synthesis in mitochondria is inhibited by chloramphenicol whereas cytoplasmic synthesis is not.
5. Mitochondrial protein synthesis employs N-formylmethionine as initiating amino acid whereas cytoplasmic synthesis employs methionine. Mitochondrial synthesis resembles bacteria.

6. Mitochondria contains a specific circular mitochondrial DNA and DNA-dependent RNA polymerase hence it can regenerate mRNA.
7. Evidences have shown that majority of mitochondrial enzymes and proteins, are synthesized in extramitochondrial ribosomes, i.e. in the cytoplasm.
8. Mitochondrial DNA codes for mitochondrial rRNA, tRNAs, leaving only a small amount for function of coding protein synthesis.
9. Mitochondrial DNA specifies the synthesis of hydrophobic protein subunits of ATPase, cytochrome $aa_3$, cytochrome b.

## Sigma Factor

- **Definition**

Transcription, initiation requires specificity protein known as a sigma (δ) factor.

- **Mode of Action**

The sigma factors bind reversibly to the catalytically active core RNA polymerase.

The core RNA polymerase interacts with sigma factors to form a holoenzyme. After the release of δ factor the core enzyme elongates an RNA chain. The majority of cellular transcription requires the primary sigma (δ) factor. In *E.coli* the primary sigma factor is referred to as $\delta^{70}$ and in *B. subtilis* $\delta^{43}$.

- **Chemical Nature**

Recently, sigma factors have been isolated from the genes encoding many bacterial and phage and their amino acid sequences have been analysed from the corresponding DNA sequences. This shows that many sigma factors have similar amino acid sequences and form a homologous protein family. Sigma factors thus may use a variety of protein structures to perform related functions in bacterial cells.

- **Biochemical Actions**

The bacterial sigma factors have the following biochemical actions:

1. Core binding
2. Activation of promoter recognition
3. DNA melting
4. Inhibition of non-specific transcription

All sigma factors have at least the 1 and 2 of these actions.

1. **Core Binding:** Different sigma factors interact with the same region of the core RNA polymerase complex. Therefore, many sigma factors may share a common protein structure that allows core binding.
2. **Promoter Recognition:** Promoter recognition can occur only when holoenzyme is formed. It seems most likely that sigma factors themselves make sequence-specific contacts with promotion. Bacterial sigma factors do not bind lightly to DNA in the absence of core RNA polymerase. As with promoter recognition, sigma factors do not show a DNA melting activity in the absence of the core RNA polymerase subunits.

   *E.coli* $\delta^{70}$ factors have been shown to reduce the affinity of the core subunits for non-promoter containing DNA promoter by about $10^4$. It has been claimed that this action is important in allowing promoter recognition against large background of non-specific DNA. In *B. subtilis*, a separate protein component, $\delta^{21}$ inhibits core-mediated transcription of non-promoter containing DNA. The $\delta^{21}$ protein binds to the *B. subtilis* core enzyme in the presence and absence of sigma factors and reduces the interaction of RNA polymerase with the DNA template. The ability to inhibit non-specific transcription has also been attributed to a specificity factor from yeast mitochondria that may be functionally similar to a sigma factor.

- **General Characteristics of Sigma Factor**

1. Most bacterial sigma factors are single polypeptide.
2. Sigma factors form a heterogeneous group of proteins.
3. All the related sigma proteins are acidic with an excess of negative charge at pH 7.0. *E.coli* $\delta^{70}$ factor is one of the most acidic proteins in *E.coli*.
4. *E.coli* $\delta^{70}$ and *B. subtilis* $\delta^{43}$ have approximately 50 per cent amino acid identity allowing for a 245-amino gap introduced into the smaller $\delta^{43}$ factor.

## QUESTION

1. Describe the mechanism or process of biosynthesis of proteins or enzyme in the cell.
2. Discuss molecular relationship between gene and protein synthesis.
3. Describe the role of DNA and RNA in protein synthesis.
4. How sequence of amino acids are regulated during protein synthesis in the cytoplasm
5. "Specific functions of proteins depends upon the number and sequence of amino acids." How it is regulated?
6. How gene controls the cytoplasmic proteins and enzymes synthesis?
7. How genetic information or code reaches the cytoplasm and how it is effected?
8. Describe the experiment done by Beadle and Tatum. What is the rationale behind it? What is the conclusion?
9. Discuss the "one-gene one-enzyme" synthesis.
10. What experiment can be done to show that ribosome as the site of protein synthesis?
11. What is "base triplet hypothesis"?
12. How does DNA communicate with cytoplasm?
13. Describe briefly the process of transcription.
14. Describe the process of translation.
15. What is the direction and rate of chain growth during polypeptide synthesis? Describe briefly the processing and route of synthesised protein.
16. Describe briefly the error of protein synthesis in senescence or aging.
17. Describe briefly the protein synthesis in the mitochondria.
18. What is sigma factor? Describe the mode of action, chemical nature and biochemical action.

# 18

# RECOMBINANT DNA TECHNOLOGY OR GENETIC ENGINEERING OR GENE TRANSFER

1. Introduction, 2. Preparation of recombinant DNA (chimeric DNA molecule) cleavage; splicing; cloning, 3. Importance of cloning, 4. DNA sequencing, 5. Libraries and library construction; probe; Blotting, 6. Hybridization techniques, 7. Importance of recombinant DNA technology, 8. Application of recombinant DNA technology, 9. Immunomodulator for immunotherapy.

## Introduction

Recombination is the exchange of genes between similar chromosomes in prokaryotic and eukaryotic organism. It occurs frequently in:

1. The transformation of bacterial cells,
2. Viral transduction,
3. The exchange or crossing-over genes during meiosis in eukaryotic germ cells.

One of the most recent and rapidly advancing frontiers of molecular biology is the *construction and new combination of genes in the laboratory*. Genetic recombination in which a new DNA molecule is formed by the breakage and reunion of DNA strand. In this process the normal genetic make-up of a cell responsible for performance of specific functions is altered by manipulating the genome to make it for new functions. It involves the splitting and combining genetic material derived from different organisms to produce *recombinant DNA* and this novel genome can then be introduced into suitable cells and *amplified* many times by the DNA synthesizing machinery of the host. Some of these inserted genes are also transcribed and translated in their new setting. This kind of biotechnology is known as *recombinant DNA technology* or *genetic engineering*.

Animals and human genes, treated as above, are made to grow in bacteria and other microorganisms for use in this technique. The genetic engineering has been made possible by the discovery of double-helical structure by Watson and Crick. It contains purine and pyrimidine bases and the sequence of these bases embodies the genetic code. When the helix splits lengthwise to replicate itself and pass on its message of heredity to the next generation of cells, the split occurs between the pair of bases. Thus the scientists have learned to read the message of gene. These advances have generated a new biotechnology industry—the genetic engineering.

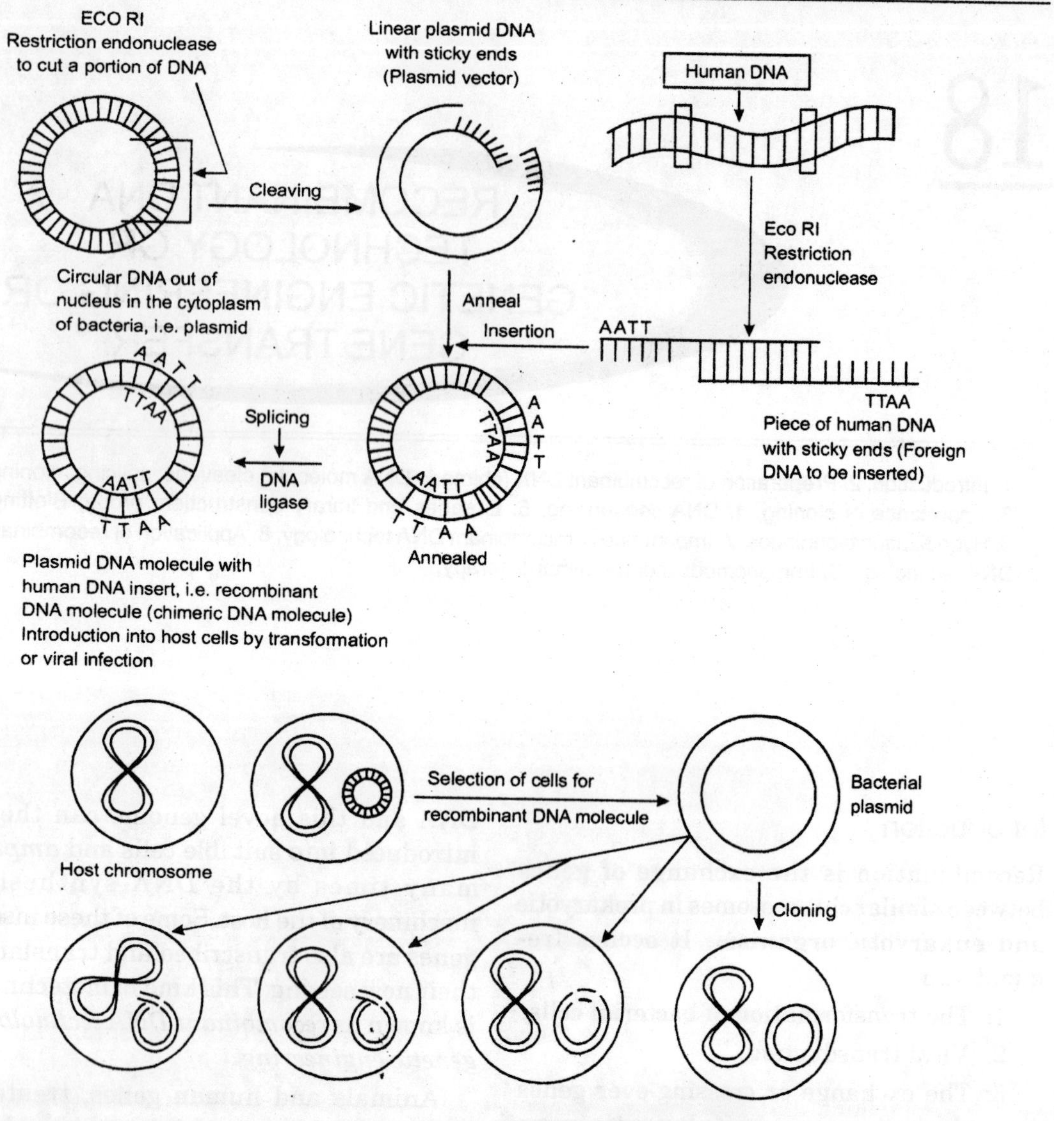


**Fig. 18.1:** Synthesis and cloning of recombinant molecules.

- **Recombinant DNA Technology**

New combination of unrelated genes can be constructed in the laboratory. The production of recombinant DNA is effected by combining (splice) fragments of genetic material obtained or derived from different organism by *gene splitting* or *cleavage*. This novel *genome* can be introduced into suitable cells and *amplified* many times by the DNA synthesizing machinery of the host. Some of these inserted genes are also transcribed and translated in their new setting. In this manner, a bacterial culture can be utilized for the autonomous replication of recombinant plasmid including the new segments of DNA. The *major steps* for cloning of DNA are given below:

CLONING (Fig. 18.1)

## I. Preparation of Recombinant (chimeric) DNA Molecule

- **Cleavage (splitting)**

1. Plasmid DNA molecules. (Most of the bacteria contains small, extra

chromosomal circular molecule of DNA present in the cytoplasm which can replicate independently of the host DNA. This extranuclear DNA is known as plasmid) can be readily spliced (unite, joining) in vitro by the restriction endonucleases, DNA ligase and other highly specific enzymes that act on DNA.

2. The vector (vehicle) in a recombinant experiment can be prepared for splicing by cleaving (splitting) it at a single specific site of the plasmid DNA (circular) molecule by restriction endonuclease (Eco RI) to form linear plasmid DNA with *sticky ends.*

   The staggered cuts made by this enzyme generate complementary single-stranded ends (cohesive ends).

3. (a) Now if the foreign DNA fragment to be inserted into this plasmid e.g., human DNA is cleaved by Eco RI restriction endonuclease into piece containing same sticky ends. The single-stranded ends of the fragment would then be complementary to those of the cut plasmid.

   (b) The essential feature of a vector is that it can replicate autonomously in an appropriate host, e.g. plasmids and $\lambda$ phage are choice vectors for cloning in *E.coli*.

4. (a) *First Method-Cohesive-end method*: The DNA fragment (e.g. human DNA) and the cut plasmid DNA can be *annealed* (toughen) and then covalently *joined by DNA ligase* by the cohesive-end method.

   (b) *Homopolymeric Tails*: A *second method* for joining or splicing two unrelated DNA molecules requires the addition of poly dA tails to both 3′ ends of one molecule and poly dT tails to both 3′ ends of the other molecule. These homopolymeric extensions are synthesised by *terminal transferase* (terminal deoxynucleotidly transferase) an enzyme that adds nucleotides to the 3′ hydroxyl group of DNA chain. This transferase, in contrast with DNA polymerase, does not take instruction from a template, and so tails made up of just one kind of nucleotide (100 residues long) can be added. The poly A tails on one DNA molecule are annealed to the poly T tails on the other. The lengths of the tails are not precisely defined and so the gaps are filled by DNA polymerase I before the strands are joined by DNA ligase.

   (c) *Chemically Synthesized Linkers*: A *third* method for joining DNA molecules combines the advantages of the first cohesive-end method with the dA-dT method, i.e. second method. A chemically synthesised linker (from 6-10 base-pairs) that is susceptible to cleavage by a restriction enzyme is covalently joined to the ends of a DNA fragment or vector. The 5′ ends of the decameric linkers and of the DNA molecule are phosphorylated by polynucleotide kinase and joined by $T_4$ ligase, which can form a covalent bond between blunt-ended (flush-ended) DNA molecules. Cohesive ends are generated when these terminal extensions are cut by an appropriate restric enzyme. Thus, cohesive ends corresponding to the specificity of a particular restric enzyme can be fashioned virtually any DNA molecule.

## II. Introduction into Host Cells

(a) **By Transformation:** Most bacterial and eukaryotic cells take naked DNA molecules from the medium. The efficiency of uptake is low (1 of $10^6$ DNA molecules), but an appreciable proportion of cells can be transformed under appropriate experimental conditions.

(b) **By Infection:** Another process is that the target cells can be infected with reassembled virions harbouring the recombinant DNA molecule. In this synthetic viral genome, a gene of interest replaces a segment of viral DNA that is not essential for replication.

## III. Selection of Cells for Cloning

Next, selection is done *which cells harbour the recombinant DNA molecule* containing the gene of interest as clone.

A clone is a large population of identical molecules, bacteria or cells that arise from a common ancestor. Cloning allows for the amplification or production of a large number of identical DNA molecules which can then be used for many purposes. Clones containing recombinant DNA are stable for at least several hundred generations. The desired clones, i.e. *cloning vector* can be selected on the basis of the presence of the vector or of the inserted gene itself, e.g.

1. Some plasmid vectors confer resistance to an antibiotic,
2. To determine which cells bind RNA that is complementary to the gene of interest or synthesise protein encoded by it.

The cloning vector may be:

(a) Bacterial plasmids
(b) Phages
(c) Cosmids

(a) **Bacterial Plasmids**

1. Bacterial plasmids are small, circular duplex DNA molecules whose natural function is to confer antibiotic resistance to the host cell (resistance to Ampicillin and Tetracycline).
2. Plasmids possess several properties which make them much advantageous as cloning vectors.
3. They are present as single or multiple copies within the bacterium, and replicate independently from the bacterial DNA.
4. Plasmids are smaller than the host chromosomes and are therefore easily separated from the chromosomes, and the desired DNA is easily removed by splitting the plasmid with the enzyme specific for restriction site into which the original piece of DNA was inserted.
5. The common plasmid vector is pBR 322.

(b) **Phages**

1. They have linear DNA molecules into which foreign DNA can be inserted at several restriction enzyme sites.
2. When the phage proceeds through the lytic cycle for production of mature, infective phage particles then the chimeric DNA is collected.

(c) **Cosmids**

1. Larger fragments of DNA can be cloned in *cosmids* which combine the best features of plasmids and phages.
2. Cosmids are plasmids that contain the DNA sequence, so-called *cos, sites,* required for packaging Lambda DNA into phage particles.
3. DNA insertion into a functional region of the vector will interfere with the action of this region.

## Importance of Cloning

1. For understanding of chromosome structure.
2. For gene expression.
3. Amplification of inserted genes.
4. Production of large amounts of DNA helped for study of DNA sequence and electron-microscopic analysis.

5. Large quantities of scarce genes and proteins are synthesised by these clones.
6. Dissection of complex genomes and analysis of the regulation of their expression.
7. Specific regions of DNA that control gene mobility, DNA replication, and transcription are being analyzed.
8. Discovery of intervening sequences in many eukaryotic genes.
9. Complex chromosomes are rapidly being mapped and dissected into units that can be manipulated and explored.
10. Large quantities of specific proteins can be produced, e.g. production of DNA ligase by *E.coli* can be increased 500-fold by this process.
11. The synthesis of eukaryotic peptides and proteins, e.g. insulin and antiviral agents such as interferon may be produced by the bacteria.
12. Gene cloning for increasing agricultural yields is also being explored.
13. Eukaryotic genes are now being introduced into eukaryotic cells as well as into bacteria, e.g. SV 40 virus has been used as vector to bring the rabbit β-globin gene into monkey-kidney cells. These infected cells synthesise amounts of rabbit β-globin protein. This experiment elucidates regulatory mechanisms in eukaryotic gene expression.
14. By this technique, some protein hormones have been produced in experimental quantities.

## IV. DNA Sequencing

The segments of specific DNA molecules obtained by recombinant DNA technology can be analyzed for their molecule sequence. The method depends upon having a large number of identical DNA molecules. This requirement is fulfilled by:

1. Recombinant DNA technology.
2. The stringent sequence specificity of the restriction endonucleases. The methods are:
   i. **Maxam and Gilbert Method:** If employs chemical methods to cleave the DNA molecules where they contain the specific nucleobases.
   ii. **Sanger's Method:** It employs specific nucleotide analog that terminate DNA strand synthesis at specific nucleobases as the strand is synthesised on purified template nucleic acid.

## Libraries and Library Construction

1. Various cloning vectors and restriction enzymes in combination allow the entire genome to be packaged into a vector and this collection of different recombinant clones is called a *library*. Libraries may be either:
   (a) *A Genomic DNA Library*: It is formed by the total DNA of a cell line or tissue. Genomic libraries are prepared by the partial digestion of total DNA with a restriction enzyme that cuts DNA frequently.
   (b) *A cDNA Library*: A cDNA is a single-stranded DNA molecule which is complementary to an mRNA molecule and is synthesised from it by the action of reverse transcriptase.

   A cDNA library is the population of mRNAs in a tissue.

   cDNA libraries are prepared by isolating all the mRNAs in a tissue and then copying these molecules into double-stranded DNA, using the reverse transcriptase and DNA polymerase.

   The protein coded by the gene introduced by recombinant

technology is synthesised in a vector called an *expression vector*. These vectors are used to detect specific cDNA molecules in libraries and to produce proteins by genetic engineering. These vectors are especially constructed to contain very active inducible promoters.

2. A human library which contains $10^6$ recombinant fragments of large size has a 99 per cent probability of being complete.

- **Probe**

1. A molecule used to detect the presence of a specific fragment of DNA or RNA in common probes are cDNA molecules.
2. Probes are generally pieces of DNA or RNA labelled with a $P^{32}$ containing nucleotide.
3. The probe should know a complementary sequence to be effective.
4. A cDNA synthesised from a specific mRNA can be used to screen either a cDNA library for a longer cDNA or a genomic library for a complementary sequence in the coding region of a gene.
5. cDNA probes are used to detect DNA fragments on Southern blot (a method for transferring DNA from an agarose gel to a nitrocellulose filter on which the DNA can be detected by a suitable probe) transfers and to detect and quantitate RNA on Northern blot (a method for transferring RNA from an agarose gel to a nitrocellulose filter on which the RNA can be detected by a suitable probe) transfers.

- **Blotting**

1. Observation of a specific DNA or RNA fragment in many "contaminating" molecules requires a number of techniques which are collectively called *blot transfer*.
2. Sometimes, if a specific base is altered and a restriction site is changed, these procedures can detect a point of mutation.
3. The Northern blot and Western blot (A method for transferring protein to a nitocellulose filter on which the protein can be detected by a suitable probe, e.g. an antibody) transfer techniques are used to size and quantitate a specific RNA and protein molecules respectively.

- **Hybridization Techniques**

1. Colony or plaque hybridization is the method by which *specific clones are identified and purified*. Bacteria are grown on colonies on an agar plate and overlaid with a nitrocellulose filter paper (Fig. 18.2).
2. Perfect matches hybridize readily and withstand high temperatures in the hybridization and washing reactions. These complexes also form in the presence of low salt concentrations.
3. Less than perfect matches do not tolerate these *stringent conditions*.
4. Gene families can be detected by varying the stringency of the hybridization and washing steps.
5. Several techniques have been developed to detect the presence of clones with specific DNA sequence inserts. The best method is being used widely by *nucleic acid hybridization*. Colonies of transformed host bacteria with recombinant clones are used to make replica plates. Then these are blotted on to nitrocellulose filter to which nucleic acid binds. The DNA of the replica blot is then denatured to make the DNA single-stranded which allows it to hybridize with single-stranded, radioactively $^{32}P$ labelled DNA or RNA probes which can then be detected by exposure to an X-ray film, or what is known as *autoradiography*. In this

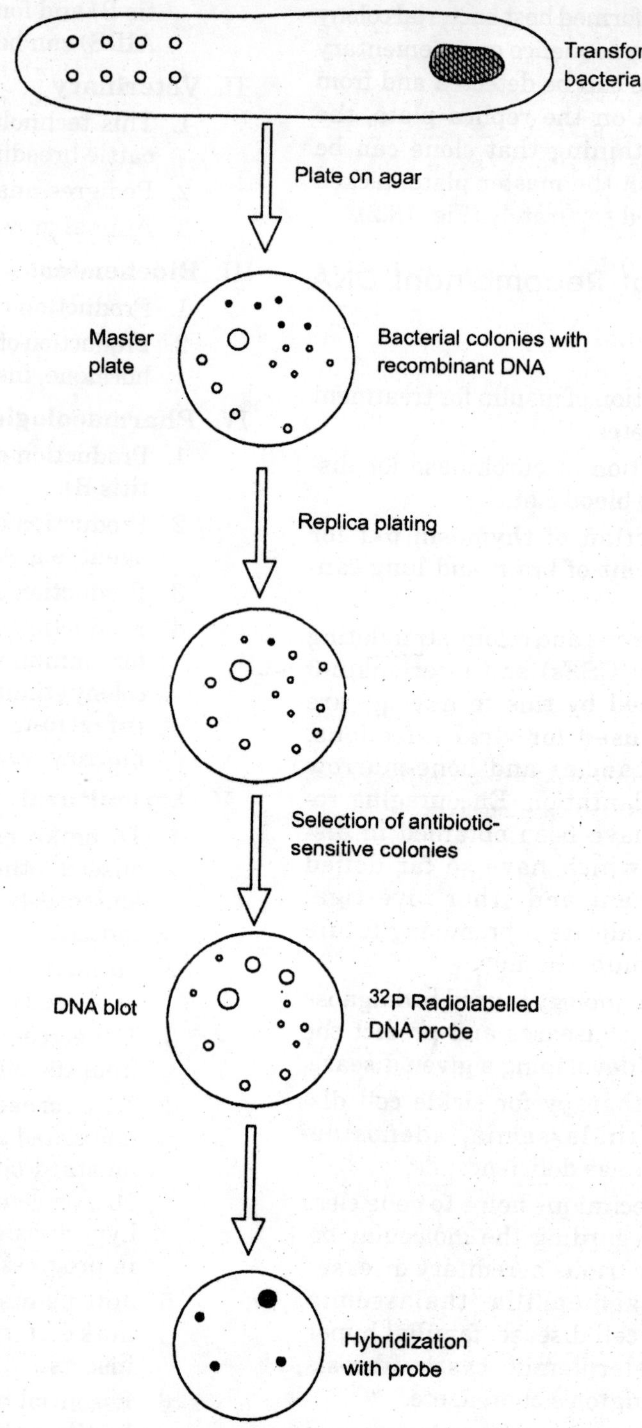


**Fig. 18.2:** Identification of recombinant DNA clones with specific DNA inserts by loss of antibiotic resistance, nucleic acid hybridization and autoradiography.

way a transformed host bacterial colony containing a sequence complementary to the probe can be detected and from its position on the replica plate, the colony containing that clone can be identified on the master plate, picked and cultured separately (Fig. 18.2).

## Importance of Recombinant DNA Technology

**I. Clinical**

1. Production of insulin for treatment of diabetes.
2. Production of eurokinase for dissolving blood clot.
3. Production of thymosin $\alpha$-1 for treatment of brain and lung cancer.
4. Interferons and colony stimulating factors (CSFs) and interleukin-3 produced by this technology are being used for viral infections, malignancies and bone-marrow transplantation. Encouraging results have been obtained in diseases which have so far defied treatment and other investigations indicate a promising future for immunotherapy.
5. The technology is used to diagnose existing diseases and predict the risk of developing a given disease.
6. Gene therapy for sickle cell disease, thalassemia, adenosine-deaminase deficiency, etc.
7. This technique helps to gene clear idea regarding the molecular basis of various hereditary diseases, e.g. haemophilia, thalassemia, sickle cell disease, familial hyperchoteslerolemia, cystic fibrosis, Huntington's chorea, etc.
8. By this technology, a large quantity of human proteins are prepared.
9. With the help of this technology, proteins for vaccines (e.g. hepatitis B) and for diagnostic tests (e.g. AIDS) can be prepared.

**II. Veterinary**

1. This technology is being used for cattle breeding.
2. Pedigree analysis.
3. Animal growth.

**III. Biochemical**

1. Production of proteins.
2. Production of hormones, e.g. growth hormone, insulin, endorphin.

**IV. Pharmacological**

1. Production of vaccines (e.g. hepatitis B).
2. Production of diagnostic chemical agent, e.g. AIDS.
3. Production of drugs.
4. Production of immune substances for immunotherapy (interferons, colony stimulating factors) for viral infection, malignancies, bone marrow transplantation etc.

**V. Agricultural**

1. To make resistant type of rice, oilseed, other crops from bacteria and insects to increase the protein content.
2. To increase the percentage of oil in oilseeds.
3. To decrease unwanted fatty acids from the oil seeds.
4. To decrease the percentage of unwanted uric acid in lipid part of mustard oil. This acid present in 50 per cent which is harmful in hypertension. The investigation is in progress.
5. Jute plants have been protected to make it resistant to harmful insects.
6. Essential amino acid is absent in lentils and lysine is absent in rice, wheat and corn. The work is in progress to increase methionine in Mung Dal and lysin in rice, wheat, etc. The work is in progress.

- **Methods**

I. **Bacterial Method:** The scientists are using "agrobacterium" for recombinant DNA technology. After isolation of $T_1$ plasmid it is combined with disease-resistant or protein-increasing gene, etc. by genetic engineering. Then after making clone, it is released in the tissue culture of any plant. The plant cells are infected by these bacteria, i.e. recombinant DNA enters the cells of the plant. Now these altered cells are used to grow plant, the latter will carry the characteristics of foreign gene.

Genetic Engineering
(Recombinant DNA Technology)
Cloning

- Introduction
  Importance
  Technology
  (a) Isolation
  (b) manipulation of DNA
  - Requires several techniques
  - Reagents
- I. Preparation of R. DNA
  A. Cleavage
  B. Splicing (Joining (a), (b), (c)
  II. Introduction to host cell
  III. Selection of cells
  IV. Cloning importance

II. **Cell-Fusion Method:** When bacterial method is difficult in case of monocot, e.g. rice and wheat the cell-fusion method is applied. Plant cells from two different plants are combined in such a way that the DNA of each plate is mixed up. And the gene having desired property is mixed up in the other cell.

III. **Gene-Gun Method:** In this method the gene is introduced with the help of a gun into the target-cell. Metal pellets having diameter of few micrometer painted with DNA and packed in a cylinder and then introduced in the nozzle of the gun kept in erect position. Then it is fired and the cylinder comes out speedily on a plate where it the cylinder becomes attached. But DNA or gene painted pellets pass through the minute holes of the plate and enters the cells in the culture media. Foreign gene combines with DNA of the cells in culture and makes a DNA having different character.

### Application of Technology

1. Gene mapping
2. Protein production
3. Molecular analysis
4. Gene therapy
5. Production of immunologic cell products for use in clinical trials.

A. **Interferons**
Viral infection
Malignancies

B. Colony-stimulating factors (CSFs)—Immunotherapy
1. Chemotherapy-induced Neutropenia.
2. Bone-marrow transplantation
3. Other uses
Acute myeloid leukemias, Leucopenia-associated AIDS.

## Application of Recombinant DNA Technology

I. **Gene Therapy**
1. Adenosine deaminase enzyme deficiency causes decrease in resis-

tance power to the toxins because of inhibition of production of "T" and "B" lymphocytes. Introduction of heathy gene responsible for the enzyme into the blood can cure the disease. It warrants further clinical appraisal.

2. **Diabetes:** Introduction of insulin-producing gene in diabetic cases are being investigated.
3. **Bone-marrow stem cells:** Bone-marrow stem cells transplantation in the bone marrow of cases of inherited blood disorders are being investigated. These stem cells will resettle and replicate there. The introduced gene will begin to direct the expression of its protein and cells products to correct the deficiency.
4. **Tissue specific effect on gene expression:** When some percentage of genes injected into a fertilized mouse ovum, they are incorporated into the genome and found in both somatic and germ cells and this type animal is called *transgenic animals* and are useful for analysis of tissue specific effect on gene expression.
5. **Transgenic technique:** The transgenic technique is being used to correct a genetic deficiency mice.
6. **Hormone production in vivo:** When fertilized ova obtained from mice having genetic hypogonadism were injected with DNA having coding sequence for secretion of GnRF (gonadotropin-releasing hormone), the gene was expressed in the hypothalamus of certain number of mice and these animals were normal and the offspring did not show any GnRH deficiency.
7. **Stem cell therapy:**

   (a) **Fetus with severe combined immunodeficiency was** successfully treated in utero with stem cells taking from bone-marrow cell of the father. The cells were infused into the fetal peritoneum under ultrasonographic guidance during gestation, at weeks 16, 17.5 and 18.5. At birth, the newborn showed a normal lymphocytic count.

   (b) **Stem cells of spermatogonia:** Recent study has shown that when stem cells of spermatogonia of rat placed in mice resulted in the mice producing mature sperms of the rat. The scientists has successfully frozen stem cells from spermatogonia to ensure unlimited source of parental cells that had the capability of spermatogenesis. This has the possibility of application in men with high reproductive risk for specific gene mutations.

   (c) **Ovarian tissue**

   1. Researches have shown the possibility to grow section of ovarian tissue under appropriate laboratory conditions and harvest a large cohort of oocytes that could be utilised for in vitro fertilisation.
   2. There is enormous development of research to preserve fertility in women undergoing chemotherapy by removing their ovaries and freezing sections of the ovarian tissue with the intention of placing the ovarian tissue back into the original donor after completion of the chemotherapy.

8. **Zygotic gene therapy:** Direct injection of genetic material at the zygotic stage, although successful

in animals but encounters many technical problems.

9. **Embryonic and fetal gene therapy:** Recent advance has been made the chances of applying gene therapy in the embryonic and fetal period.

10. **Pregnant mice and intravenous DNA:** Gene transfer and expression in progeny after intravenous injection of DNA in pregnant mice are being investigated. It involves transferring a foreign gene, complexed with a promoter substance, across the placenta and into embryos by a simple injection into the tail of the maternal host on the day 9.5 post-conception. The gene could be picked up in the DNA of the maternal liver and in the DNA of the treated embryos and newborn mice up to 14 days after treatment, in their liver, lung, heart, thyroid, brain, spleen, pancreas, kidneys and small intestines. The progeny did not show any congenital abnormality when they were born.

## II. Prenatal Diagnosis

1. Prenatal diagnosis can be done if the genetic lesion is known and a specific probe is available.
2. A fetus with restriction pettern AA does not have sick cell disease, nor is it a carrier but fetus with SS pattern will develop the disease and probes are available for this type of analysis of many genetic disease.

## III. Gene structure mapping and Functions

1. Specific genes to distinct chromosomes are localised by this technique and thus to define a map of the human genome. This process produces useful information in the definition of human disease.

   Somatic cell hybridization and in situ hybridization are two techniques are used for this.
2. In situ hybridization, a radioactive probe is added to a metaphase spread of chromosomes on a glass slide. The exact area of hybridization is localized by layering photographic emulsion over the slide and after exposure lining up the grains with some histologic identification of the chromosomes some of the human genes are localised by this procedure.
3. Genes having code for proteins with similar functions can be located on separate chromosomes.
4. Genes responsible for growth hormone and prolaction have been located.
5. In many hereditary diseases, which are due to protein deficiency, the genes involved have been located including X chromosome-linked conditions.

## IV. Pedigree Analysis

1. For cattle breeding, pedigree analysis is applied to a number of genetic diseases and is most useful in those caused by deletions and insertions or the rarer instances in which a restriction endonuclease cleavage site is affected.
2. Incubation of DNA from normal (AA), heterozygous (AS), and homozygous (SS) individuals results in three different patterns on southern blot transfer.

## V. Protein Production

1. This technology has two advantages regarding protein production over the conventional methods:
   (a) Large amounts of protein are produced

(b) It can produce human protein hormones like growth hormone, insulin (Fig. 18.1)

2. It also produces protein materials for:

(a) Diagnostic purpose, e.g. AIDS in human.

(b) Diagnostic purposes in animal diseases.

(c) Preventive purposes, e.g. hepatitis B vaccine, etc.

VI. **Therapeutic—Immunomodulator for Immunotherapy**

*Interferons* (IFNs) interleukins and *colony stimulating factors* (CSFs) are produced within one's body in response to various stimuli. These immunologic cell products have been isolated and synthesized in the laboratory using recombinant DNA technology are being used in clinical trials called immunomodulator indicating a promising future for immunotherapy.

A. **Interferons:** The interferons are a family of proteins produced and secreted by leucocytes, fibroblast and epithelial cells. The major subtypes on the basis of antigenicity, biologic and chemical properties are alfa beta and gamma. The human interferon alfa is commercially available as alfa-n3, alfa-2a and alfa 2b which are of recombinant DNA origin and exist as single interferon subtype preparations.

It is indicated in:

i. Viral infections
ii. Malignances
iii. Tried in other diseases like expophytic genital wart, anal wart, multiple sclerosis, AIDs and certain angiomatous disorders.

B. **Colony Stimulating Factors (CSFs):** When cell culture techniques for WBCs were being carried out, it was recognized that certain substances permit growth of cell colonies, known as colony stimulating factors (CSFs) they appeared to act as intracellular signals to activate their target cells. Several such factors are now recognized, and are identified by

**Table 18.1:** Proteins produced biosynthetically by Recombinant DNA technology

| Proteins | Used in disease |
|---|---|
| 1. Insulin | Insulin-dependent diabetes mellitus |
| 2. Growth hormone | Short stature due to growth-hormone deficiency |
| 3. Factor VIII | Haemophilia A |
| 4. Factor IX | Haemophilia B |
| 5. Interferon | Infections, cancer (?) |
| 6. $L_1$-Antitrypsin | Emphysema due to $L_1$-antitrypsin deficiency (?) |
| 7. Somatostatin | Excess growth |
| 8. Vaccines | Hepatitis B, malaria, other tropical diseases (?) |

(?)—Efficacy not yet established

the cells whose growth and development they control. CSFs are complex glycoproteins that can now be produced by recombinant DNA technology. The first CSFs to be produced by genetic engineering and most widely used are GM-CSF (granulocyte-macrophage CSF) and its gene is located on chromosome 17.

These are indicated in:

i. Chemotherapy-induced neutropenia
ii. Bone-marrow transplantation
iii. Myelodysplastic syndromes
iv. Other diseases with limited application—acute myeloid leukemias, leukopenia associated with AIDs.

C. **Interleukins:** These are family of proteins (lymphokines, produced by lymphocytes. IL-2 is the best studied of these proteins which was formerly known as T-cell growth promoting factor. It can stimulate growth of certain cells called lymphokines-activated killer (LAK) cells. IL-2 affect the natural killer cells, T-lymphocytes, and tumor-infiltrating lymphocytes within the immune system. They have the ability to recognise and destroy tumor cells without damaging normal cells. It is used in:

i. Tumors resistant in conventional treatment
ii. Renal cell carcinoma
iii. Malignant melanoma.

### • Method of Production of Immunomodulator

Genetic engineering has employed bacteria to produce immunomodulator. The following steps are used:

1. Human DNA is cleaved by restriction endonuclease.
2. Cut piece of DNA for immuno- modulator is synthesised.
3. After purifying the synthesised gene is inserted into the bacterial DNA for splicing causing to produce immunomodulator.
4. The bacteria is then multiplied by culture for multiplication.
5. Bacterium then produces the immuno modulator and it is extracted in large-scale (Fig. 18.3).

## VII. Molecular Analysis of Disease

### A. Gene variations causing disease

1. Genetic diseases were due to point mutations that resulted in an impaired protein.
2. β-globin gene is located in a cluster on chromosome 11. Defect of production of β-globin results in a variety of diseases and is due to many types of lesions in and around the gene for β-globin.

### B. Normal gene variations

1. *Polymorphism* occurs once in every 500 nucleotides or about $10^7$ times per genome.
2. There are deletions and insertions of DNA as well as single substitutions.
3. In normal individuals, these alterations occur in concoding regions of DNA or at sites that cause no change in function of the encoded protein.
4. The polymorphism of DNA structure may be associated with certain diseases.

### C. Deletions, insertion, and rearrangement of DNA

1. Disease is caused by the deletion of a critical piece of DNA, the rearrangement of DNA, the rearrangement of DNA within a gene, or the insertion of a

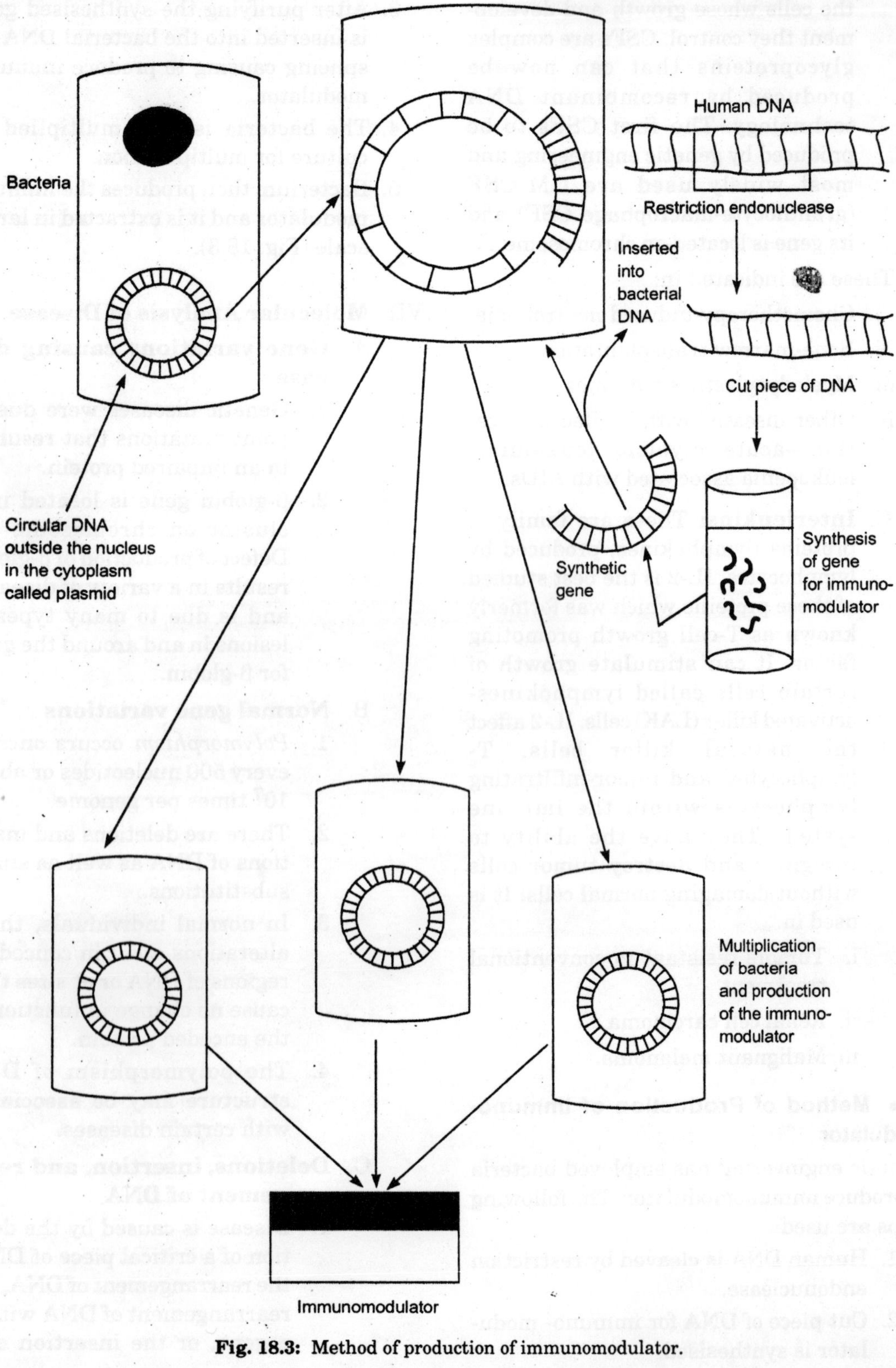


**Fig. 18.3:** Method of production of immunomodulator.

piece of DNA within a coding or regulatory region can all cause changes in gene expression.

2. Pieces of DNA can move from one place to another within a genome on the study of bacteria, viruses, yeast, and fruit flies.
3. Deletions or insertions of DNA larger than 50 bp can often be detected by the southern blotting procedure.
4. Deletions in the alpha-globin cluster, located on the chromosome 16, cause alpha thalassemia.

D. **Restriction Fragment Length Polymorphism (RFLP)**

1. Inherited differences in the pattern of restriction is known as RFLP.
2. **This results from single base changes, e.g. sickle cell disease or from deletions or insertions of DNA into a restriction fragment, e.g. thalassemia and are proving to be a useful diagnostic measure.**
3. **RFLPs are inherited and they segregate in a Mendelian fashion.**
4. **RFLPs may disrupt the function of the gene or may have no biological importance.**
5. **These can be used to establish linkage groups, which in turn, by the process of *chromosome walking*, will define the disease locus.**
6. **In chromosome walking, a fragment representing one end of a long piece of DNA is used to isolate another that overlaps but extends the first.**

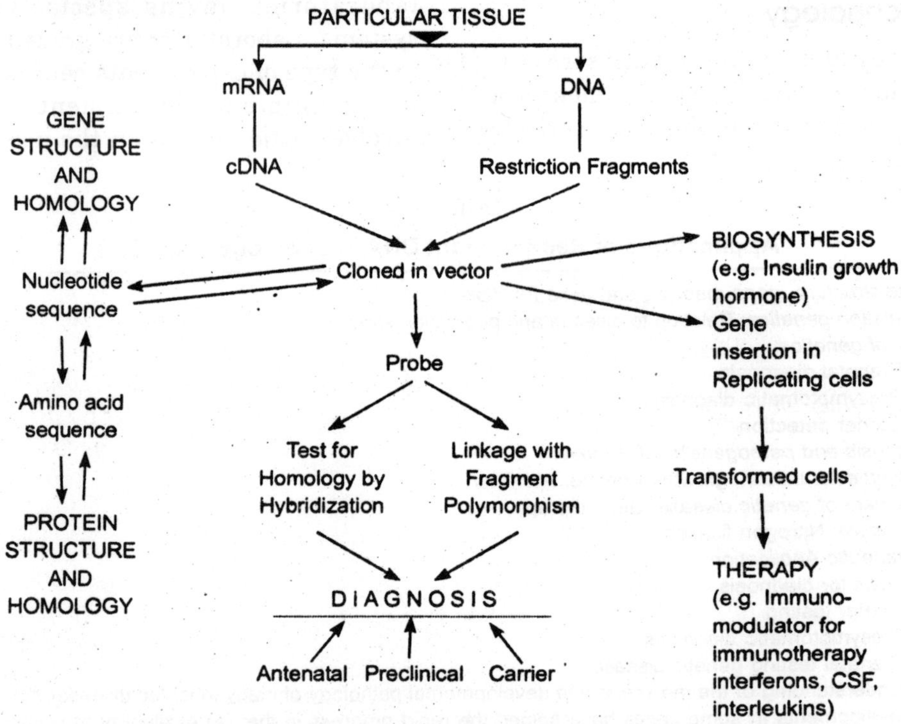


**Fig. 18.4:** Some of the application of DNA technology in medical genetics.

7. X-linked disorders such as Duchenne type muscular dystrophy will be defined using RFLPs.
8. 20 per cent of the defined RFLPs are on the X-chromosome.
9. The defect that causes polycystic kidney is linked to the alpha-globin locus on chromosome 16.

VIII. **Population Genetics:** Relation to disease and population.

IX. **Clinical Genetics:**
1. Prenatal diagnosis
2. Presymptomatic diagnosis
3. Carrier detection

X. **Diagnosis and pathogenesis of disease**

XI. **Agriculture:** Nitrogen fixation.

## Biological Hazards of Recombinant DNA Technology

1. Engineered microbial strains, if released into the environment, could cause for more problems than the benefit.
2. Danger of producing transgenic organisms which could contain genes for cancer susceptibility or be resistant to all known antibiotic.
3. One of the host organisms which is commonly used in the experiments is *E.coli,* which is normal bowel commensal.

The consensus of scientific opinion that these dangers have been very much exaggerated. Various Authorities like Genetic Manipulation Advisory Group in UK and the National Institutes of Health in USA have laid down very careful guidelines for the use of techniques.

- **Techniques of limitation of Biohazard**

I. Physical containment method
II. Biological containment method.

**I. Physical containment method**

The subject is to make sure that any microorganisms of potential danger are contained by use of specially designed laboratories having special ventilator systems. Laboratories are graded according to the type of experiments being conducted, from minimum containment (CI) which involves little more than the use of careful

SUMMARY

**Applications of Recombinant DNA Technology** (Fig. 18.4)

1. *Gene structure, gene mapping* and *gene function.*
2. *Population genetics*: Relation to disease and population structure.
3. *Clinical genetics*:
   a. Prenatal diagnosis
   b. Presymptomatic diagnosis
   c. Carrier detection
4. *Diagnosis and pathogenesis of disease.*
5. *Biosynthesis*: Insulin, growth hormone, interferon.
6. *Treatment of genetic disease:* Gene therapy.
7. *Agriculture*: Nitrogen fixation.
8. *Therapeutic Application*:
   a. Tests for diagnosis
   b. Carrier testing
   c. Presymptomatic diagnosis
   d. Prenatal testing genetic disease
   e. Understanding of the molecular and developmental pathology of many inherited diseases

These developments in some cases have helped the rapid progress in the: (a) Availability of biosynthetic gene products, (b) Planning of strategies for gene therapy of certain inherited diseases.

techniques to maximum containment (CIV) where containment is of an exceptionally high-order. The high level of containment is used in laboratories involved in biological warfare.

II. **Biological containment**

Host organisms used in recombinant DNA work have been attenuated or crippled, so that they cannot survive outside the confines of specialised laboratory culture conditions. This is achieved by careful selection of particular mutants which led, for example, to the production of the *E.coli*. K 12 strain 1776 (after the year of American independence) which requires many complex growth factors if it is to survive outside the laboratory.

At the same time, complementary work has produced safer vectors. The combination of physical containment and the use of attenuated organisms should reduce the risk of any possible hazard.

## QUESTION

1. What is recombinant DNA technology? Describe briefly the procedure with diagram.
2. What do you mean by cloning? How it is done? What are the suitable vectors for cloning? Which is the best? Why?
3. State briefly the importance of cloning? Why cloning is necessary? How long it is stable?
4. What are the methods for splicing foreign DNA molecule?
5. How DNA sequencing is done?
6. What do you mean by library and library construction?
7. Define—Probe, Blotting, Hybridization.
8. Describe briefly the importance of genetic engineering with special reference to clinical, veterinary, biochemical, pharmacological.
9. What do you mean by gene therapy? Describe the progress made on the therapy.
10. How genetic engineering has been employed for production for immunomodulator?

# 19

# HUMAN CELL CLONING AND THERAPEUTIC

*Hwang,* a veterinary medical researcher and *Moon,* an obstetrician, these two scientists in South Korea, faculty members at Seoul National University had successfully cloned a human embryo and extracted a stem cell line from it.

They collected 242 egg cells from which they succeeded in cloning 30 blastocysts—early stage embryos containing around 100 cells. From those cells, they harvested one colony of stem cells. This is an important step in the development of new therapies so that patients can look forward to having transplants of cells exactly matched with their needs.

Scientist have used therapeutic cloning to partially cure laboratory mice with an immune system disease. But attempts at cloning a human embryo in the stem cell quest have failed until now. Embryonic stem cells which can form any tissue in the human body, have long been considered one of the most exciting areas of research in modern medicine as they provide the raw material from which spare parts might be grown.

The prospect of therapeutic cloning is that collecting those master cells from a clone of the patient to be treated. The cells would then be genetically identified to the person who needs them and would not be rejected by the immune system.

The scientists had taken the genetic material of an adult cell from a female volunteer and transplanted it into an egg (oocyte) from the same female. From the egg the nucleus had been removed (enucleated). The egg was then coaxed into dividing as if it were an embryo, reaching the blastocysts stage after a few days of growth.

The stem cells were then isolated from the blastocyst and started to divide into all standard cell types seen at this of embryonic development. When transplanted into mice, the stem cells differentiated into still more specific cell types, offering further proof that they are *pluripotent* or capable of growing into any type of tissue.

These scientists put their success down to a finely *honed technique* in which DNA was gently squeezed from eggs rather than sucked out with a pipette. This more sensitive procedure appears to have protected the DNA from damage which is common problem with cloning.

The use of eggs from the adult cell donor may have also enhanced the prospect of success owing to their genetic compatibility. While stem cells have been successfully cultured before, the South Korean scientist is the first to produce them using therapeutic cloning which has tremendous medical potentiality. Because these stem cells carry the nuclear genome of the individual after differentiation they can be expected to be transplanted without immune rejection for treatment of degenerative disorders and it will be very much useful in transplantation medicine with special reference to Alzheimer's disease, Parkinson's disease and various other muscular disorders.

- **Advantages**

1. It offers real hope for new therapies.
2. Tissue for transplantation would not be rejected by the patient's immune system.

   The major problem with conventional transplant techniques such as in case of Parkinson's disease, the idea would be to produce a line of stem cells from a cloned blastocyst that is genetically identical to the patient, then differentiate these cells to form dopamine neurons, the type of brain cell that malfunctions in this condition. This would then be transplanted into the patient's brain with little chance of rejection to improve or even restore function.

- **Discussion**

Though the technique has been lauded by the experts, but it is also cautioned that considerable work still has to be done and any patient trials of new therapies are probably at least few years away. The potential embryonic stem cells is enormous but the scientists still must overcome significant scientific hurdles.

The cell and tissue transplantation and gene therapy are still emerging.

- **Sequence of events in procedure**

Genetic material of an adult somatic cell from a female volunteer (somatic cell biopsy)

↓

Transplanted it into an *egg cell* (donor oocyte) from the same woman from which the nucleus had been removed (i.e. enucleated donor oocyte) i.e. *nuclear transfer*

↓

The egg was then coaxed into dividing as if it were in *embryo*

↓

Reaching the *blastocyst* stage after a few days of growth (i.e. 6 days embryo)

↓

The stem cells were then isolated from the blastocyst

↓

Started to divide into all the standard cell types seen at the stage of embryonic development

↓

When transplanted into mice the stem cells differentiated into still more specific cell types, offering further proof that they are "pluripotent" or capable of growing into any type of tissue (Fig 19.1).

Human therapeutic cloning opens the door to a host of therapies technologies and it may be years yet before embryonic stem cells can be used in transplantation medicine.

This work also make it possible to learn how to reprogramme the human genome to an embryonic state. This will likely accelerate the development of alternative ways of reprogramming human cells which could in the future diminish the need to use human eggs for this purpose.

This breakthrough has created alarming reaction in the religious groups as the culling of stem cells from cloned or normal blastocyst involved the destruction of an early stage embryo.

It has also raised controversial point among the opposition group that all forms of human cloning if are allowed for therapeutic cloning may create a *slippery slope* leading to cloned babies.

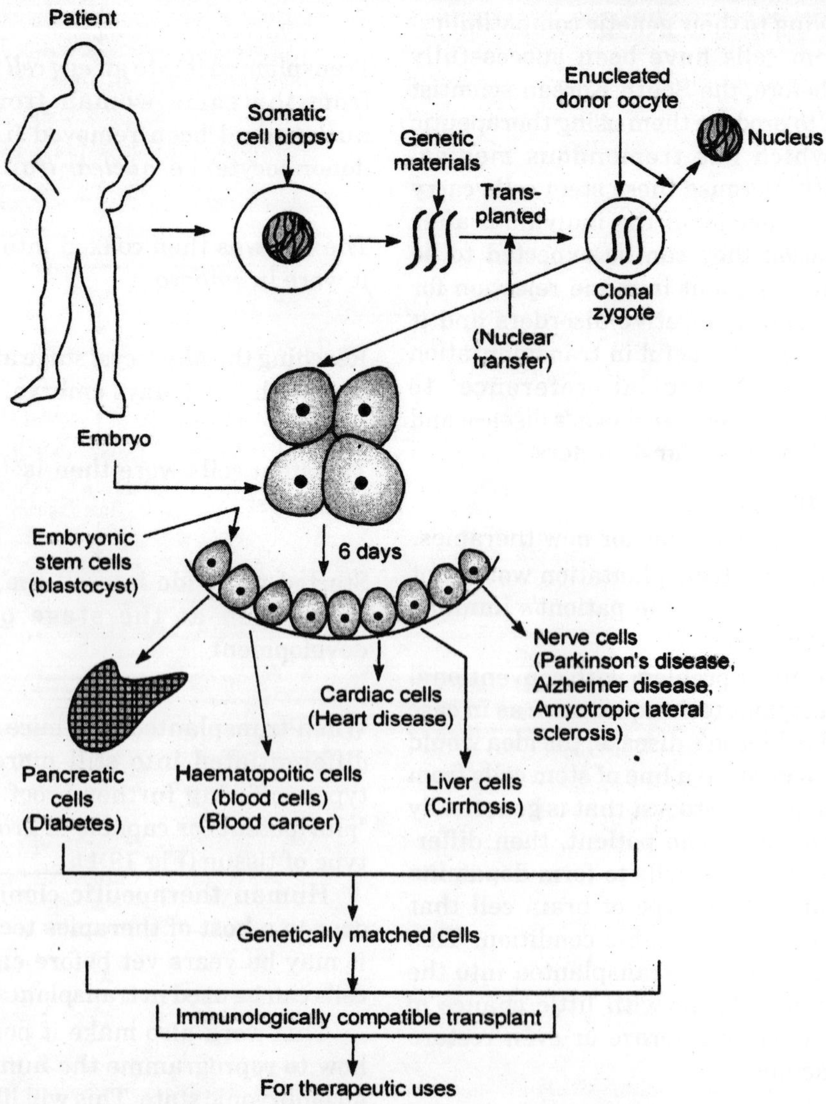


**Fig. 19.1:** Human therapeutic cloning opens the door to a host of therapies.

- **Scientific, ethical, legal and medical aspects**

1. **Britain:** Granted its first license for human cloning, first nation to authorise the technique to produce stem cells for medical research. British regulations allow the embryo to develop for no more than 14 days. No other European country has granted cloning.
2. **South Korea:** Scientists there announced, they had become the first in the world to successfully clone a human embryo and extract stem cells. But the country has not yet issued license for the research.
3. **Japan:** Government Science Council has voted to adopt recommendations that would permit limited cloning of human embryo for research.
4. **USA:** Prohibits any kind of embryo cloning and has lobied strongly against it, and funding for stem cells research is restricted there.

5. **UN:** The United Nations, will consider proposals for a global ban on all forms of cloning. More than 60 National Science Academies worldwide have argued strongly against global ban, calling instead for a ban on reproductive cloning while research into therapeutic cloning is allowed because any attempt to clone a human being would be highly dangerous and wrong, and therefore, all reproductive cloning should be banned.

## QUESTION

1. Describe the technique of human cell cloning. Who are the scientists for this pioneer work?
2. Describe the human therapeutic cell cloning.
3. Discuss the advantages and disadvantages of human therapeutic cell cloning. How the different countries have reacted to this breakthrough research?

# 20 HUMAN GENOME PROJECT

### What is Genome?

Genome is a complete set of hereditary factors as contained in the haploid assortment of chromosomes.

A genome therefore contains genetic codes from both parents of the individual. There are about 100,000 odd genes in a genome and 4550 out of them have been identified so far. Out of these about 1500 have been properly located on the respective chromosome.

It is also defined as:

1. Complete set of genes characteristics of a species.
2. Total DNA carried by a gamete.
3. The complete DNA sequence of an organism containing its complete genetic information.

### What is Human Genome Project?

Considering the future of genetics and medicine, the emergence of the Human Genome Project (HGP) is realised. It is an organized and coordinated effort to map and sequence all of the human DNA (i.e., base-by-base reading of the entire 3 billion nucleotides long human genome) within few years. A project of this type, with specific scientific goals and milestones, and differs from the convention biomedical research format. This project is being executed worldwide by many laboratories and institutions in collaboration with great enthusiasm.

### Goal or Objectives of Human Genome Project

The general impression of the public regarding objectives of human genome project is *to sequence all* $3 \times 10^9$ *base pairs of the human genome.* There are *five* main objectives, the above is one of them. The other four objectives are:

1. Preparation of *human genome maps* and *mapping of human-inherited diseases,* i.e. *genetic map.*
2. *Physical maps* (Development of new DNA technologies).
3. *Sequencing* of the human genome.
4. Development of *Informatics*.

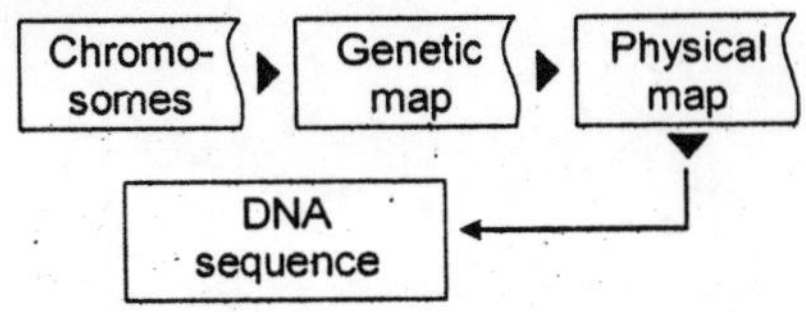


1. **Genetic Map—preparation of human genome project and mapping of human-inherited diseases:** Genetic mapping means the

exact position of gene occupied in a chromosome.

The information is available regarding the power of linkage analysis and positional cloning for identification of genes that predispose to human disease, even without prior knowledge of the genes. These methods initially developed for Mendelian single gene diseases and more recently extended to polygenic conditions and it would be successful if a robust map of the human genome is available and that is the first step in any positional cloning project. A genetic map with highly informative markers spaced on the average 2-5 centimorgans (cM) apart was thus felt to be a very high priority for the Human Genome Project. The genetic maps initially involved the production of fairly low level resolution index, skeleton or framework maps, which were based on polymorphic variable number di-, tri-, and tetranucleotide tandem repeats spaced approximately 10 cM intervals throughout the genome. The most recent high-level resolution maps have polymorphic markers which are on average spaced at intervals of less than 1 cM.

The technology of marker production shifted rapidly from restriction fragment length polymorphisms (RFLPs) to simple sequence repeat (SSR) microsatellites because of the highly informative nature of the latter, and their easy adaptation to the polymerase chain reaction. More than 5000 highly informative markers had been placed on the map, resulting in an average density 4 to 8 times better than the originally planned. With this large array of genetic markers, nearly any human phenotype that has a hereditary component should be mappable, provided a sufficient number of families are available for analysis.

2. **Physical Maps:** The mapping information from these genetic maps has been integrated with high resolution physical maps.

One of the important objectives of human genome project is to construct a physical map of 30,000 evenly distributed landmarks, distanced at an average space of 100 kilobases (1 kilobase = 100 bases). Recently published map of human genome is composed of 5264 microsatellite markers each separated by 0.7 centimorgan (cM) of map distance (1 centimorgan = 1 percent recombination frequency). This is at present the best available marker density on human genome. Dense genetic maps make possible the identification of genes for single-gene disorders and the dissection of polygenic traits.

The information for the physical maps comes from a variety of sources using a number of different techniques which include:

1. Fluorescent in situ hybridization (FISH).
2. Somatic cell and radiation hybrid (RH).
3. Cosmid contigs or YAC (yeast artificial chromosomes).
4. Sequence-tagged site (STS) content.
5. Clone-based fingerprint mapping (FM).

Best way is to use vector systems that allows the cloning of very large pieces of human DNA in order to generate physical maps.

Yeast artificial chromosomes (YACs) permit the cloning of pieces of DNA from 100,000 up to 1,000,000 bp, and have been the standard currency for the physical maps despite the technical problems.

The sequence-tagged site (STS) has become the common currency of physical

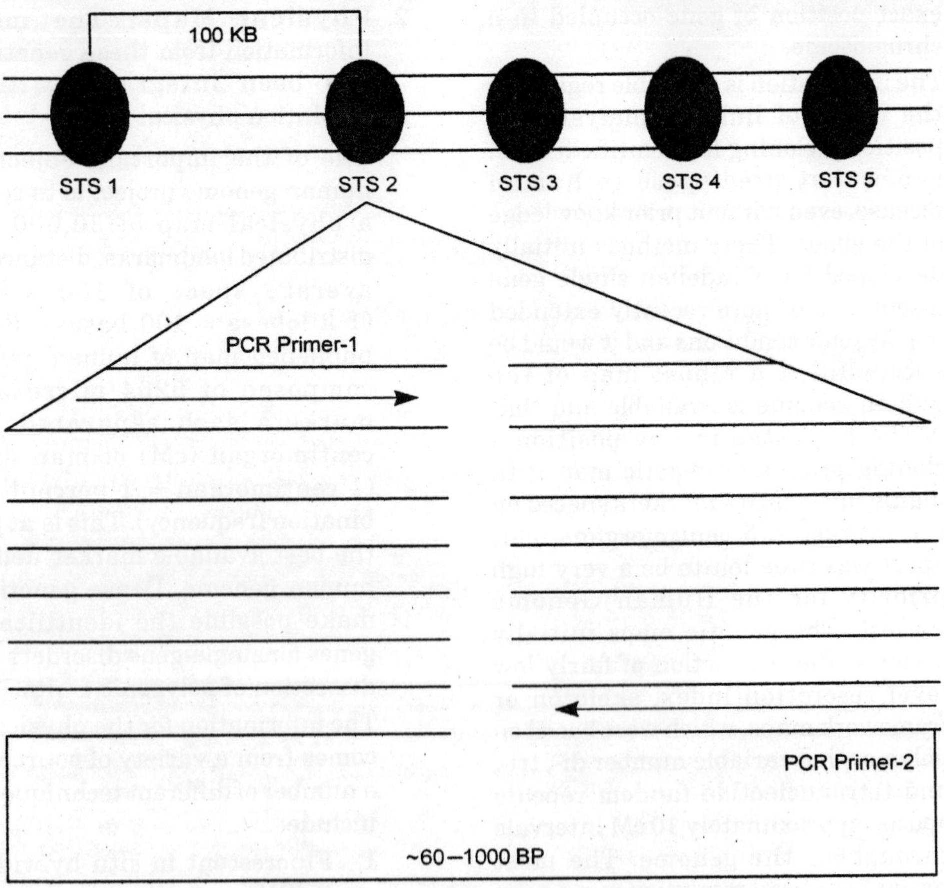


**Fig. 20.1:** The concept of STS.

mapping. These STSs can be used to readily construct a set of overlapping clones for any region of interest, provided that the STSs are sufficiently closely spaced. The spacing of the STSs needs to be an average several fold closer together than the average insert size of the cloned fragments that will be assembled into the map. The STS is a unique location in the human genome, defined by two oligonucleotide primers. These primers can be used in the polymerase chain reaction to amplify that specific region of DNA. Although vector systems will change in future, the sequence defining the STS is a basic property of the genome and thus the currency for physical mapping should be quite durable (Fig. 20.1).

Access to the detailed information from these high resolution genetic map and physical map allows individual research teams, often interested in a specific or particular inherited disease or group of diseases, to rapidly and precisely localize or map a disease gene to a specific region of a chromosome.

The presence of CpG islands, exon trapping and expressed sequences tags (ESTs) can be used to identify coding sequences in the region involved. These along with the known genes or ESTs which map to the region can then serve as positional candidate genes for analysis to determine whether they are responsible for the inherited disorders.

- **Development of New DNA Technologies**

The second major goal is the development of new DNA technologies for human genome project, e.g. the technology for DNA sequencing was very time-consuming, laborious and expensive. The development of automated genome sequencing technology, along with the use of fluorescence labelling, has helped transform the ease and cost of large-scale DNA sequencing projects. In future, developments in the use of DNA 'chips' technology are likely to further revolutionize the ability to produce human genome sequence data.

- **Sequencing of the Entire Human Genome**

This is main focus of the human genome project to obtain the sequence of all 3 billion bp of human DNA. Initially it was thought to be a difficult job but due to automation, optimization of all of the steps involved, and microscaling, it now appears certain that this effort will succeed.

Doubt arises if the complete human sequence is known, whether it is possible to predict the location of genes. Recognising that, less than 5% of genomic DNA is involved in coding for protein, it is essential to have available computer algorithm that are capable of recognizing the coding information (the exon) from a large stretch of uncharted DNA. Exons do have distinguishing characteristics they are preceded by a splice acceptor, they in general contain an open reading frame and they are terminated by a splice donor. However these characteristics are not sufficiently obvious to pick out coding regions from a large stretch of DNA by simple visual inspection. Recent sophisticated neural net approaches have allowed the development of computer algorithms that appear quite powerful in identifying exons of 100 bp or larger in a region of DNA never previously studied. Output of a computer algorithm called GRAIL for a sequence of 17,000 bp of DNA from chromosome 16. This region contains part of the gene for polycystic kidney disease and the agreement between the computer prediction and the actual location of the exons in quite good. When it was known that exons have great value in sequencing, interest has been arisen for sequencing cDNA. Already some sectors have determined partial sequences of tens of thousands of cDNAs and effort is in progress of place all these in their proper on the location in the physical map, using the 3′ untranslated regions of each cDNA (known as ESTs, or expressed sequence tags) as STSs in the physical mapping strategy described above. The combination of information from cDNA and genomic sequencing is expected to be very powerful. Some have been argued that the genomic sequence is superfluous, since the cDNAs will contain the coding region. However this assumes that it will be possible to identify all human cDNAs, which may be difficult for those that are of low abundance or expressed only in tissues that are inaccessible for study. Furthermore, sequencing the cDNAs will not provide crucial information about regulatory sequences that determine which genes are expressed in which tissues.

- **Development of Informatics**

The another goal of the human genome project is known as informatics—the establishment of facilities for:

(a) Collection (c) Analysis
(b) Storage (d) Communication

the data from the project. It is essential to have the information from the human genome project.

The information is received from large number of electronic databases available on the World Wide Web (WWW) on the internet. These encompass:

1. Protein and DNA sequence database, e.g. Genbank and EMB.
2. Databases of genetic maps for humans, e.g. GDB, Genethod, CEPH, CHLC and the Whitehead Institute sites.
3. Databases of Genetic Map for other species e.g. Mouse Genome Database and the *caenorhabditis elegans* database.

4. Linkage Analysis Programmes, e.g. Rockefellor University Website.
5. The catalogue of inherited diseases in humans (online Mendelian inheritance in Man or OMIM).

The latter was first set up as a computer-based information system by Victor Mckusik long ago.

- **Model Organism Genome Project**

In addition of Human Genome Project there are separate genome projects for a number of other species, it is known as *model organisms*. Because the complete genomic sequence will still present many puzzles. Interpretation of the sequence, and experimental application of that knowledge, will be greatly benefitted by a comparison with other model organisms more amenable to research analysis.

**Table: 20.1** : These organisms were carefully chosen based on the availability of a large amount of genetic and phenotypic information. The organisms are (Table 20.1).

| Prokaryote | Eukaryote |
|---|---|
| A. Bacterias<br>(a) Escherichia coli<br>(b) Haemophilus influenzae | (a) Saccharomyces cerevisiae (yeast)<br>(b) Caenorhabditis elegans (flat worm)<br>(c) Drosophila melanogaster<br>(d) Mus muscularis (mouse)<br>(e) Rattus norvegicus (rat)<br>(f) Fugu rubripes ruripes (puffer fish) |

Though the model organism project of different species is not essential but it has vital importance in the human genome project for two reasons:

(a) The availability of the total genome sequence of yeast has revealed it to contain a total of approximately 6000 genes, only some 2300 of which had been previously identified from traditional genetic studies in yeast. In addition of the 3700 new genes identified by the sequencing project, only approximately 2300 share significant sequence homology with any known genes from any other organisms, i.e. they are novel genes whose functions are unknown. Mapping of the human homologues of these genes identified from these and other model organisms, such as the fruitfly or mouse provides new candidate genes for inherited diseases in humans.

(b) It provides the means to follow the expression of genes and the function of their protein products in normal development as well as their dysfunction in inherited diseases. The ability to introduce targeted mutations in specific genes along with the production of transgenic animals, e.g. in the mouse, allows the production of animal models to study the patho-developmental basis for human inherited disorders as well as serve as a test system for the safety and efficacy of gene therapy.

- **Identification of a Diseased Gene**

There are two procedures:

**1. Functional Cloning**

The RNA or protein product of a gene is explored first and then an already available clone in the library is analyzed for the same function (Fig. 20.2).

- **Comparative Genomic Hybridization**

This is a modification of reverse painting which is used widely in cancer genetics *to detect regions of allele loss and gene amplification*. Tumour or test DNA is labelled with a green paint and control normal DNA is labelled with a red paint. The two samples are mixed and hybridized to normal metaphase chromosomes. If the test sample contains more DNA from a particular

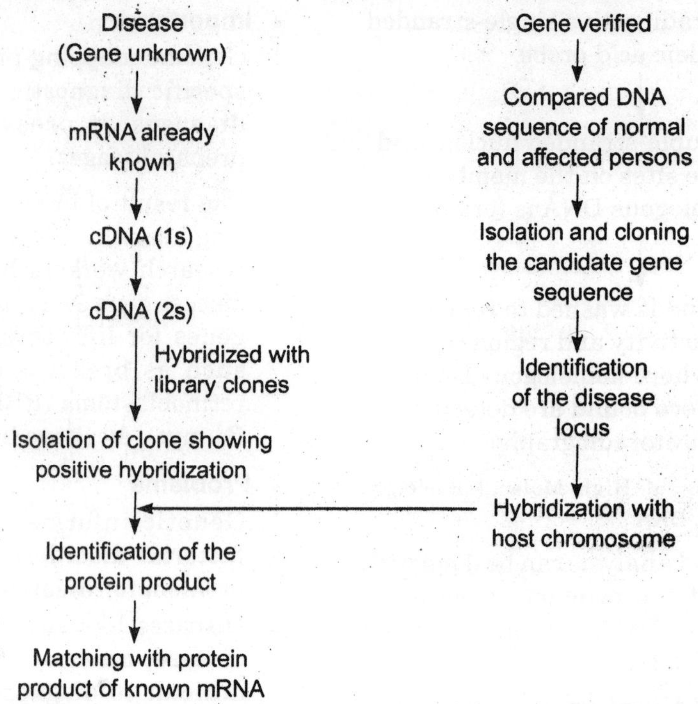


**Fig. 20.2:** Flow chart showing the process of functional cloning.

chromosome region than the control sample, then the region is identified by an increase in the green to red fluorescence ratio. Similarly, a deletion in the test sample is identified by a reduction in the green to red fluorescence ratio.

- **Flow Cytometry**

Because of their differing size and DNA composition, chromosomes bind different amounts of fluorescent dyes, some of which bind specifically to GC ('gene-rich') sequences. This property of differential binding allows chromosomes to be separated out by the process of *flow cytometry* or *fluorescent activated cell sorting (FACS)*. This involves staining metaphase chromosomes with a fluorescent DNA-binding dye and then projecting them in a fine jet of droplets across a focussed laser beam which excites the chromosomes to fluorescence. The fluorescence intensity is measured by a photomultiplier and the results are analysed by a computer which draws up a distribution histogram of chromosomes size. This is referred to a *flow karyotype.* Flow cytometry can be used to analyse an individual's chromosomes but its clinical application is severely limited at present by its expense and by the relatively poor separation achieved for certain chromosomes, most noticeably those in the C group. It finds greater application in separating out preparations of single chromosome for the construction of chromosome-specific DNA libraries and in the manufacture of chromosome paints for FISH.

Cells of peripheral eukaryocyte or fetal cell
↓ Digested with restricted enzyme
Series of reproducing fragments
↓
Separated by electrophoresis agarose gells
↓ Transferred
To a membrane that binds DNA
↓ Soaked
↓

In a solution of radioactive single-stranded nucleic acid probe

↓

Probe forms double-stranded nucleic acid complex at the sites on the membrane where homologous DNA is formed

↓

The membrane is washed to remove unbound radioactivity and regions on the membrane where homologous DNA sequences were bound are detected using autoradiography

**Fig. 20.3:** Isolation of High Molecular Weight Genomic DNA

By this method analysis can be done with a tiny portion of the primary structure of human genome DNA taken from an individual (Fig. 20.3).

## 2. Positional Cloning

At first a broad region is marked on the chromosome that has shown substantial evidence of lodging the gene. The second step is to narrow down the region by contig mapping and determine the function of the gene after its isolation (Fig. 20.4).

Mendelian disorder with clear mode of inheritance

↓

DNA markers cosegregation with disease locus identified

↓

Firm genetic linkage to the closest marker established

↓ Chromosome walking and jumping

Candidate sequence identified

↓

Gene cloned

**Fig. 20.4:** Flow chart of positional cloning.

- **Importance**

1. Genome mapping provides quick and specific diagnostic tests for genetic diseases at presymptomatic and prenatal stages.
2. The result of human genome sequencing has achieved a good returns. The research workers have isolated new disease gene every month. In this way, genes for life- threatening diseases such as breast cancer (C-erb $B_2$), retinoblastoma (RBI), neuroblastoma (N-myc), etc. have been identified.

- **Problems**

1. Genetic information about multi-factorial disorders present greatest problem to research workers.
2. Diseases like susceptibility to heart disease, cancer, infections are more difficult to diagnose and treat than those caused by single-gene defect (monogenic disorders).

- **Genome Database**

The problem is how to acquire, manage, analyze, interpret and disseminate these data. A perfect database can provide answer of many valuable queries such as:

(a) Specific location and developmental stage of gene products.
(b) Expression of what genes along specific differentiation pathways.
(c) To know in which order do the gene products appear or disappear. If these are established, the information resource for gene expression will give fundamental information for elaborating the function of 50,000 to 100,000 genes and for determine the reason of heritable mutations and epigenetic modifications lead to genetic disease. At present, there are three types of database storing different kinds of information:

1. Gene bank—For functionally cloned mRNA and genes.
2. Genome database—For chromosome mapping information.
3. Protein information resource—For protein sequence and structure. (Fig. 20.5).

- **Future consequences**

1. The Human Genome Project can be considered as building infrastructure for the field of genetics and accelerated the pace of gene discovery.
2. As a consequence of this effort, the flow of molecular medicine discoveries will be occurring in great abundance (Fig. 20.4).
3. The map and sequence information provided by the Human Genome Project, coupled with the collection of DNA from families in which a particular disorder is occurring at high frequency, should allow the identification of susceptibility genes for almost all possible heritable phenotypes, using the principles of positional cloning.
4. The advent of the Human Genome Project greatly accelerates the ability to identify human disease genes by the positional cloning strategy. After a gene has been identified, preventive medicine strategies can be designed for those individuals found to be at high risk. In the long run, however, the greatest promise of gene discovery will be the development of new and more effective therapies.
5. Those gene discoveries will lead quickly to the possibility of presymptomatic testing. In situations in which a single Mendelian gene with high penetrance is present, the testing may predict a high risk of future illness for the individual or their offspring.
6. In polygenic conditions, however, or those with relatively low penetrance, testing will not provide a negative or affirmative answer but rather an increased and decreased relative risk. For many disorders, the information about risk will be useful, as it will allow design of an individual program of medical surveillance and lifestyle planning to reduce that risk.

- **Ethics, Legal and Social Aspects**

The effort representing the largest research investment in ethics in the history of humankind is commonly known as the ELSI (i.e. Ethical, Legal and Social Implications) Program.

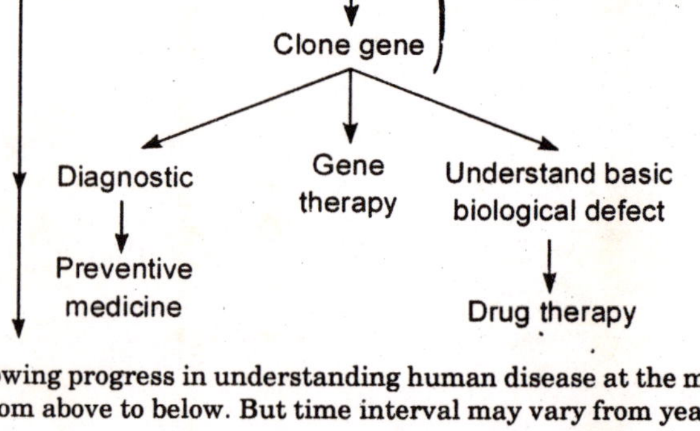


**Fig. 20.5:** Flow chart showing progress in understanding human disease at the molecular level follows a predictable pattern from above to below. But time interval may vary from years to decades.

A number of ELSI issues have been pointed in the early years of the HGP and targeted for research, deliberation and legislative initiatives such as:

1. Fairness in the use of genetic information because it may carry considerable value for the individual and could be used to discriminate against individuals.
2. A health insurance system can allow the insurer to discontinue coverage for persons found to have preexisting condition.
3. Intensive efforts are being made, accelerated by the ELSI Program to rectify this potential for discrimination through the legislation. Already, the passage of the Health Insurance Protability and Accountability Act in other countries has provided important protections for individuals in the group health insurance market but the loopholes must be closed. The urgency of doing this grows with new gene discovery.
4. The potential for genetic discrimination in employment has represented a major problem. But some progress has been made by Equal Employment Opportunities Commission (EEOC).
5. Privacy is to be maintained while genetic information about individuals can be passed to others without their consent.
6. Regarding safety and efficacy of genetic test of consumers, they wish to be reassured that tests are safe and accurate, and also are made available in a timely fashion if the information can be of potential value.

In the United States, some groups have taken interest in this issue, e.g. the American Society of Human Genetics, the National Society of Genetic Counsellors, Federal Agencies (Food and Drug Administration, Health Care Financing Administration) etc. In spite of these facts some questions need proper explanation such as:

1. Is the genetic test mandatory or not?
2. At what circumstances or condition it should be done?
3. Whether the quality or standard of testing maintains the norms properly?
4. Who are authorised to receive or supply or disclose the information.
5. From therapeutic point, which tests are to be considered?
6. If the following agencies or organisations (e.g. court, employer, insurance companies, etc.) ask for test result, whether the information will be communicated or not?
7. How many can afford the expenses for these tests?
8. Is counselling essential before taking decision for testing as genetic information mestly based on probability?

## QUESTION

1. What is genome? What is Human Genome Project?
2. What are the goals or objectives of Human Genome Project? State briefly.
3. What do you mean by mapping of genes? What are different types of maps?
4. What are the techniques for physical map? What are the steps for physically mapping in Human Genome Project?
5. State briefly the concept of STS.
6. State briefly the sequencing of the entire Human Genome.
7. What is the working principle of informatics in the Human Genome Project?
8. Describe briefly the concept of STS (sequence-tagged site).
9. What is microsatellite marker? What are the techniques and sources of information for physical map?
10. Write briefly about genetic map.
11. Write short notes on:
    GRAIL, model organism, genome, genetic map, physical map, RFLP, microsatellite markers, FISH, YAC, STS, Genome Database.
12. What are the model organisms? How does it help for Human Genome Project?
13. What are the procedures for identification of a disease gene?
14. What do you know about Genome Database?
15. What are the future consequences of genome project?
16. Dicuss briefly the ethical, legal and social implications of genome projects.

# 21 INHERITANCE OF BLOOD GROUPS

1. A, B, AB, O, M, N, Inheritance of Rh system, 2. Haemoglobin and Haemoglobinopathies-sickle cell Anemia, Thalassaemia.

Landsteiner discovered in 1900 that, there is existence of blood groups and a vast number of antigens have been detected on human blood cells of which about 10–15% form well-defined system and only 1–2% play a significant role in blood transfusion.

**Human Blood Antigens** (membrane antigens).

Human blood antigens are:

## I. RBC Related

At least 400 antigens have been identified on RBC surface and about 20 well-defined blood group systems have been demonstrated in Europeans.

- **Membrane Antigens**

Antigens are present on the external surface of the red cell membrane. The antigens of different individual differ in their serologic reactivity as a result of minor structural differences between individuals in the carbohydrates, proteins, and lipids of the membrane. Antigens stemming from structural differences controlled by genes at the same locus (alleles) or closely linked loci are classified together as a blood group system. Over 20 different blood group systems containing about 400 antigens have been identified to date.

Before blood transfusion, the antigenic composition of two blood groups of stems the ABO system and the Rh system must be investigated. The red blood cells to be infused (donor cells) should be of the same type in these two systems as the red blood cells of the recipient.

The four phenotypes of the ABO system (A, B, AB, and O) result from small differences in the terminal sugars of carbohydrate chains of trace amounts of fucose-containing glycolipids in the red ceil membrane.

## I. RBC Related

Important blood groups are:

1. ABO
2. Rh blood group
3. MNS's
4. Lutheran

However out of the above *only ABO and Rh are important for blood banking.* Other are *less important because the antigens are weak or because corresponding antibodies* do not actually occur in serum.

## II. WBC Related

1. HLA typing
2. RBC group antigen in small amounts

**III. Platelet Related**

1. HLA typing
2. Platelet-specific matching
3. ABO matching
4. Rh typing

**ABO Blood Groups**

The ABO system consists of 4 main groups:

1. A  3. AB
2. B  4. O

These are determined by the presence or absence of A and B antigens on the RBC.

## Inheritance of the ABO Group

The A and B antigens, like other blood group antigens, are the expression of genes inherited from the previous generation. If an antigen is demonstrated, the gene controlling it must have been inherited from one or both of the parents, and this gene can be passed to the next generation. The genes A, B and O are *alleles,* i.e. any one of the three may occupy the ABO *locus* on each of the pair of *chromosomes* responsible for this system. It is not known which chromosome pair it is. If the chromosome inherited from the father carried the gene A and the chromosome from the mother carried the gene B, the child will be of the *genotype* AB and his or her red cells will have both A and B antigens. Persons who inherit O genes from both parents belong to group O. The O gene is an *amorph,* i.e. it does not produce a detectable antigen. Group O cells are recognised by the absence of A or B antigen when the O gene is inherited with A, only the A gene expresses itself; red cells from an individual of the genotype AO will react in the same way as cells from someone of the genotype AA in tests will anti-A. Similarly, it cannot be differentiated BO from BB in tests with anti-B. The symbols A and B denote *phenotypes,* whereas AA, BO, etc. are *genotypes.*

When the same gene has been inherited from each parent, for example OO, the individual is said to be *homozygous* for that gene. When different genes are inherited, as is evident in the case of AB, the individual is *heterozygous.* The genotype of someone of the phenotype A (or B) may be deduced from a study of his family; the person who is homozygous AA will pass an A gene to each of his children while the heterozygous AO individual may pass either an A or an O gene to his offspring. As for all chromosomes, chance alone determines which of the pair ends up in the *zygote* from which the new individual result.

Since each of the pair of chromosomes of an AB person carries either an A or a B gene, an AB parent cannot normally have a group O child. By the same reasoning, someone who is group O cannot have an AB child because the child must inherit one O gene of the parent, and the gene from the other parent could not transmit both A and B.

**Table 21.1:** Inheritance of classical blood group (A,B,C)

| If person receives gene | Blood group | Gene-type |
|---|---|---|
| 1. A + A or A + O | A | AA or AO |
| 2. B + B or B + O | B | BB or BO |
| 3. A + B | AB | AB |
| 4. O + O | O | OO |

- **Inheritance and medicolegal value**

It must be borne in mind that the child's blood may not be set in its true ABO type until as late as one year after birth. It follows that:

**Table 21.2:** Inheritance and medico-legal test

| If the baby having this group | Parents must have given | If mother was | Father could not have been |
|---|---|---|---|
| 1. O | O + O | Does not matter | AB |
| 2. AB | A + B | " | O |
| 3. A | A + O or A + A | B or O | B or O |
| 4. B | B + O or B + B | A or O | A or O |

The *medicolegal value* of blood grouping tests is greatly made *more valuable if the MN and the various Rhesus factors* are also studied. The M and N factors depend on two minor blood genes. It is very rare indeed for these to elicit an output of agglutinins when injected into man and they may therefore be ignored in carrying out transfusions. They are however antigenic to rabbits and agglutinating sera can be prepared by injecting human M or N cells into rabbits.

Each person carries two of the genes of the M and N group, i.e. M + M (=M), N + N (=N), or M + N (=MN). As a result

**Table 21.3:** Inheritance of MN blood group

| If a baby's supplementary group is | Parents must have given it | If mother's supplementary group was | Father could not have been |
|---|---|---|---|
| 1. M | M + M | Does not matter | N |
| 2. N | N + N | Does not matter | M |
| 3. MN | M + N | N | N |

It is fact that blood grouping tests can never prove that any suspected person is the actual father; they can only show that he could not possibly have been the father or that he like many others might have been.

## Rh Blood Groups

- **Rh Blood Groups**

Rh or Rhesus system was so named because the original antibody was raised by injecting RBC of Rhesus monkey into rabbits and guinea-pig.

- **Antigens**

For most clinical purposes, it is sufficient to determine whether a subject is Rhesus positive or Rhesus negative. This division is made by testing the RBCs with the commonest type of Rhesus antibody, anti-D. But in the transfusion centres, ideally the RBCs of Rhesus negative donors are further tested for presence of C and E antigens.

- **Antibodies**

According to Fisher's classification 7 different antigens are described. *Antibodies against all the Rhesus antigens except 'd'* have been described, e.g. anti-D, anti-C and anti-E, etc. D-negative individuals are relatively easily stimulated to form immune anti-D antibodies. Observation shows if the immunized rabbits serum is tested against *human* red cells, agglutination occurs in 85 per cent of white men; these people are called $Rh^+$(positive) and their serum contain no Rh antibody. No agglutination occurs in 15 per cent; these are called $Rh^-$(negative) and their serum also contains no Rh antibody except in certain cases discussed below:

We have seen above that individuals whose red cells contains no D agglutinogen, anti-D agglutinins are not naturally present in the serum, but the production of anti-D may be achieved by:

1. Transfusion of an $Rh^-$ individual with $Rh^+$ blood.
2. By the presence of $Rh^-$ fetus in an $Rh^-$ mother, in the latter case the titre of anti-D is not likely to be high, unless the woman has undergone one or more pregnancies.

A serum containing anti-D acting *in vitro* may agglutinate $Rh^+$ cells suspended in saline (complete agglutinations). In other cases no such agglutination is demonstrable *in vitro,* although the transfused $Rh^+$ Cells may produce a severe reaction *in vivo.* In such cases, the presence of agglutinin may be made manifest by allowing them to act *in vitro* on $Rh^+$ cells suspended not saline but in ox albumin (20 per cent). Such anti-D is known as an incomplete antibody.

- **Antigens and Antibodies Other Than D**

In addition to the antigen, which distinguishes Rh positive and Rh negative persons, there are other antigens in the Rh system. The major ones are C, c, E and e

Antibodies may be produced to any of these antigens by the persons, whose red cells lack them. As with anti-D, the antibodies to the other Rh antigens may cause transfusion reactions and haemolytic disease of the newborn. Although Rh antibodies other than anti-D are usually a response to red cells, they have been reported occasionally in patients who have had neither transfusions not pregnancies. Anit-C, anti-c, anti-E and anti-e are available as reagents to test red cells for the corresponding antigens.

### • Effects of Rh Antigen-Antibody Interaction *in vivo*

Every tests must be done to assure that a patient whose serum contains antibody is never transfused with red cells which possess the corresponding antigen. Thus, tests for red cell antigens (grouping and typing) and test for antibodies (screening and crossmatching) are vital procedures prior to transfusion. If antigen and antibody come into contact *in vivo* a transfusion reaction may result.

### • Transfusion Reactions

When antibody attaches to antigen on the surface of red cells *in vivo,* the red cells are no longer normal and they are labelled for destruction. Rh antibodies rarely cause lysis of the cells but they attach to the cells and sensitize them. Sensitized cells are recognized by the spleen as damaged cells which must be removed from the circulation. If a large number of cells, such as a unit of Rh positive blood, is accidentally transfused into an Rh negative patient who has anti-D, the spleen may be over-taxed. The toxic products of red cell destruction may also affect the functioning of the liver, kidneys and heart. The symptoms of a transfusion reaction experienced by the patient may be mild or the reaction may be of such severity as to cause death.

Early symptoms are chills, fever, and low back pain. They may be followed by chest pain, fall in blood pressure and shock, thought to be caused by substances such as histamine released during the course of the reaction. The low back pain is indicative of renal damage. This may result from haemoglobin and other substances released when red cells are lysed.

These tragic consequences can usually be prevented by performing antibody-detection tests to reveal the presence of antibody prior to transfusion. When antibody is detected in the serum of a patient, it must be identified. The selected donor is then tested to confirm that the antigen corresponding to the patient's antibody is absent from his cells. The serum of the patient is cross-matched with the cells of the donor to assure that they are *compatible.*

### • Rh Factor and Blood Transfusion

The rules which should be followed during blood transfusion pertaining Rh factor:

1. To avoid sensitization it is essential that Rh as well as ABO testing should be carried out.
2. NO $Rh^-$ female at any age before the menopause should ever be given an $Rh^+$ blood transfusion which should be avoided. If she is $Rh^-$, she becomes sensitized by the injected $Rh^+$ blood and forms anti-D; she is likely to destroy, subsequently, any $Rh^+$ fetus with which she becomes pregnant and transfusion may make her permanently childless.
3. Any $Rh^-$ female who has given birth to a child suffering from haemolytic disease is undoubtedly immunized to D and her serum contains anti-D. Likewise, any $Rh^-$ person who is given $Rh^+$ blood may become sensitized and the serum may contain anti-D. Anyone whose serum contains anti-D will show all the changes attributable to mismatched transfusion if they are transfused with a blood, otherwise compatible, but containing D.

### • Rh Blood Group and Haemolytic Disease of the Newborn (HDN)

The child of an $Rh^-$ mother (genotype dd) and an $Rh^+$ father (genotype DD) must be

$Rh^+$ (Dd) if the $Rh^+$ father is Dd the offspring may be $Rh^+$ (Dd) or $Rh^-$ (dd). If the mother is $Rh^-$ and the fetus is $Rh^+$; serious complications may occur: Cells containing D may pass across the placenta from the fetus to the mother; the latter responding by forming anti-D which returns to the fetus circulation and tends to destroy the fetal red cells. The degree of damage done to the fetus depends on the magnitude of the maternal anti-D response and the ability of maternal Rh agglutinins to cross the placenta (Fig. 21.1)

Generally, no harm is done during the first pregnancy; but serious results may occur in the second or later pregnancies depending on the degree of sensitivity of the mother. If the mother has been immunized previously by an $Rh^+$ transfusion at any time even in childhood, a dangerously high response may occur during a first pregnancy. But it should be remembered that in most cases, agglutinins are not formed and the great majority of matings between an $Rh^-$ mother and an $Rh^+$ father result in normal offspring; presumably fetal red cells do not usually cross the placental barrier and so maternal agglutinins are not evoked.

- **Effects of Anti-D on the Fetus**

The changes in the fetus may be termed haemolytic disease of the newborn (HDN) because they are due to the destruction of the red cells by maternal anti-D. The major clinical syndromes are:

**I. Hydrops fetalis:** The fetus is grossly oedematous; it either dies in the uterus, or if born prematurely or at term, it dies within a few hours.

**II. Haemolytic Disease of the New born (HDN) or Icterus Gravis Neonatarum:**

1. The infant is born at term, it is jaundiced (haemolytic jaundice) or becomes so within 24 hours.

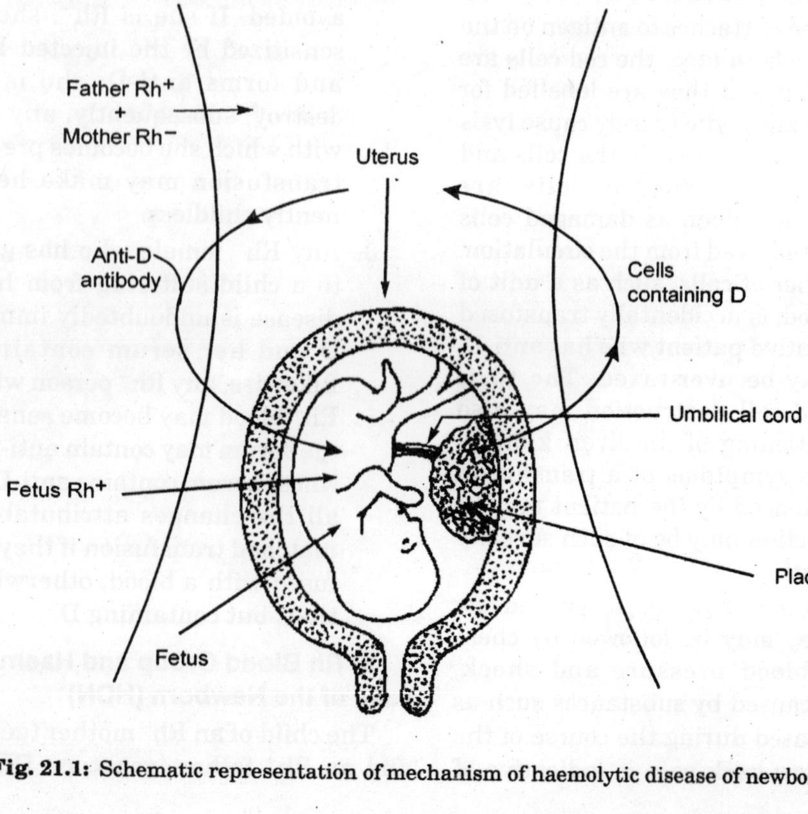


**Fig. 21.1:** Schematic representation of mechanism of haemolytic disease of newborn.

2. There may be no anemia at birth because excessive red cell destruction is more or less compensated for by an intense normoblastic response of the red bone marrow, associated with a high reticulocyte count and the presence at many nucleated red cells in the circulating blood (erythroblastaemia, erythroblastosis fetalis).
3. The rate of red cell destruction by anti-D is maximum at birth and so anemia may develop in the first few days.
4. There may be severe neurological lesions involving the *basal ganglia* especially; they secondarily become stained bright yellow with bile pigments, known as *kernicterus* or *bilirubin encephalopathy.*
5. The liver may also be severely damaged and death may occur from liver failure.
6. Free anti-D (derived from the mother) is present in the infant's blood for at least one week after birth and continues to destroy the infant's red cells, though at a diminishing rate, all this time.
7. Alloimmunization of pregnancy and haemolytic disease of the newborn are not confined to the Rh blood group system. The potential for alloimmunization of a mother can exist whenever any blood group incompatibility exists, and thus antibodies may be produced to almost any of the cell antigens or even the white cell antigens.

## Inheritance of Rh System

There are several varieties of Rh antigen and of Rh antibody; the commonest Rh antigen is called D, and its antibody is called anti-D. Blood group antigens are the result of the action of genes which are present in the chromosomes. The gene corresponding to the antigen D is also called D, when D is absent from the chromosome, its place is occupied by the alternate form (allelomorph) called d. An Rh gene is inherited from both the father and the mother. If gene D is carried by both sperm and ovum the resulting composition (genotype) of the offspring is DD; if the gametes carry D and d respectively the result is Dd; if both gametes carry d, the result is dd. DD (called homozygous) and Dd (called heterozygous) are both $Rh^+$; dd (called homozygous) is $Rh^-$ of the 85% $Rh^+$ English people examined, 35% were DD and 48% were Dd; the remaining 2% of the $Rh^+$ people had some other genotype containing D. In the case of a homozygous father of genotype DD, all the sperms contain D; with a heterozygous father of gene Dd, half the sperms containd and half d. These factors are of importance in relation to the genotype of the child and the likelihood of development of haemolytic disease (Table 4.3).

**Table 21.4:** Inheritance pattern of the Rh-group to the children

| Father | Mother | Children |
|---|---|---|
| $Rh^+ / Rh^+$ | $Rh^+ / Rh^+$ | $Rh^+$ (100%) |
| $Rh^+ / Rh^+$ | $Rh^+ / Rh^-$ | $Rh^+ / Rh^+$ (50%)<br>$Rh^+ / Rh^-$ (50%) |
| $Rh^+ / Rh^-$ | $Rh^+ / Rh^-$ | $Rh^+ / Rh^-$ (50%)<br>$Rh^+ / Rh^+$ (25%)<br>$Rh^- / Rh^-$ (25%) |
| $Rh^+ / Rh^-$ | $Rh^- / Rh^-$ | $Rh^+ / Rh^-$ (50%)<br>(incompatibility of blood might occur)<br>$Rh^- / Rh^-$ (50%) |
| $Rh^+ / Rh^+$ | $Rh^- / Rh^-$ | $Rh^- / Rh^-$ (100%)<br>(incompatibility of blood might occur) |
| $Rh^- / Rh^-$ | $Rh^- / Rh^-$ | $Rh^- / Rh^-$ (100%) |

- **Technique of Rh Grouping**

I. **Slide Test**

(A) 1 drop of 40% cell suspension of unknown cell + 1 drop anti-D sera.

(B) Cell suspension as in (A) + 1 drop of 22% bovine albumin (as control). Tilt back and forth and see for agglutination.

II. **Tube Test**

(A) 1 drop of 5% unknown cell suspension + 1 drop Anti-D sera.

(B) 1 drop of 5% unknown cell suspension + 1 drop of 22% bovine albumin.

Centrifuge at 100 rpm for 1 minute watch for agglutination both macroscopically and microscopically.

- **Observation and Inference**

1. If agglutination in (A) and no agglutination in (B) = $Rh^+$.
2. Agglutination in both (A) and (B) = Repeat again by retesting.
3. If agglutination is negative in both (A) and (B) = $Rh^-$.

- **Confirmation**

It should be confirmed in the following way:

(a) Take 2 tubes
One drop of 5% RBC suspension of unknown cell + 1 drop Anti-D serum.

(b) Same as in A + 1 drop of 22% bovine albumin.

(c) Centrifuge at 1000 rpm for 1 minute.

(d) Incubate at 37ºC for 15 minutes.

(e) Fill with saline, centrifuge at 1500–2000 rpm for 1/2 minute. Decant and repeat 3–4 times.

(f) Add 1 drop AHG to both (A) + (B).

(g) Centrifuge at 1000 rpm for 1 minute.

(h) See for agglutination.

**Comment:** Ideally any blood negative for D antigen must also be tested for the presence of C and E antigens.

## QUESTION

1. What do you mean by blood group? What are different types? What is the mechanism of blood group reaction?
2. State briefly the inheritance of ABO blood group. What is the importance of blood group?
3. What do you know about Rh blood group?
4. State briefly the effects of antigen-antibody reaction in blood group Rh
5. What do you know about haemolytic disease of New born pertaining to Rh blood group.
6. State briefly the inheritance of Rh blood group
7. Describe briefly the technique of Rh grouping

# 22

# GENETICS IN FORENSIC MEDICINE

1. Blood-typing in legal medicine, 2. DNA fingerprint technique, 3. RNA fingerprinting.

## Blood-typing in Legal Medicine

Determination of blood grouping and their inheritance according to Mendelian fashion has become an important legal procedure in case of disputed paternity. Including all the blood types, as many as 50 common blood group genes can be determined for each person by blood grouping procedure. After determining the mother's gene and child's gene, many of the father's gene and his corresponding blood factors are then known immediately. Because any gene present in a child but not present in the mother must be present in the father. If a suspected man misses any one of the blood factors he is not the accused person. As for example, a female gives birth to a child and accuses one well-known prestigious man being the father. The mother belongs to group A and the baby to group O. If the accused person belongs group AB, he cannot be a father. The culprit must be A, B, O.

If the accused belongs to A, B, O, this alone would not be sufficient to accuse him because many persons belong to these groups. Hence other evidences would have to be obtained, e.g. other blood groups such as the MN system to corroborate that supplied by the ABO system. But it can be very useful in positively eliminating a suspect.

The M, N or MN blood group system is similar to the ABO system except that there are only two alleles, M and N, instead of three. A person belonging to group M has the genotype MM, a person belonging to group N is genotypically NN, and a person belonging to group MN is genotypically MN.

- **Use of MN Groups**

It does not have importance in blood transfusion but have got medicolegal importance for disputed paternity.

- **Limitation**

There are exceptions due to:

1. Mistakes in classifying the blood type of the persons concerned.
2. Illegitimacy.
3. Mutation.

- **Importance of Blood Group Determination**

1. Blood transfusion
2. For study of certain blood diseases
3. Paternity test

4. In forensic medicine
5. To study the racial science (ethnological studies)
6. Anthropological studies

## DNA Fingerprint Technique

- **Introduction**

Beside the traditional method of identifying the fingerprint at the site of crime, a new method has been developed and is increasingly in use which helps to identify the source of biological material with almost cent per cent precision. The identification is done be *genetic fingerprint* or *DNA fingerprint technique* which is unique to each and every individual *except in the case of identical twins.* This technique was discovered by Alec Jeffrey of Leicester University.

- **Basis**

The 46 chromosomes (23 pairs) of a cell contain a long chain of DNA (about 3 feet long) which are the blueprint of an individual. The contents of DNA determine the characteristics of an individual, e.g. colour of the skin, type of hair, nails and also some of the hereditary diseases. The two strands of DNA are joined by purine and pyrimidine bases like adenine, guanine, cytosine and thymine. These four bases are chained in a wide variety of combination to form the DNA chain. These bases carry the hereditary characters. As DNA chains of no two persons can be identical, this explains why no two persons look exactly similar in the world except identical twins. The basic features like nose, eyes, hands, feet are common to all humans but *DNA is not identical in all humans* and the chance of occurrence of that particular pattern of DNA is one in 10 billion.

The use of hypervariable regions (HVRs) of human DNA consisting of some tandem repeats of short nucleotide sequences of 15–30 bp in length has opened a new chapter in forensic science for the identification of individuals. The length of DNA is variable in this region in the population. DNA probes are made from this region and are hybridised with the southern blots of human DNA. The pattern obtained from this hybridisation is very individual specific which is known as *DNA Fingerprint.*

- **Biological Samples**

An important feature of this technique is that no matter what biological sample is obtained—hair, bit of skin, blood, sperm—the results are not going to be different because *all the cells in the body contain the same set of DNA.*

- **Procedure**

1. **Isolation of DNA:** The cells are isolated and burst open to release DNA chains.
2. **Cleavage:** The DNA chains of different individuals have sections that are identical as well as dissimilar. Using restriction enzymes, the DNA chain cleaved at a specific site. This results in DNA fragment called restriction fragment length polymorphism (RFLP) of different lengths which contain both the highly personalised segments and also a lot of repetitive stretches the function of which is not exactly understood.

**Restriction Fragment Length Polymorphism (RFLP):** With the help of restriction enzymes, a map can be established in DNA, nearly similar to genetic map is called *restriction map.* Restriction map differs from the genetic map in the following respects:

A. Any sequence of DNA can be mapped whether any mutation occurs or not; but the genetic map locates the sites of mutations.

B. The restriction map or sites can identify if there is any base alterations or mutations.

The difference in the restriction map between the two species is called *Restriction Fragment Length Polymorphism* (RFLP).

RFLP can be used:

(a) As a specific marker in the identification between the species.
(b) To draw the phytogenetic relationship between the species.
(c) To show Mendelian segregation.
(d) For identification of many genetic diseases:
   i. Identification of the altered gene in many genetic diseases and advancement of technology of synthesizing and cloning genes causes correction of the genetic defect by introducing the normal gene and this is known as *gene therapy*.
   ii. Introduction of normal β-globin gene to the thalassemia patient for correction of genetic defect.
(e) Due to advancement of RFLP, it is being used to identify individuals which is known as *DNA Finger-printing*.

   Using restriction enzymes, the DNA chain, cleaved at a specific site. This results in DNA fragments called restriction fragments length polymorphism (RFLP) of different lengths which contain both the highly personalised segments and also lot of repetitive stretches the function of which is not exactly understood.

3. **Separation and Arrangement of DNA:** Using gel electrophoresis technique, the fragments of DNA of different sizes are separated and arranged according to size.
4. **Identification of Type of Segment:** The DNA samples now contain highly individualised fingerprints, but the types of segment have to be identified. The double strands are separated into halves by using chemicals, the other half strands (created in the lab) are used to identify the kind of half chain that already exists in the sample. The lab-made DNA bits are tagged with radioisotope so that they can be detected by an X-ray film. After several days of exposure, the film is developed and shows black bands where the lab-made DNA bits have joined to the original sample. As we saw earlier that the bits in the original are almost exclusive to an individual.

   The accuracy with which it can predict the source of biological material is almost cent per cent. The advantage is that just any biological sample.

- **Application**

This method is very much useful for forensic scientists to identify the criminals, murderers, disputed paternity, immigration cases, identification of individuals.

- **RNA Finger Printing**

RNA finger prints are based on the analysis of two-dimensional gels after partial digestion with ribonucleases. There are different RNA bases which have certain specificity in digesting the RNA. Two dimensional electrophoresis is a special technique to separate complex nucleic acid mixtures which cannot be clearly shown in one-dimensional gel. Polyacrylamide gels are used and the shift in the electrophoretic conditions in the two-dimensional gels is made by changing the acrylamide concentration, change from non-denaturing to denaturing or by changing the pH.

- **Application**

1. To note the sequence similarities between different species of RNA.
2. For the analysis of tRNA genes and their transcripts.
3. To study the viral genomes.

## QUESTION

1. **What are the genetical methods used in forensic medicine? Describe the blood-typing method and its authenticity as a test.**
2. **What is DNA fingerprint technology? What materials can be used for this method. Describe in brief the DNA fingerprint technology. Is it an authentic method?**

**PART-IV**

# BIOCHEMICAL GENETICS— INBORN ERROR IN METABOLISM

# 23

# BIOCHEMICAL GENETICS—INBORN ERROR IN METABOLISM

1. Introduction, 2. Method of study, 3. Hereditary anemias, 4. Carbohydrate metabolism, 5. Lipid metabolism, 6. Protein metabolism, 7. Nucleic acid metabolism.

## Introduction

Biochemical reactions are catalyzed by the enzymes. Enzymes proteins are synthe-sized in the cell according to genetic code which regulates the type of enzyme and their sequence of amino acids in the molecule. Biochemical reactions in the metabolic pathway is altered either due to absence of enzymes or defective enzyme formation due to genetic error. According to *"one gene one polypeptide"* or *"one gene one enzyme"* or more accurately *"one cistron one polypeptide" hypothesis* every gene is responsible for synthesis of an enzyme, so if there is defect in one gene, the enzyme concerned would be absent or defective. These are manifested by metabolic diseases and are hereditary and tend to occur only in closely-related family groups in man. The field of biochemical genetics was initiated by the physician A.E. Garrod in 1902. He identified only a small number of such metabolic diseases, one of which is alkaptonuria, the "black urine" anomaly due to a rare recessive gene.

- **Definition**

Error in metabolism for absence of enzyme or defective enzyme due to genetic defect is known as *inborn error in metabolism.*

Such genetic defects, if they are not lethal often result in the accumulation and excretion of the substrate of the defective enzyme.

## Method of Study

1. The abnormal excretion of *homogentisic acid* in the urine is due to one such genetic defect in human patients called *alkaptonuria.* Since excretion of this acid is increased by feeding phenylalanine or tyrosine but not by other amino acids, it was concluded that homogentisic acid is formed as an intermediate during the catabolism of phenylalanine or tyrosine. It was concluded that homogentisic acid is the intermediary product of phenylalanine and tyrosine metabolism, and it is not further catabolized like normal man due to defective enzyme *"homogentisic acid 1, 2 dioxygenase"*.
2. **Beadle and Tatum experiment** (see protein synthesis): Far more useful for

study of metabolism are specific genetic defects induced in micro organisms by various mutagenic agents, such as X-Ray radiation. Mutant microorganisms deficient in a specific pathway or enzyme can literally be produced at will and have become powerful tools for studying intermediary metabolism and by this method "one gene one enzyme hypothesis" was postulated by Beadle and Tatum using "*Neurospora crassa*".

3. **Cross-feeding experiment:** Such sets of mutants can be used in another way to identify intermediates in a biochemical sequence. For example,

Wild type A $\xrightarrow{E_1}$ B $\xrightarrow{E_2}$ C $\xrightarrow{E_3}$ D $\xrightarrow{E_4}$ Arg

Mutant IA $\xrightarrow{E_1}$ B $\xrightarrow{E_2}$ C $\xrightarrow{E_3}$ D $\xrightarrow{E_4}$ Arg

Mutant IIA $\xrightarrow{E_1}$ B $\xrightarrow{E_2}$ C $\xrightarrow{E_3}$ D· $\xrightarrow{E_4}$ Arg

Mutant IIIA $\xrightarrow{E_1}$ B $\xrightarrow{E_2}$ C $\xrightarrow{E_3}$ D $\xrightarrow{E_4}$ Arg

When mutant I is grown in the presence of small, limiting amounts of arginine, the blocked precursor D will accumulate in the growth medium since it cannot be converted into arginine. After removal of mutant I cells by filtration, this medium will support the growth of mutant II in the absence of arginine since it supplies the precursor D. The filtrate of mutant II, however will not support the growth of mutant I. This type of experiment is known as *"cross-feeding"* experiment. Mutants which are defective in a biosynthetic pathway but which revert to normal growth when they have been provided with the normal product of the pathway are called *"auxotrophic mutants"*.

4. **Inducible enzyme (β-galactosidase) experiment (Monod and Jacob):** Mutants can also be used to analyze catabolic pathways. For example, wild type *E.coli* cells are able to grow on lactose, glucose or galactose as carbon source. In some *E.coli* mutants the capacity to grow on lactose is lost, but the cells are still able to grow on the other sugars. The genetic defect in this case involves the enzyme β-galactosidase which can hydrolyze lactose to its components glucose and galactose.

**Table 23.1:** Example of inborn errors in metabolism

| Name | Defective enzyme or process |
|---|---|
| **Some genetic defects in man affecting amino acid metabolism** | |
| 1. Albinism | Tyrosine 3-monooxygenase |
| 2. Alcaptonuria | Homogentisic acid 1, 2-dioxygenase |
| 3. Phenylketonuria | Phenylalanine 4-aminooxygenase |
| 4. Cystinosis | Storage and/or release of cystine from lysosomes |
| 5. Cystinuria | Renal and intestinal excretion of cystine |
| 6. Homocystinuria | Cystathionine β-synthase |
| 7. Histidinemia | Histidine ammonia-lyase |
| **Affecting carbohydrate metabolism** | |
| 8. Galactosemia | Galactokinase or Galactose-I-phosphate uridyl transferase or uridine diphosphogalactose-4-epimerase |

*contd.*

| | |
|---|---|
| **Affecting lipid metabolism** | |
| 9. Gaucher's disease | Glucocerebrosidase |
| 10. Krabbe's disease | Galactocerebrosidase |
| 11. Niemann-Pick disease | Sphingomyelinase |
| 12. Farber's disease | Ceramidase |
| **Affecting lipoprotein metabolism** | |
| 13. Abetaliporoteinemia | Apoprotein B synthesis |
| 14. Wolman's disease (cholesteryl ester storage disease) | Cholesteryl ester hydrolase |
| **Affecting porphyrin metabolism** | |
| 15. Hereditary coproporphyria | Coproporphyrinogen oxidase |

## I. Hereditary Anemias

1. **Sickle-cell anemia**
   (See Lethal genes.)
2. **Hereditary spherocytosis:** The red cells are very small in size, and they are *spherical* in shape rather than being biconcave discs. These cells cannot be compressed because they do not have the normal loose bag-like cell membrane structure of the biconcave discs. Therefore, on passing through the splenic pulp they are easily ruptured by even slight compression.
3. **Thalassemia (Cooley's anemia or Mediterranean anemia):** Thalassemia or Cooley's anemia or Mediterranean anemia, is another hereditary type of anemia in which *the cells are unable to synthesize adequate amounts of the polypeptide chains required to form the globin part of haemoglobin.* Therefore, haemoglobin synthesis is greatly depressed. The hemopoietic system responds by producing tremendous numbers of new RBC but with little haemoglobin and also cells that are small and have fragile membranes. Here again, the cells are easily ruptured upon passing through the tissues. In thalassemia, mutation of regulator genes repress the synthesis of a particular type of peptide chain of globin, with a compensatory rise in the synthesis of its other peptide chains. The repressed peptide chains are replaced in globin molecules by the other peptide chains producing abnormal haemoglobins. Thalassemia are of seven types:

- **Types of Thalassemia**

1. α-thalassemia
2. β-thalassemia (with haemoglobin A production).
3. β-thalassemia (with no haemoglobin A production)
4. β-δ-thalassemia
5. δ-thalassemia
6. Haemoglobin-Lepore thalassemia
7. Thalassemia-like state.

1. **α-Thalassemia:** In this case, repression of α-chain produces abnormal haemoglobin deficient in α-chain; e.g. *Haemoglobin H* appears in RBC at the adult stage, and Haemoglobin Barts occurs in infancy stage. Absence of α-chain in Hb (having β, γ or δ) alters $O_2$ affinity but $PCO_2$ is not affected. The amount of $HbA_2$ is subnormal.
2. **β-Thalassemia (with haemoglobin A production):** The rate of β chain

translation is reduced and the amount of $HbA_2$ and HbF rise in the RBC. HbS and HbC are also present in high amount simultaneously, carry abnormal β-chain, show the characteristic of HbS-β-Thalassemia and HbC-β-Thalassemia.

3. **β-Thalassemia** (with no haemoglobin production): Severe anemia, haemoglobin consists of F and $A_2$ only.
4. **β-δ-Thalassemia:** Severe anemia with haemoglobin F only.
5. **δ-Thalassemia:** Reduced haemoglobin $A_2$ level.
6. **Haemoglobin-Lepore Thalassemia:** Severe anemia with haemoglobin Lepore and F in homozygous state and haemoglobin Lepore present and normal level of $A_2$ haemoglobin in heterozygous state.
7. **Thalassemia-like state:** Clinical picture of thalassemia but normal levels of haemoglobin $A_2$ and F in homozygous state.

- **Genetic**

1. The disease of thalassemia is controlled by an allele—C or β thal or α thal which in *homozygous condition* produces the severe thalassemia but in *heterozygous condition* produces mild form of the disease.
2. This allele is most prevalent in equatorial region.
3. But the effects of both types of thalassemia gene, i.e. β thal and α thal are due to the combined effects of defective globin production and discrepancy in globin-chain with *inclusion body* formation, membrane damage and premature red-cell destruction. (Inclusion body—Absence of sufficient γ or β chains, the excess α-chains precipitate forming inclusion body.)

- **Defect in Molecular Level**

1. Partial or total deficit of globin synthesis could arise from either:
   (a) A reduced amount of mRNA for a particular chain, or
   (b) From the production of an abnormal mRNA, such as the mechanism of chain assembly, i.e. initiation, translation and termination is retarded.
2. It might follow an alteration in a codon such that chain initiation is not effective.
3. It might be shortage of tRNA which is required to insert given amino acids due to alter codon anywhere along the mRNA stand.
4. It might be that the chain releasing mechanism is abnormal.

## II. Carbohydrate Metabolism

1. **Essential Pentosuria:** It is rare hereditary disease where large amount *L-Xylulose* appear in the urine. It occurs in males of Jewish persons. It is caused due to inherited *deficiency* of the NADPH-dependent *L-Xylulose dehydrogenase* (Fig. 23.1) responsible for reducing L-Xylulose to xylitol in the liver during metabolism through pentose phosphate pathway.
2. **Fructosuria:** Essential fructosuria is a rare congenital disorder in which there is the deficiency of fructokinase and characterized by inability to utilize fructose completely. Insulin does not act in this case. Fructosuria are of two types:
   (a) The metabolism of other carbohydrates is undisturbed and there are no clinical symptoms (Fig. 23.2).
   (b) The rise in blood fructose is accompanied by a sharp drop in the blood glucose concentration with severe symptoms of hypoglycemia. This is characterized by a rise in Fructose-1-phosphate level in tissues and urinary elimination of fructose. Fructose-1-phosphate

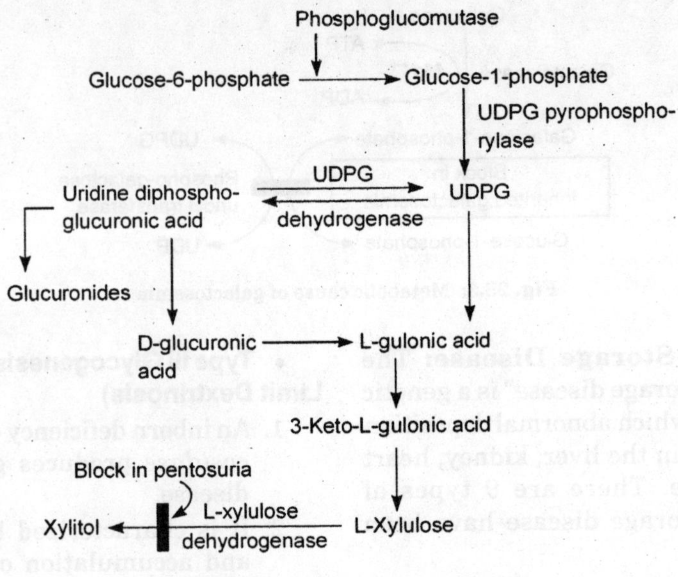


**Fig. 23.1:** Metabolic cause of pentosuria.

cannot be converted to Fructose-1-6-diphosphate due to an inborn failure to synthesize *aldolase B* in the adult liver and the condition is known as *hereditary fructose intolerance*.

3. **Galactosemia:** Galactosemia, an inherited metabolic disorder due to inborn deficiency of *phosphogalactose uridyl transferase*. The inability of the infant to metabolize galactose raises the plasma galactose level and galactose-1-phosphate in RBC, brain and liver, resulting in growth failure, damages of liver and central nervous system, cataract of lens and urinary excretion of galactose (galactosuria) (Fig. 23.3). The symptoms are controlled by eliminating lactose and galactose from the diet. Symptoms may also subside with age because of gradual rise in hepatic UDP-galactose pyrophosphorylase activity initiates galactose metabolism.

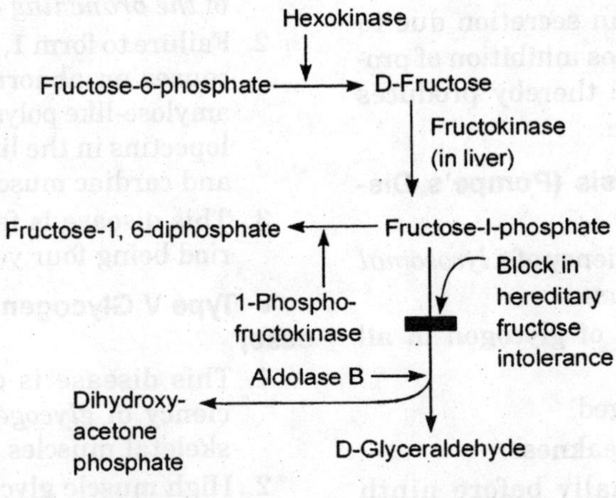


**Fig. 23.2:** Metabolic cause of fructosuria.

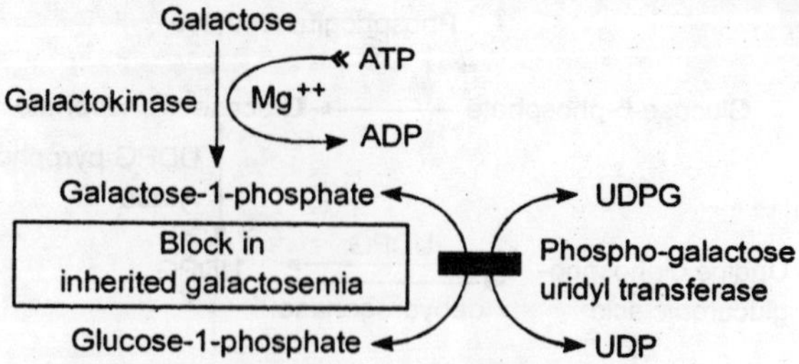


**Fig. 23.3:** Metabolic cause of galactosemia.

4. **Glycogen Storage Disease:** The "glycogen storage disease" is a genetic disorder in which abnormal deposition of glycogen in the liver, kidney, heart and muscle. There are 9 types of glycogen storage disease have been identified.

- **Type I Glycogenesis (von Gierke's Disease)**

1. This type of Glycogenesis is due to genetic mutations causing *glucose-6-phosphatase deficiency* for which glycogen cannot be broken down to liberate glucose and glucose-6-phosphate and promotes glycogen synthesis.
2. Children develops hypoglycemia.
3. Failure of glycogenolysis and hyperglycemic effect of epinephrine.
4. Utilization of fat predominantly as a source of energy leading to lipemia, acidemia and ketosis.
5. Fatty infiltration of the liver.
6. Inhibition of insulin secretion due to hypoglycemia causes inhibition of protein synthesis and thereby produces cessation of growth.

- **Type II Glycogenesis (Pompe's Disease)**

1. It is due to the deficiency of a *lysosomal enzyme, acid maltase.*
2. Excessive amount of glycogen in all tissues.
3. The heart is enlarged.
4. Extreme muscle weakness.
5. Death occurs usually before ninth month of life.

- **Type III Glycogenesis (Cori's Disease, Limit Dextrinosis)**

1. An inborn deficiency of *amylo-1, 6-glucosidase* produces glycogen-storage disease.
2. It is characterized by hypoglycemia and accumulation of "limit dextrin" type polysaccharides (limit dextrinosis) due to failure of hydrolysis of 1, 6-glycosidic linkages during glycogenolysis.
3. Accumulation of limit dextrin in liver and muscle. This substance produces by the action of phosphorylase on glycogen.
4. Debrancher enzyme is deficient in this disease.
5. Patients survive well into adult life.

- **Type IV Glycogenesis (Andersen's Disease, Amylotinosis)**

1. This disease is due to genetic mutations producing the inborn deficiency of *the branching enzyme in the liver.*
2. Failure to form 1, 6-glycosidic linkages causes an abnormal accumulation of amylose-like polysaccharides, i.e. amylopectins in the liver, skeletal muscles and cardiac muscle.
3. This disease is fatal and survival period being four years.

- **Type V Glycogenesis (McAldle's Disease)**

1. This disease is due to genetic deficiency of *glycogen phosphorylase* in skeletal muscles.
2. High muscle glycogen, reduced glycolysis, low blood lactate, failure of

epinephrine ore, severe exercise in increasing glycolysis and blood lactate, and reduced capacity for muscular work.

3. Cramps occurs in muscles after moderate exercise.
4. The exercised muscles can metabolize fructose.

- **Type VI Glycogenesis (Her's Disease)**

1. This type is due to inborn deficiency of *glycogen phosphorylase*.
2. This causes failure of glycogenolysis, higher liver glycogen, hypoglycemia, ketosis and a failure of the hyperglycemic response to epinephrine.

- **Type VII Glycogenesis**

1. This is due to genetic deficiency of **phosphofructokinase** isozyme in muscles.
2. Characterized by reduced muscle glycolysis and high muscle glycogen.
3. But glycolysis in RBC and hepatic phosphofructokinase are unaffected as these tissues possess a different phosphofructokinase isozyme.
4. There is marked accumulation of glucose-6-phosphate and fructose-6-phosphate and a marked diminution of fructose-1, 6-diphosphate.

- **Type VIII Glycogenesis**

1. This is caused by a genetic deficiency of *dephosphorylase kinase* in the liver.
2. Failure of hepatic glycogen phosphorylase activity, high liver glycogen, and hypoglycemia may result.

- **Type IX Glycogenesis**

1. This type of disorder is due to genetic mutations producing deficiency of *glycogen synthetase* in the liver.
2. Characterized by low hepatic glycogen.

## Hurler's Syndrome (Mucopolysaccharidosis I or Gargoylism)

1. This is due to genetic deficiency. *α-L-iduronidase* produces excessive accumulation of acid mucopolysaccharides, e.g. dermatan and keratan sulphate in lysosomal and Golgi vesicles of hepatic and connective tissue cells.
2. There is mental retardation, skeletal abnormalities, high urinary elimination of low molecular weight mucopolysaccharides, grotesque facial features, corneal opacity and very short lifespan.
3. This disorder is due to defect in catabolism rather than to excessive synthesis.
4. Deafness of a conductive type.
5. The voice is harsh and behaviour is apelike.
6. Dermatan sulfate and heparan sulfate are present in urine and tissues.
7. Dilatation of the ventricles.
8. Cardiovascular findings are prominents.

## Hunter's Disease or Syndrome (Mucopolysaccharidosis II)

1. A genetic deficiency of iduronate sulfatase produces excessive accumulation of keratan and dermatan sulfate in the liver and connective tissues.
2. Characterised by mental retardation and bone deformities; but no corneal opacity.
3. This disorder is also due to defect in catabolism rather than to excessive synthesis.
4. Presence of dermatan sulfate and heparin sulfate in urine and tissues.
5. Dwarfing and abnormal facies are present.
6. Cardiovascular changes, obstruction and restriction of chest movement.
7. Death frequently occurs from cardiac troubles at an early age.

## III. Lipid Metabolism

### Sphingolipidosis

These are inborn error of sphingolipid metabolism due to genetic deficiency of degradative enzymes and characterized by the accumulation of intermediary products whose further catabolism depends on the blocked enzyme. Most of them are inherited as *autosomal recessive traits,* but *Fabry's disease* is an *X-linked disorder,* maternally transmitted.

## Gaucher's Disease

In this disorder, the glucocerebroside content of the reticuloendothelial cells of liver, bone marrow, spleen, brain is very high. In the glucocerebroside molecule, the Kerasin is characterized by glucose replacing galactose. The concentration of cerebrosides are much higher in medullated than in nonmedullated nerve fibres. The disease is caused by the deficiency of *glucocerebrosidase* (*glucosyl seremide hydrolase*).

The symptoms are:

1. The size of the spleen is increased.
2. There are signs of leukopenia and thrombocytopenia.
3. The liver is enlarged and the marrow cavity is widened.
4. Eyes show a yellow-brown wedge-shaped elevation.
5. In early infantile form, the disease starts early and death occurs before attaining the age of 3 years.
6. Neurological symptoms, mental retardation are usually present.
7. There is loss of bone minerals.

## Neimann-Pick Disease

In this disorder, excessive accumulation of sphingomyelin in spleen, brain and liver. This is due to the deficiency of *sphingomyelinase.*

1. The hereditary disease is found in infancy and death occurs within the first two years of life.
2. The clinical findings are:
   (a) Enlarged liver and spleen (splenomegaly).
   (b) Mental retardation.
   (c) Anemia and leukocytosis.
   (d) Nervous system is affected.
   (e) Cherry-red spot may be seen over the retina.
   (f) Cholesterol deposits in the tissues are increased.

## Tay-Sachs Disease

This disorder is characterized by the increased accumulation of gangliosides ($GM_2$) in brain and spleen. The defect is due to the deficiency of the enzyme *Hexosaminidase*. *A* in tissues. The disease is characterized by:

1. Demyelination, cerebral degeneration, mental retardation, severe motor deterioration, blindness.
2. A cherry red spot appears in the muscular region of the eye within the first year of life.
3. At about two years of age, the circumference of the head becomes 50 per cent greater than normal.
4. There is repeated respiratory-tract infection of the person and the patient expires at the third and fourth years.

## Fabry's Disease

In this disease, accumulation of large amount of ceramide trihexoside take place in the kidney, intestine and lymph glands. This is due to the deficiency of the enzyme *ceramide trihexosidase.*

The clinical findings are:

1. Kidney failure and male patients generally die due to progressive renal failure in the fourth or fifth decade of life.
2. Cardiac enlargement and oedema of the extremities.
3. Skin rashes.
4. Corneal opacities and vascular dilatation.
5. Some patients suffer from excessive pain in the joints.
6. No mental retardation.

### Refsum's Disease

This disease causes accumulation of large amounts of phytanic acid (3, 7, 1, 15-tetramyl hexadecanoic acid). This is due to the deficiency of the enzyme *phytanic acid oxidase.*

The clinical findings are:

1. The early symptoms are signs of chronic polyneuropathy with distal muscular atrophy.
2. Deep tendon reflexes are weak or absent.
3. Severe pain in the knees.
4. Night blindness and narrowing of the visual fields.
5. Deafness and anosmia.
6. Cardiac involvement may lead to tachycardia.
7. The cerebrospinal fluid protein is always increased while the cell count is normal.

### Krabbe's Disease

This disorder causes accumulation of galactocerebroside in the liver and white matter. This is due to the deficiency of *galactocerebrosidase* which catalyzes the hydrolysis of galactocerebroside to form ceramide and galactose and the compound is the important component of myelin.

The clinical manifestions are:

1. Severe mental retardation in infants.
2. Total absence of myelin in the central nervous system which is replaced by gliosis and "globoid bodies" appear in the white matter.
3. Diagnosis by determination of the galactocerebrosidase in the leukocytes.

### Farber's Disease

This disease shows accumulation of ceramides in brain, mental retardation, skin lesions, bone malformation. This is due to the deficiency of the enzyme *ceramidase*.

### $GM_1$ Gangliosidosis

This disease is due to the deficiency of the enzyme $GM_1$ β-galactosidase. This is manifested by accumulation of $GM_1$ gangliosides in viscera and brain, bone malformation, liver enlargement.

### Lactosyl ceramidosis

This disease is due to the deficiency of the enzyme *ceramide lactosidase.* This is manifested by the accumulation of lactosyl ceramides in liver, bone marrow and brain, cerebral degeneration, liver enlargement.

### Metachromatic Leucodystrophy

This disease is due to the deficiency of the enzyme *sulfatidase.* This is manifested by accumulation of sulfatides in white matter, demyelination, failure of mental growth, psychopathic disorders.

## Protein Metabolism

### Albinism

- **Definition**

Lack of pigment melanin in the body due to defective tyrosine metabolism for genetic defect. Hence it is an inborn error of metabolism (see Fig. 23.4).

- **Characteristics**

Occasionally among human beings and also in other animals, individuals are found without black pigment melanin in the skin, iris, retina, choroid and hairs. Such individuals are known as *"albinos"* and the phenomenon is known as *"albinism"*.

- **Causes of Albinism**

It has been proved that albinism results from a failure on the part of the amino acid tyrosine to become metabolized to melanin—a black pigment (Fig. 21.4). Absence of melanin results the condition known as albinism. Melanin-protein complexes remain stored as pigments or black granules in the *"melanocyte"* cell of skin, iris, hairs, choroid and substantia nigra, etc. Synthesis of melanin occurs in the melanocytes by a series of enzymatic actions (Fig. 23.4).

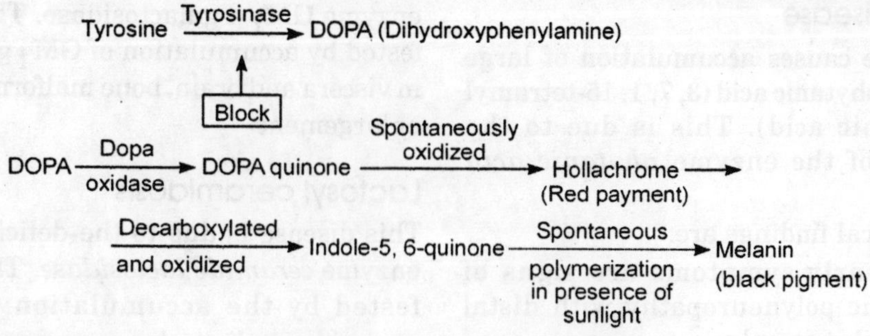


**Fig. 23.4:** Biosynthesis of melanin.

Albinism is recognized as hereditary and is under the control of an autosomal recessive gene "a". The trait appears in genotypes bearing "aa" or in other words it appears in a person who has homozygous recessive genotype.

### Albinism as a model of inborn errors of metabolism

Albinism together with Alkaptonuria (a condition in which the urine of the patient blackens upon exposure to air and cartilage becomes black and hardened. It has been established that a biosynthetic product homogentisic acid fails to become metabolized further for want of an enzyme "Homogentisic acid 1,2 dioxygenase") are cases first recorded as inborn error in metabolism due to genetic defect by Garridm 1909. He proved these traits behave genetically recessive alleles and become expressive in persons with homozygous recessive genotype. From these studies, Garrod postulated that genes regulate and control the presence and working of enzymes. Thus albinism is a useful model for study of inborn error in metabolism. Garrod's views were not accepted in his time but present-day knowledge has supported the view that genes regulate the synthesis of enzymes and thereby control metabolic reactions. Garrod's study has provided an answer to the question—how the genes function to elaborate a phenotype and helped to establish *"one gene one enzyme hypothesis"*.

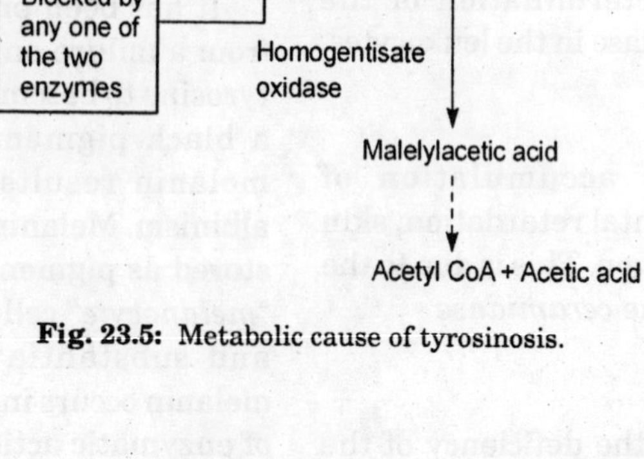


**Fig. 23.5:** Metabolic cause of tyrosinosis.

## Tyrosinosis

1. This is due to the deficiency of the enzyme either of hepatic *p-hydroxyphenylpyruvate hydroxylase* or of *tyrosine transaminase* (Fig. 23.5).
2. The patient excretes large quantities (1.5-3 g/d) of tyrosine in the urine.
3. Diet rich in tyrosine causes excretion of other p-hydroxyphenyl acids, including 3, 4-dihydroxyphenylalanine (dopa) and p-hydroxyphenyl lactic acid.

## Tyrosinemia

1. Plasma tyrosine levels are elevated above normal.
2. The common clinical findings are: hepatosplenomegaly, nodular cirrhosis of the liver, abnormalities of tyrosine and methionine metabolism, p-hydroxyphenyllactic aciduria, multiple defects in renal tubular reabsorption, rickets, hyperphosphaturia and proteinuria, and amino aciduria. The enzyme defect appears to be hereditary, its exact nature is unknown. Treatment with a diet, low in tyrosine and phenylalanine improves renal function and possibly also retards degenerative liver changes.

## Phenylketonuria

1. Phenylketonuria is an inherited disorder of phenylalanine metabolism. The frequence of the disease 1 : 10,000. This is due to the deficiency of the enzyme phenylalanine hydroxylase, component I (Fig. 23.6). The patient is therefore unable to convert phenylalanine to tyrosine and as a result, an alternative catabolites of phenylalanine which includes:
   (a) Phenylpyruvic acid (product of deamination of phenylalanine),
   (b) Phenyllactic acid (reduction product of phenylpyruvic acid),
   (c) Phenylacetic acid (decarboxylation and oxidation product of phenylpyruvic acid)

   Phenylacetic acid is conjugated in the liver with glutamine and excreted as conjugate phenylacetylglutamine.
2. In infants and children, there is mental retardation. Further deterioration of mental performance can be pre-

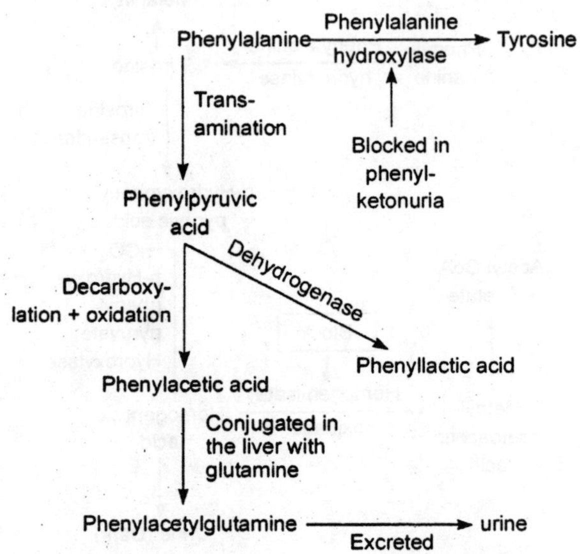


**Fig. 23.6:** Metabolic cause of phenylketonuria.

vented if they are maintained on a diet of low levels of phenylalanine and the diet can be terminated at 6 years of age, when high concentrations of phenylalanine no longer are injurious to the brain. The presence of the Ketoacid phenylpyruvic acid in the urine gives the disease its name—phenylketonuria.

3. Patients tend to have a deficiency of serotonin.
4. Impairement of melanin synthesis due to accumulation of phenylalanine.
5. Phenylpyruvate in the urine can be detected by a simple biochemical spot test, definite diagnosis requires determination of elevated plasma phenylalanine levels.

## Alkaptonuria

1. The disease alkaptonuria has historical interest because it was first observed by Garrod and it formed the basis of Garrod's ideas concerning inherited metabolic disorders.
2. This is characterized by the excretion of homogentisic acid resulting dark urine on standing in air (Fig. 23.7).
3. This is due to the deficiency of the enzyme *homogentisic acid oxidase.*
4. This condition is often found in infancy and 600 cases have been reported.
5. The incidence of alkaptonuria is 2-5 per million live birth.
6. This condition is present at birth and persists throughout life.
7. Late in the disease, there is generalized pigmentation of connective tissues gives rise to a condition known as *ochronosis* and a form of arthritis.

## Inborn Error of Branched Amino Acids Metabolism (Valine, Leucine, Isoleucine)

### (A). Hypervalinemia

1. This disease is characterized by high plasma levels of valine.
2. The patient is unable to transaminate valine to α-keto-isovalerate.
3. Transamination of leucine and isoleucine is unimpaired.
4. In hypervalinemia, feeding a diet low

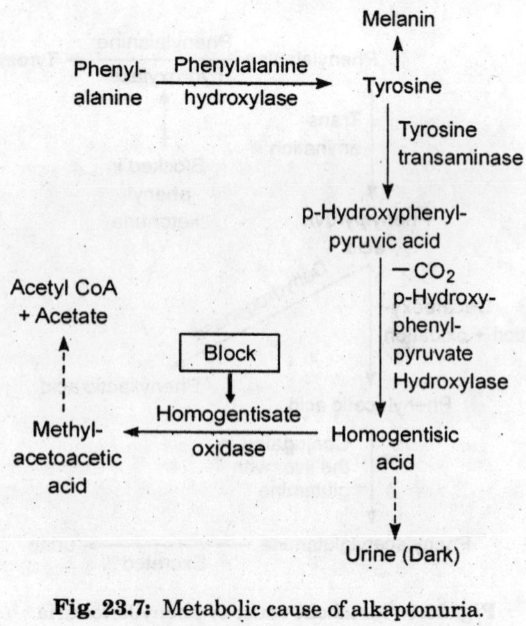


**Fig. 23.7:** Metabolic cause of alkaptonuria.

in valine prevented vomiting, improved weight gain, and reduced hyperkinesis.

(B). **Maple Syrup Urine Disease**

1. This hereditary disease is characterized by the odour of urine, which resembles that of maple syrup or burnt sugar.
2. This disease is characterized by the absence of the enzyme *α-ketoacid decarboxylase* which required for oxidative decarboxylation of the keto acids derived from the branched chain amino acids valine, leucine and isoleucine.
3. As a result, these ketoacids are accumulated in the blood and excreted in the urine, and it is also termed *as branched chain ketonuria.*
4. The affected newborn infant initially appears normal but the characteristic signs are observed by the end of the first week of extrauterine life.
5. The infant is difficult to feed and may vomit. The patient may show lethargy.
6. Extensive brain damage occurs in surviving children and recognized by convulsions, attacks of flaccidity and apnoea.
7. The patient may be treated with amino acid mixtures low in the above three amino acids.
8. Death may occur by the end of the first year of life without treatment.

## Tryptophan Metabolism (Hartnup's Disease)

1. This is a hereditary abnormality in metabolism of tryptophan which is characterized by a pellagra-like skin rash, intermittent cerebellar ataxia and mental deterioration.
2. The urine of patient contains increased amounts of indole-acetic acid and tryptophan.
3. Abnormal patterns of indole acid excretion were found in the urine from the mentally ill.

## Glycine Metabolism

(A) **Glycinuria**

1. This is a rare disease of glycine metabolism, has so far been observed in one family.
2. It is characterized by excess urinary excretion of glycine (glycinuria) associated with a tendency of formation of oxalate renal stone, although the amount of oxalate excreted in the urine is normal.
3. Glycinuria appears to be inherited as a dominant, possibly X-linked, trait.
4. The plasma content of glycine is normal while the urinary excretion of glycine ranges from 600-1000 mg/d.
5. Glycinuria is attributed to a defect in renal tubular transport of glycine whereby decreased reabsorption of glycine by the renal tubule permits the amino acid to escape in the urine.

(B) **Primary Oxaluria**

1. It is characterized by continuous high urinary excretion of oxalate unrelated to dietary intake of oxalate.
2. There is progressive bilateral calcium oxalate urolithiasis, nephrocalcinosis, and recurrent infection of the urinary tract.
3. Death occurs in childhood or early adult life from renal failure or hypertension.
4. The excess oxalate is apparently of endogenous origin, possibly from glycine, which may be deaminated to form glyoxalate, a precursor of oxalate.

5. This is considered to be a disorder of glyoxylate metabolism associated with failure to convert glyoxylate to formate or to glycine by transamination resulting the errors glyoxylate oxidation to oxalate.
6. Glycine transaminase deficiency, together with some impairment of oxidation of glyoxylate to formate, may be the reason for the inherited metabolic disease hyperoxaluria.

## Sulfur-Containing Amino Acids
## Cystine Metabolism

**(A) Cystinuria**

1. This inherited disorder is characterized by urinary excretion of cystine is 20-30 times normal.
2. Excretion of lysine, arginine and ornithine is significantly increased.
3. This disorder is considered to be due to a renal transport defect. The greatly increased excretion of lysine, arginine, and ornithine as well as cystine in urine of cystinuric patients indicate a defect in the renal absorption of these amino acids.
4. Cystine is relatively insoluble, in cystinuria patients if may precipitate in the kidney tubules and form cystine calculi.
5. Apart from cystine which is the principal sulfur-containing amino acid in the urine of the cystinuria patient, another sulfur-containing amino acid, e.g. mixed disfulfide of L-cysteine and L-homocysteine which are more soluble than cystine.
6. There may be an intestinal transport defect for these amino acids. Failure of concentration of cystine and lysine in the cells of jejunal mucosa has heen observed by biopsy.
7. Cystinuria term is a misnomer, so that cystine-tysinuria term may be preferred.

**(B) Cystinosis (Cystine storage disease)**

1. Cystinuria is different from cystinosis. Cystinosis is also inherited, cystine crystal is deposited in many tissues and organs particularly in the reticuloendothelial system of the body.
2. It is usually accompanied by a generalized amino aciduria in which all amino acids are considerably increased in the urine.
3. These patients usually die at an early age with all of the manifestations of acute renal failure.

**(C) Homocystinuria**

1. The incidence of this inherited disease for defective catabolism of methionine is about 1 : 160,000 births.
2. Homocystine up to 300 mg/d, together with S-adenosylmethionine in some cases, is excreted in the urine, and plasma methionine levels are elevated.
3. Associated manifestations are occurrence of thromboses, osteoporosis, dislocated lenses in the eyes, and frequently mental retardation.
4. Two forms of the disease are known:
   (a) Vitamin $B_6$-responsive form,
   (b) Vitamin $B_6$-unresponsive form.
5. Feeding a diet low in methionine and high in cystine effectively prevents pathologic changes if initiated early in life.

6. This is due to impairment of the enzyme *cystathionine synthetase.*

## Histidine Metabolism

### (A) Histidinemia

1. This is an inherited disease of histidine metabolism. In addition to increased levels of histidine in blood and urine, there is also increased excretion of imidazole-pyruvic acid.
2. Speech development may be retarded.
3. The metabolic block of histidine is due to the insufficient activity of liver histidase which impairs the conversion of histidine to urocanic acid.
4. The histidinemia patients are treated well with a diet of protein hydrolysate free from histidine.

## Nucleic Acid Metabolism

### (A) Lesch-Nyan Syndrome

1. This is an inherited X-linked recessive disorder due to complete *HGPRTase* (Hypoxanthine-guanine phosphoribosyl transferase) enzyme deficiency.
2. It is characterized by cerebral palsy with choreoathetosis and spasticity, a bizarre syndrome of self-mutilation, and severe overproduction of hyperuricemia. There is usually an associated uric acid lithiasis.
3. The mothers of these children, who are heterozygous and mosaic for the HGPRTase deficiency, frequently exhibit overproduction of hyperuricemia but without any neurologic manifestations.
4. There are male patients with partial deficiencies of HGPRTase attributable to a different mutation of the same gene.

### (B) Hereditary Xanthinuria

1. This is rare genetic disorder due to the deficiency of xanthine oxidase which leads to the diminished level of blood uric acid (1 mg/100 ml).
2. The urine contains more xanthine with lesser hypoxanthine.
3. Urinary calculi of xanthine may be produced.

### (C) Orotic Aciduria

1. This inherited disorder causes the excessive production of orotic acid due to the deficiency of orotate phosphoribosyl transferase.
2. The urinary excretion of large amounts of pyrimidine nucleotide is observed.
3. The urine becomes cloudy on cooling with the deposition of needle-shaped crystals of orotic acid.
4. Affected children develop a severe megaloblastic anemia with physical and mental retardation.
5. Uridine administration improves this condition significantly.

**PART-V**

# DIFFERENT FACTORS AND GENETICS

# 24 PHYSIOLOGICAL GENETICS

1. Introduction, 2. Genic control of pigments in flowering plants, 3. Genic control of antigens, 4. Genic control of protein structure, 5. Genic control of metabolic pattern.

## Introduction

**In higher plants and animals the differentiating characters do not form directly from genes but result from a chain of developmental processes, initiated by one or more genes, through interactions with events controlled by other genes and factors of internal and external environment to the final phenotype. But how the genotype is causally related to the developed character? By what means do the thousands of individual genes of the fertilized egg cell produce the phenotype of the multicellular individual?**

**Two closely related aspects of these questions may be distinguished. One concerns the mechanisms of by which the hereditary characters achieve their final form. The other concerns the mechanisms by which genes influence the reactions that give rise to specific substances. These reactions occur within the cell during metabolism; subsequently they influence the phenotype. Both the problems are in the domain of *physiological genetics*, i.e. it includes both**

1. *Developmental genetics or pheno-genetics.*
2. *Genic control of metabolism or metabolic genetics.*

**There is no sharp distinction between the two but both are concerned with ontogeny, they deal with it at different levels; therefore, different methods and materials are employed in the two kinds of studies.**

**Experiments with Acetabularia is the best example which illustrate the general rule that genic control over metabolic and developmental processes is exercised by influences emanating from the nucleus and carried into effect by reactions in the cytoplasm (see next chapter).**

## I. Genic control of Pigments in Flowering Plants

**The variations in flower colours in sweet peas, stocks, primroses, and other plants had attracted the geneticist and a number of genes affecting flower colour have been detected. Subsequently the pigments were analysed by the chemists and the gene-influence reactors responsible for the colours have been established. Most of the work is based on water-soluble sap pigments of two**

groups, the *anthocyanins* and the *anthoxanthines.* The former are responsible for the various shades or mixtures of red and blue (pinks, purples, magentas, lavenders), and yellow and ivory colours are due to the related group of anthoxanthin. The anthocyanins and anthoxanthins all contain similar ring structures derived from the condensation of sugars. Known genes for flower-colour differences influence all the following components; presence of pigment and copigment, degree of oxidation and reduction, position of methyl groups, relative acidity or alkalinity, and local distribution of the pigments.

The production of these pigments is clearly under the control of specific genes that determine the arrangement of elementary radicals or atoms in a molecule. The relation of gene to pigment character is indirect, probably through modification of synthetic steps carried out by the enzymes.

## II. Genic Control of Antigens

A direct relation between a gene and an antigen has already been established pertaining to blood group (A, B, O) antigens on the surface of RBC. In persons of blood group AB, both alleles, $I^A$ and $I^B$, express themselves, each responsible for its own specific antigen. In case of persons having MN blood group, two different antigens, each due to an allele of the same gene, can be demonstrated as in heterozygotes for alleles at the Rh. This indicates a one to one correspondence between genotype and phenotype.

## III. Genic Control of Protein Structure

A complex molecule like proteins are formed from several amino acids held together by peptide bonds, are essential composition of all cells of the body. They occur in an enormous variety of specific forms. Individuals within a species as well as members of different species are often distinguishable by the specific forms of the proteins they contain and are able to synthesize.

A direct relation between specific chemical differences in protein molecules have been demonstrated i.e. study of the difference between normal and abnormal haemoglobin in persons carrying the gene of sickle-cell anemia. Differences in the three phenotypes—normal, sickle-cell trait and sickle-cell anemia—were found to be associated with the three genotypes si/si, Si/si and Si/Si respectively.

Horowitz had shown the difference in the form of an enzyme due to different alleles of one gene. He observed in the bread mold Neurospora two forms of the same enzyme, tyrosinase, differing only in stability toward heat. The difference shows simple Mendelian inheritance.

## IV. Genic Control of Metabolic Pattern

Gene-controlled metabolic processes are those concerned with the synthesis of amino acids, the purine and pyrimidine bases of nucleic acids and of some of the vitamins. Among microorganism, many mutants have been found which are incapable of carrying out the synthesis of certain of the above compounds that are necessary for growth. Such mutants can grow only when supplied with a particular substance in the culture medium, these mutants are called "nutritional mutants" or according to requirement of a substance, e.g. "thymine-requiring". Study of bread mold Neurospora has supplied many analyzed cases of nutritional mutants and serve to illustrate the methods and results obtained generally in microorganisms. The method of detection of biochemical mutants in Neurospora has been shown experimentally by George Beadle and Edward Tatum.

## QUESTION

1. Describe the genic control of pigments in flowering plants.
2. Describe the genic control of antigens.
3. Describe the genic control of protein structure.
4. Describe the genic control of metabolic pattern.

# 25

# ENVIRONMENTAL GENETICS

1. Role of environment on gene action, 2. Examples—Fur colour of rabbit, gene action of *E.coli*.

## Role of Environment on Gene Action

The experiments can be done with plants and animals to observe whether the variation is due primarily to the genetic constitution or to environmental differences. A species of plant growing at the sea level may have larger leaves and taller stems than individuals of the same species growing on a mountain. If the two varieties are collected, planted in the same situation and grown through one or two generations and still show the differences of leaves and stem size, it may be assumed that the differences are genetically controlled. If the two varieties, after growing in the same environment, produce offspring which are similar, the original variations must have been due to the environmental differences. The phenotype of any organism is a result of the interaction of a genotype with an environment, both are absolutely necessary. A person, with all the physical, physiological and mental traits, is a product of growth and development brought about by a certain genotype in a certain succession of environment. Every person is the product of his genotype and of his life experience. Phenotypic differences between individual that have developed in similar environment can be ascribed to differences in genotypes.

Environment means the immediate chemical surroundings of the individual cell which encompasses the general factors like temperature, light, etc. The influences of these factors are describe below with examples.

- **Examples**

1. **Development of Fur Colour of Himalayan Rabbit:** There is a true-breeding variety of rabbit called Himalayan rabbit in which the pink eyes and the pattern of black feet, ears and tail and white body which might be thought that the pattern is under purely genetical control but a simple experiment causes the deletion of the idea. This pattern is transmitted faithfully to the descendants. If fur is pulled out from the white parts and the animal is kept in a cold room, the new fur comes in black rather than white. Again if fur is pulled out from the black parts and the part is kept warm—by a bandage—the new white fur comes

**Fig. 25.1:** The influence of temperature on hair colour of Himalayan rabbit. The white hair on a small area of the back was plucked and the animal was exposed in a cold room. The black hair grew on the area. Hair in this area was white when exposed to warm condition.

rather than black. It seems at first that the black-and-white pattern is itself inherited, but the experiment shows that what is really inherited is the ability of certain parts to form pigment or not to form it, depending on the particular temperature getting in that part at a specific time (Fig. 25.1).

2. In plants, chlorophyll develops in presence of light, flowers if the day length is right and the temperature suitable. Usually, environmental influences tend to be cruder than internal ones. They affect the total appearance or of a particular feature rather than the subtler aspects of differentiation.
3. Experiments on gene, action in *Escherichia coli* showed that lactose itself can turn on the switch leading to synthesis of the enzyme, i.e. a chemical molecule which is not itself part of the nuclear material can influence gene action. This fact reveals that development is controlled by a complex interaction between the nucleus, cytoplasm and environment. If the nucleus prepares the genetic plan for production of a highly differentiated multicellular organism from a fertilized egg, the cytoplasm and environment determine the way the information would be used.

Often we are encountered by question whether intelligence is predominantly due to the genes inherited or the environmental conditions of home and school. One source of answer in the nature, i.e. nurture controversy is the study of identical twins.

## QUESTION

1. Describe briefly the role of environment on gene action with suitable examples.

# 26 DEVELOPMENTAL GENETICS

1. Genes and control of development, 2. Role of nucleus, 3. Hammerling's experiment, 4. Cytoplasmic role in the development, 5. Role of DNA or genetic code during development, 6. Role of nucleolus in the development of ribosomes.

## The Genes and Control of Development

During development of an organism from fertilized egg to an adult, i.e. from as a single cell to a complex organism, which involves an increase in size as well as gradual addition of visible complexity when cell differential proceeds in the embryo. But it requires highly ordered sequence of events in which the different cells interact with each other and differentiate in proper time and place. For this events, there must be some elaborate mechanism by which the processes proceed for development and to know this mechanism the separate branch has come out known as *developmental genetics*. In this branch, the differences seen arise within members of the same cell lineage and are traceable to the same set of genes in the fertilized egg.

- **The Role of Nucleus**

1. If the nucleus is removed from the fertilized egg, the enucleated egg dies and revives if the nucleus is replaced within a short time. It indicates nucleus is important in directing the development of the fertilized egg into an embryo.
2. The nucleus is necessary for cell division.
3. It determines the type of structure of the cell.

The role of nucleus was demonstrated by Hammerling, J., using marine algae *Acetabularis* because of advantages like it is large single cell, body has a stalk about 40 mm in length, head, base of root-like structure and the nucleus is situated at the base of the stalk. In *A.mediterranea* the head is cup-shaped, but in another closely related species, *A. Crenulata* which consists of a circlet of finger-like processes.

## Hammerling's Experiment

**I. Experiment-1**

He cut through the stalk of *A. Mediterranea* so as to separate the head from the base. He found that the top half without nucleus died and the bottom half with the nucleus regenerated a new cup-shaped head. It was opined that some kind of chemical substance might be passing from the nucleus into the cytoplasm of the stalk stimulating the formation of a new head (Fig. 26.1, Exp. 1).

**II. Experiment-2**

To verify the above opinion, Hammerling cut across the stalk in two places, just below the head and just above the base. The isolated stalk was then placed in sea water and after a certain period it regenerated a new head. When the regeneration was completed the new head was removed. It was observed that the isolated stalk failed to regenerate another head. It was assumed that the supposed substance produced by the nucleus is present in the stalk which is enough for the formation of new head but not sufficient for formation of further heads (Fig. 26.1, Exp. 2).

**I. Experiment-1**

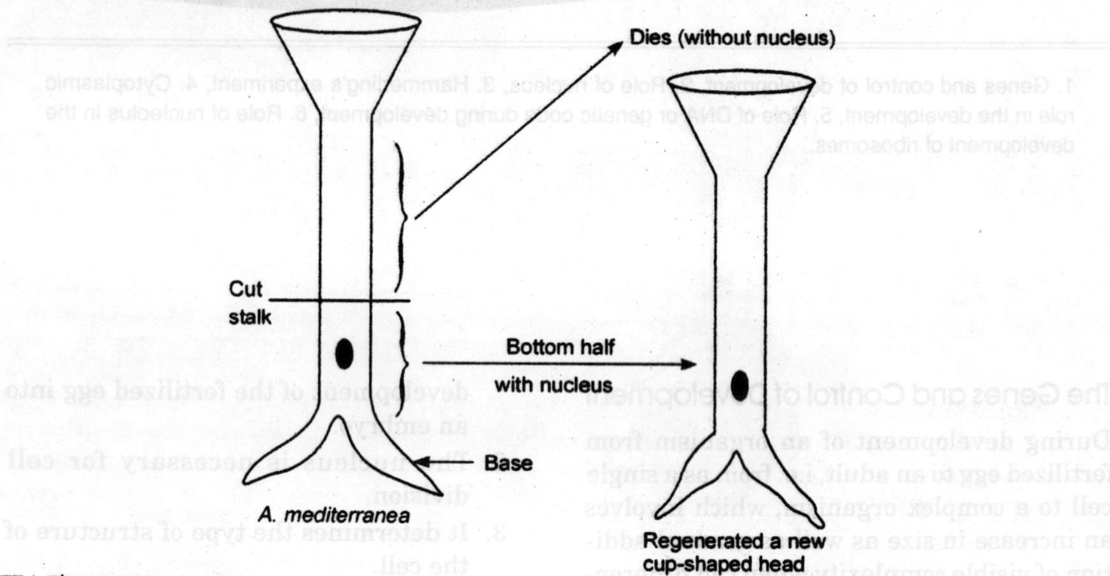


**Fig. 26.1: Hammerling's Experiment No. 1 and 2.**

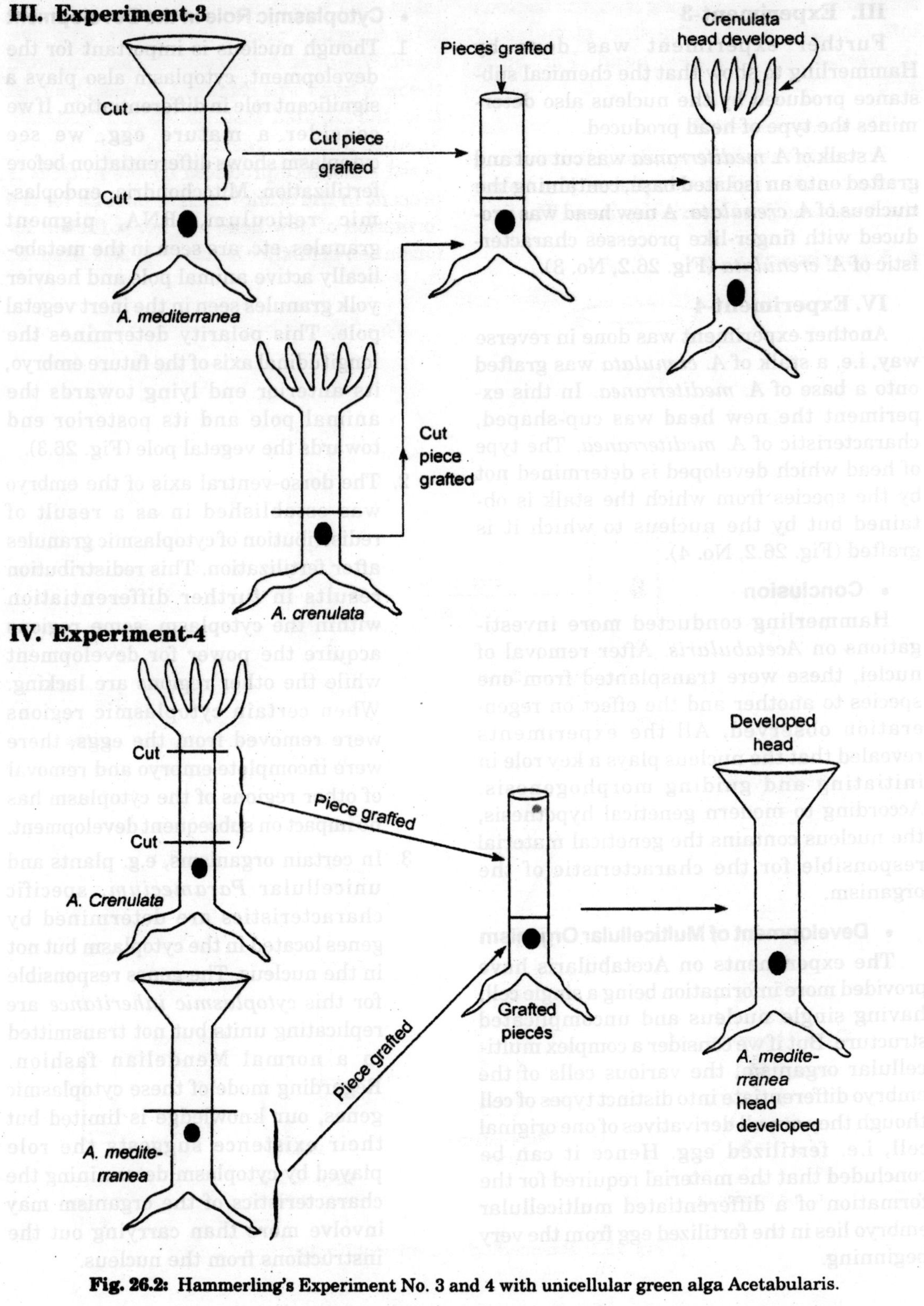


**Fig. 26.2:** Hammerling's Experiment No. 3 and 4 with unicellular green alga Acetabularis.

### III. Experiment-3

Further experiment was done by Hammerling to show that the chemical substance produced by the nucleus also determines the type of head produced.

A stalk of *A. mediterranea* was cut out and grafted onto an isolated base, containing the nucleus of *A. crenulata.* A new head was produced with finger-like processes characteristic of *A. crenulata* (Fig. 26.2, No. 3).

### IV. Experiment-4

Another experiment was done in reverse way, i.e. a stalk of *A. crenulata* was grafted onto a base of *A. mediterranea.* In this experiment the new head was cup-shaped, characteristic of *A. mediterranea.* The type of head which developed is determined not by the species from which the stalk is obtained but by the nucleus to which it is grafted (Fig. 26.2, No. 4).

- **Conclusion**

Hammerling conducted more investigations on *Acetabularis.* After removal of nuclei, these were transplanted from one species to another and the effect on regeneration observed. All the experiments revealed that the nucleus plays a key role in initiating and guiding morphogenesis. According to modern genetical hypothesis, the nucleus contains the genetical material responsible for the characteristic of the organism.

- **Development of Multicellular Organism**

The experiments on Acetabularis have provided more information being a single cell, having single nucleus and uncomplicated structure. But if we consider a complex multicellular organism, the various cells of the embryo differentiate into distinct types of cell though they are all derivatives of one original cell, i.e. fertilized egg. Hence it can be concluded that the material required for the formation of a differentiated multicellular embryo lies in the fertilized egg from the very beginning.

- **Cytoplasmic Role in the Development**

1. Though nucleus is important for the development, cytoplasm also plays a significant role in differentiation. If we consider a mature egg, we see cytoplasm shows differentiation before fertilization. Mitochondria, endoplasmic reticulum, RNA, pigment granules, etc. are seen in the metabolically active animal pole and heavier yolk granules seen in the inert vegetal pole. This polarity determines the longitudinal axis of the future embryo, its anterior end lying towards the animal pole and its posterior end towards the vegetal pole (Fig. 26.3).
2. The dorso-ventral axis of the embryo was established in as a result of redistribution of cytoplasmic granules after fertilization. This redistribution results in further differentiation within the cytoplasm, some regions acquire the power for development while the other regions are lacking. When certain cytoplasmic regions were removed from the eggs, there were incomplete embryo and removal of other regions of the cytoplasm has no impact on subsequent development.
3. In certain organisms, e.g. plants and unicellular *Paramecium,* specific characteristics are determined by genes located in the cytoplasm but not in the nucleus. The genes responsible for this *cytoplasmic inheritance* are replicating units but not transmitted in a normal Mendelian fashion. Regarding mode of these cytoplasmic genes, our knowledge is limited but their existence suggests the role played by cytoplasm determining the characteristics of the organism may involve more than carrying out the instructions from the nucleus.

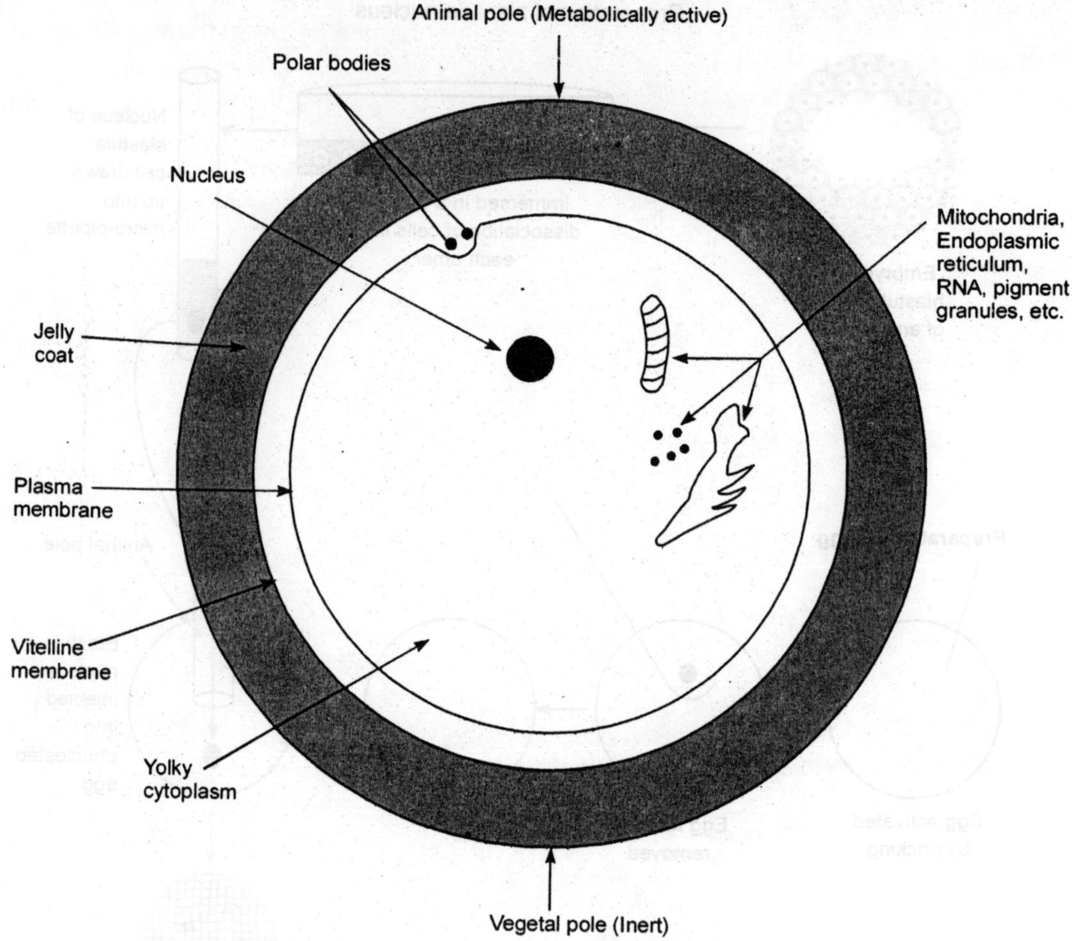


**Fig. 26.3:** Diagram of a generalized animal egg cell.

## Role of DNA in Development or Role of Genetic Code during Development

The cells of the embryo are highly specialized in structure and function. These cells, develop into epithelium or muscle or nerve though they are derived from single cell i.e. fertilized egg. But the characteristics of the cell are determined by its genetic code, i.e. by the sequence of nucleotides in the DNA present in the nucleus. The DNA replicates in correct way before cell division and the cells in adult stage get identical genetic constitutions, but the cells differ from each other. This may happen either of the two-ways:

1. Replication of DNA is correct but changes occur in the genetic code during the progress of development.
2. Alternatively, the genetic code remains same but the different parts of it are used in different cells at the different stages of development.

To prove that the genetic code remains unchanged during development, an experiment was done on amphibians by Brigg and King's (Fig. 26.4).

An embryo was taken at the late blastula stage and immersed it in a fluid which is not injurious to the cells. This fluid helps to dissociate the cells from each other. Then the cell was enucleated, i.e. the nucleus is removed by means of micro-pipette. The

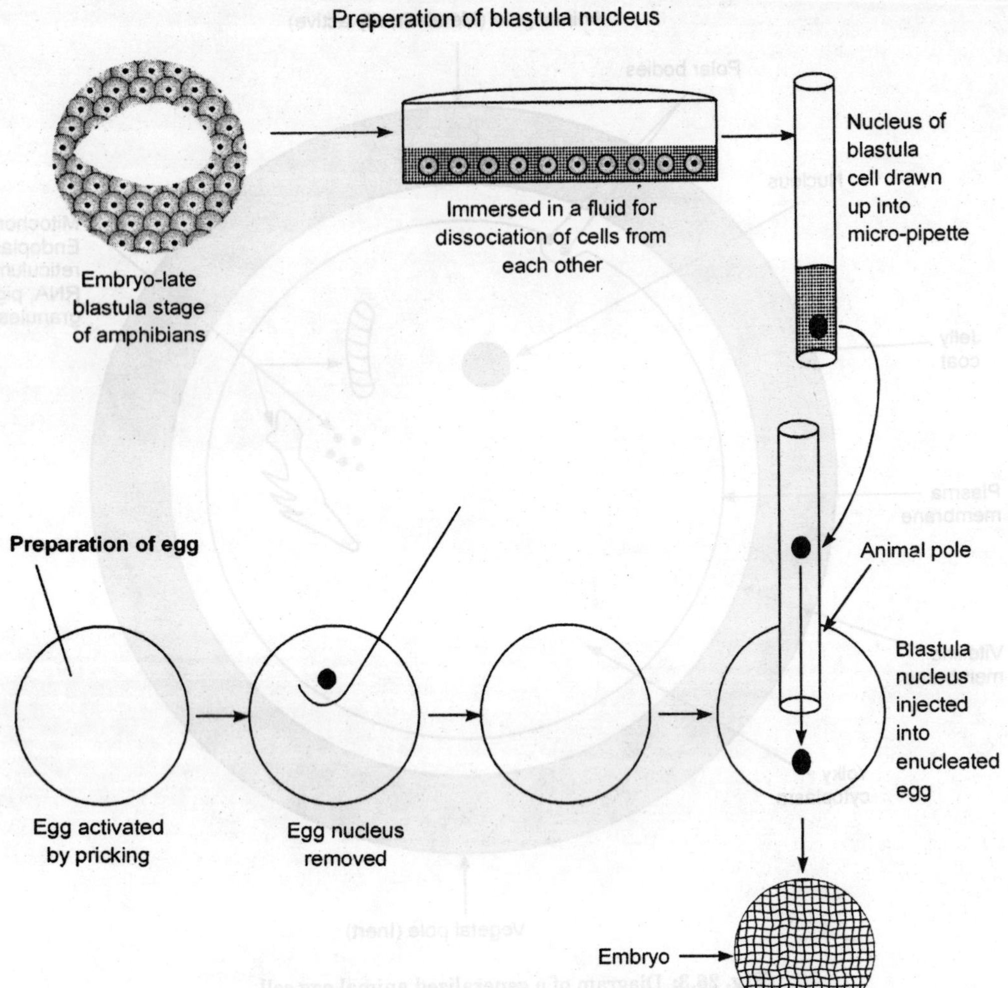


**Fig. 26.4:** Briggs and King's method of transplanting nuclei in amphibians.

micro-pipette would be such that its bore is enough to allow the nucleus only but not the cell. Then an egg cell is prepared to receive the nucleus. At first, the cell is activated by pricking its surface, or applying a brief electric shock, and its nucleus is taken out by means of a fine glass needle. The last method makes the extraction of nucleus easier because the nucleus lies close to the edge of the egg cell. Then the donor nucleus is implanted into the enucleated egg by injecting it into the cytoplasm of the egg near the animal pole and it is allowed to develop. It has been found by repeated experiments that the nuclei of late blastulae stage are capable to sustain the embryo even sometimes it may develop to adult. From the above experiment it can be concluded that up to the late blastula stage no irreversible changes occur in the genetic information contained within the nucleus.

But there might be nuclear changes in the genetic material after the blastula stage.

To get the information, repeated experiments were done with nuclei from later stages. But the following problems were encountered during the experiments:

1. It is difficult to take nucleus due to small size of the cell.
2. The results are variable.
3. Different species of amphibia give different results.
4. Nuclei taken from late gastrulae give normal tadpoles and adults.
5. Nuclei from tumour cells of an adult were found to be capable of producing normal tadpoles. These results are common in the African clawed toad, *Xenopus laevis* used in these experiments.

In case of common frog *Rana pipiens* nuclei taken from grastulae, produce a high percentage of abnormal embryos and these abnormalities are stable and heritable which indicates that they are caused by changes in the genetic material within the nuclei.

- **Conclusion**

Nucleus taken from a differentiated cell can produce normal development and it can be explained that the genetic material in the nucleus is unaltered at that time. The formation of abnormal embryos indicates that in some cases the genetic constituents of the nucleus alter, at least during the later stages of development. Both mechanisms are correct, differentiation is brought about partly *by changes occurring in the genetic code* and partly by different parts of the code is used at different times during the progress of development.

## Role of Nucleolus in the Development—Ribosome Formation

- **Structure of Ribosome**

It is known that synthesis of proteins takes place on the ribosomes. There must be a control system for the formation of the ribosomes. Chemical analysis has shown that ribosome consists of a nucleo-protein in which protein is conjugated with one type of RNA called ribosomal RNA (rRNA).

Ribosomal RNA constitutes about 60 per cent of the ribosome, the remainder of 40 per cent contains 50 different types of proteins and enzymes needed for protein synthesis. The ribosome is the physical and chemical structure on which protein synthesis occurs with the help of transfer RNA (tRNA) and messenger RNA (mRNA). The ribosomes are composed of two physical subunits called 40S and 60S particles because of their sedimentation rates in the ultracentrifuge. mRNA and tRNA first complex with 40S particles and 60S particles provide enzymes for promoting peptide linkage during protein synthesis, hence ribosome is called *protein manufacturing plant.*

- **Formation of Ribosome**

From *DNA template,* the rRNA is *synthesized* and *collected* in the *nucleolus,* the dense body normally found in the nucleus of all animal and plant cells. It has been observed that abnormal cells lacking nucleoli fail to manufacture ribosomal RNA and are thus unable to make ribosomes. The rRNA is *processed in the* nucleolus and *combines* with *ribosomal proteins* to form *granular condensation products* to form *ribosomes.* These ribosomes are then released from the nucleolus, and they migrate through the large *"pores"* of the nuclear membrane to almost all parts of the cytoplasm. Hence, chemically it is ribonucleoprotein (RNA + Protein). Ribosomes are found in *free form* or *bound to the endoplasmic reticulum.* Many of the ribosomes are associated in bead-like strings called *"polyribosomes"* or *"polysomes"* which are formed by the attachment of a number of ribosomes to a single molecule of mRNA during protein synthesis (Fig. 26.5).

- **Mechanism of Nucleolar Performance**

It was observed specific regions of certain chromosomes are always embedded in the nucleolus. It was suggested that the genes in these specific regions of the chromosomes are responsible for bringing about the synthesis of ribosomal RNA. An experiment was done to verify the suggestion.

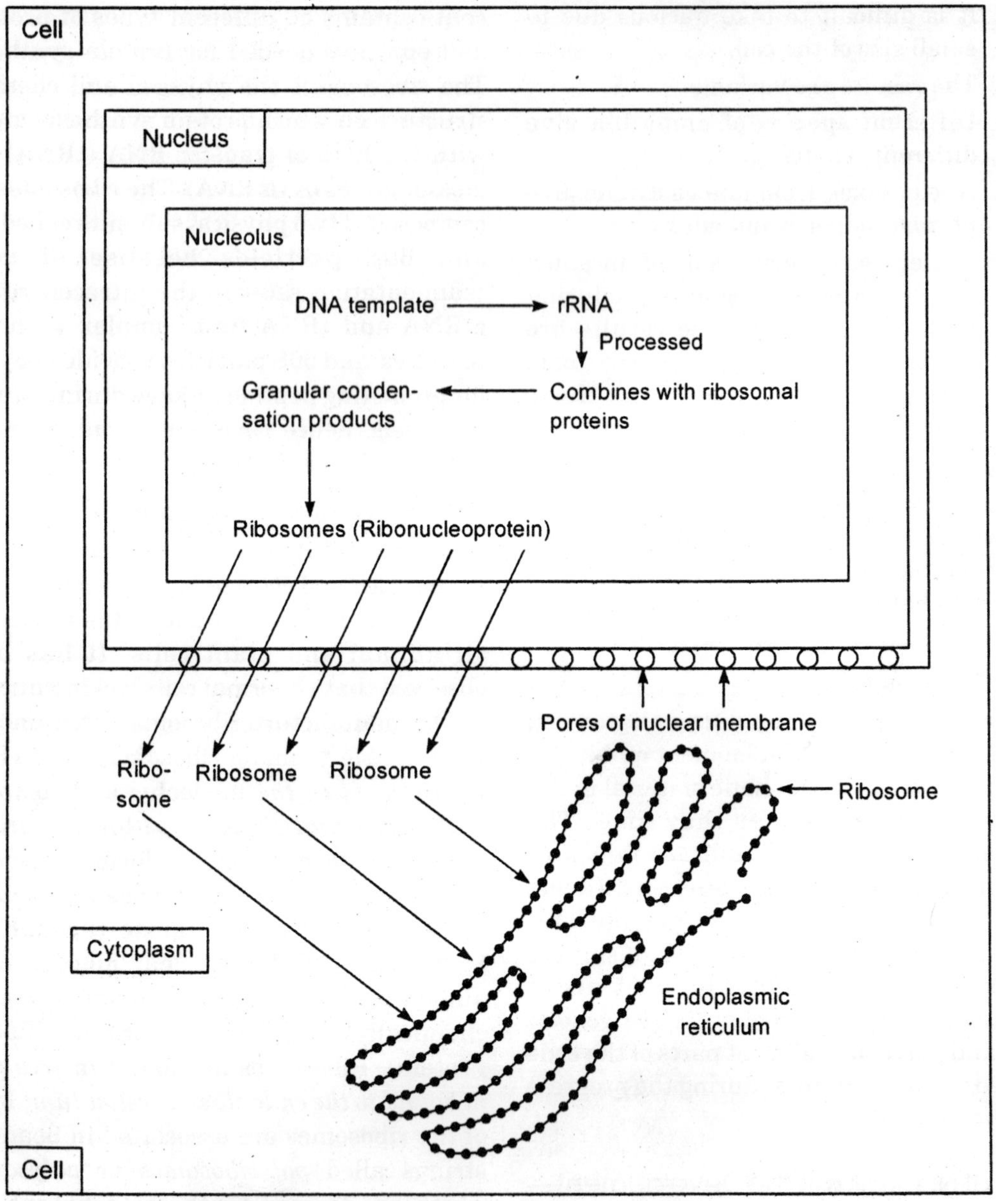


**Fig. 26.5:** Role of nucleolus in the development—Formation of ribosomes.

**From the tadpoles of the African clawed toad, *Xenopus laevis*, DNA was extracted from the cells. The advantage of tadpole cell is that some of the tadpole cells had nucleoli, others were abnormal and without nucleoli. Then DNA was hybridized, i.e. the double helix was split by chemical treatment into single strands and mixed with radioactive labelled ribosomal RNA. This radioactive rRNA proceeded to bind to the particular part of the DNA where the sequence of bases corresponded to that of the ribosomal RNA. Estimation was done regarding the quantity of bound ribosomal RNA by washing away any unbound RNA and measured the amount of radioactivity. The observation was that the**

tadpole cells without nucleoli did not show any bound rRNA; DNA from tadpole cells having nucleoli gave a level of radioactivity which revealed that over a thousand genes were responsible for coding for the ribosomal RNA.

The part of the nucleolus containing these genes known as *nucleolar organizer region.* It is believed that once the DNA has been transcribed into ribosomal RNA in the nucleolar organizer region, the ribosomal RNA is conjugated with protein in another part of the nucleolus to form a nucleoprotein chain. The nucleoprotein chains so produced then coil up and pass via the nuclear membrane pores into the cytoplasm and are assembled into functional ribosomes.

## QUESTION

1. How will you show experimentally the role of nucleus in the development? What conclusion can be drawn?
2. Describe the cytoplasmic role in the development.
3. Describe the role of DNA or genetic code during development.
4. Role of nucleolus in the development of ribosomes.

# 27

# POPULATION GENETICS

1. What is population genetics?, 2. Effect of natural selection on genes, 3. Genetic death, 4. Polymorphism, 5. Influence of natural selection on population genetics, 6. Gene frequency and factors affecting it, 7. Genetic equilibrium, 8. Genetic drift, 9. Hardy-Weinberg law.

## What is Population Genetics?

In the previous chapters, the discussion was made on the behaviour of genes in individuals and in families. Mendel, Morgan and others have discovered the rules of heredity transmission which paved the pathway for different observations, e.g.

1. Prediction of the characteristics of the progeny of some marriages and crosses.
2. In the organism like Drosophila, Corn and Neurospora which are well-studied, the strains of desired characteristics can be produced easily.
3. Prediction of probable blood groups of individuals and certain other traits in children can be made if the characters of the mother and the father are known.

But we encounter the difficulties when we discuss the heredity in population but not in individuals or families. This field of genetics is known as *population genetics*.

The study of population genetics tries to give informations, as for example:

1. Mutants induced by exposure to radiations and the effects on the particular community in various generations;
2. Frequencies of various blood groups among the inhabitants of the various centres of the world;
3. Whether the mutants induced by radiations or the blood-group genes, become more or less frequent with time or retain their frequencies unchanged;
4. Preparation or divising any method of freeing a population of harmful genes or of increasing the frequencies of useful ones.

As organic evolution is a change in the genetic composition of populations, population genetics also tries hard to throw light on the mechanisms of evolution.

Genetics is more related particularly in Mendelian populations of sexually reproducing and cross-fertilizing organism. Hence, it can be stated that a *Mendelian population* is a community of interbreeding individuals. The best example of most extensive Mendelian population is the species. Mendelian population includes the potential mates or of individuals related by bonds of marriage, parentage or common descent. Human population is a vast Mendelian population divided into:

(a) Subordinate Mendelian Population, or,
(b) Races.

These are further subdivided into several groupings such as:

(a) Linguistic,
(b) National,
(c) Religious,
(d) Economic, etc.

Mendelian population does not include the following:

1. People having blood group A, or having blue eyes or criminals or fat or lean people because they do not form mating groups.
2. The organisms reproduce exclusively by asexual means or by self-fertilization.

## Effect of Natural Selection on Genes

Natural selection is very much concerned with the continuance or elimination of genes carried by the individuals than with survival or death of the individuals.

Most mutants which arise in natural or artificial population are harmful or better to say complete lethals. The mutation process creates these harmful mutants and introduces them into gene pools of populations. This accumulation of harmful mutants of hereditary diseases is resisted by natural selection.

Darwin's theory of natural selection propounded by the fact that only a part and often a very small part of the progeny of any species surviye and become parents of the next generation. The different parents produce different numbers of surviving offspring. The population of a species may include genetic variants. These variants are more or less well-adapted to survive in a given environment. The better-adapted variants make a greater proportion of the survivals than the less well-adapted ones. The incidence of better-adapted increases and the lesser-adapted decreases in the succeeding generations. *Artificial selection* is a process similar to natural selection excluding the variants that leave the greater progeny are chosen by man rather than by the general environment. The differential mortality is of special interest from the evolutionary point of view because:

1. if it occurs before the individual has had a chance to reproduce, it eliminates the harmful genes from the population.
2. death is not the only way to achieve this but there should be any process which encourages the transmission of favourable genes and resists the transmission of harmful genes to help the evolutionary progress.

- **Genetic Death**

In the past, the evolutionist stated that natural selection means, "survival of the fittest" in the "struggle for existence" which is a misnomer. The accurate statement is that carriers of different genotypes transmit their genes to the succeeding generations at different rates. A harmful mutant is opposed by natural selection, that means that the mutant allele is perpetuated from generation to generation less efficiently than the normal allele of the same gene. It is certain that a harmful mutant will be able to remain in the population for a smaller number of generations than a neutral mutant. Muller stated that a harmful mutant will sooner or later be eliminated by *"genetic death"*. It is caused not by differential mortality but by differential fertility. As fer example, if an animal having unfavourable characteristics is prevented from breeding, either by natural infertility or losing for a mate, then this individual's genes will be prevented from going through to the next generation. Without involvement of death of the animal, it is must effective to eliminate its genes from the population, known as *"genetic death"*. It is not the fact that harmful genes are always quickly weeded out of the population by natural selection. It is seen that a gene which transmits unfavourable characteristics in the homozygous state as in *sickle-cell anemia* which is rare but confers an advantage in the heterozygous state as in case of *sickle-cell trait* which is milder. The high incidence of this trait caused by heterozygote being more resistant to *malaria* and this advantage ensures the survival of the sickle-cell gene. Owing to this heterozygote advantage, individual organism and populations always harbour a certain number of harmful genes which is known as the *"genetic load"*.

## Polymorphism

- **Definition**

Existence within a particular species of two or more distinct types of individual. Literally the meaning of the word is "having many forms".

**Occurrence:** The peppered moth and *Cepaea* both exhibit the phenomenon of *polymorphism.*

- **Genetic Polymorphism**

Different types may result from differences of environment. But in genetic polymorphism, the different forms called *morphs* have a genetic basis for which distinctive characteristics are transmitted in a Mendelian fashion. Usually the different morphs are controlled by a series of alternative alleles, any one of which may occur at a particular locus on the chromosome, thereby giving a variety of different genotypes i.e. the genes responsible for the different morphs belong to one allelic set.

### Types of Polymorphism

Two types of polymorphism are recognized:

A. **Unstable or Transient:** In this case, a genetically controlled morph spreads through the population and its numbers compared to other morphs, will be changing and later on it may become uniform, the best example is peppered moth.

B. **Stable or Balanced Polymorphism:** In this case, the different morphs i.e. alleles responsible for them occur in fairly constant proportions within a given population whose members are freely interbreeding. To a greater or lesser extent, all of them are selected for or against by natural selection. This type is obtained by a balance of relative advantages and disadvantages. Each morph is favoured by natural selection in the particular conditions which occur within a varied environment. The best examples of stable polymorphism are:

   1. Land snail *Cepaea,*
   2. Sickle-cell condition in man,
   3. Human blood groups.

### Differences in Various Forms

Sometimes in polymorphism, the various forms are having insignificant differences which is difficult to assess of its importance in natural selection. These differences are associated with other more important differences such as viability, reproductive efficiency and the like, on which natural selection can operate. At a glance it would appear that having blood group is not advantageous but it has been observed that person belonging to blood groups A and AB are more susceptible to smallpox than the persons having other blood groups. Again, blood group A is said to be associated with cancer of the stomach and group O with duodenal ulcers.

### Two or More Types in Life Cycle (Special Case of Polymorphism)

A special case of polymorphism is prevalent among some organism in which two or more types of individual occur in their course of life cycle. As for example,

1. Plants like mosses and ferns where a sexually-reproducing gametophyte alternates with an asexually-reproducing sporophyte,
2. Coelenterates, e.g. *Obelia* in which an asexually-reproducing polyp alternates with a sexually-reproducing medusa,
3. *Obelia* and its more complex variety, e.g. "Portuguese-man-of-war" exhibits another type of polymorphism in that they are colonial animals, the colony being composed of several different types of interdependent individuals each with its particular job to do such as feeding, defence or reproduction,
4. In social insects, e.g. bees and ants in which there are several distinct castes within the community.

But these are not concerned with the genetic polymorphism.

## Influence of Natural Selection on Population Genetics

The forces of natural selection though act on individual genes but it has great influence on population for an effective long-term evolutionary change *Deme*—a deme is more or less genetically isolated unit of population, e.g.

1. Deer on the Kaibab Plateau in Arizona,
2. Carp in a pond,
3. Beech trees in a corpse,
4. Paramecium in a ditch, etc.

Though there may be movement of individuals to and from an individual deme, a deme usually perpetuates itself by interbreeding of the individuals within it. It is considered as a genetic unit and the impact on it in evolutionary terms depends on the genes it contains.

- **Gene Pool**

The sum total of all the different genes in the population that means genetic constitution of a deme which is known as *gene pool.* Some of the features are:

1. In *migration,* the interchange of individuals, and hybridization with genetically different population may also increase or decrease the frequencies of some gene alleles in the gene pool.
2. Most or all genes are exposed to greater or lesser risk of undergoing mutation. These mutants after entering the gene pool of a population may persist in the gene pool for many generations after their origin, even indefinitely as the Hardy-Weinberg principle stated.

- **Gene Frequency**

The frequency of any given gene in a population, relative to all its alleles at the same locus, is known as the *gene frequency.* It was proved mathematically by Hardy and Weinberg and stated that if there are no disruptive influences e.g., mutation, or selection, the frequency of genes in a population remains constant generation and after generation which is known as *Hardy-Weinberg principle.*

- **Genetic Equilibrium**

There is continuous movement of genes known as *gene flow* within the population as a result of breeding maintaining constant gene frequencies and this stability is called *genetic equilibrium.*

During this time, a deme remains unaltered in its characteristics and evolution does not occur. Evolution is the change in the genetic composition of populations. If the gene frequencies of all genes in all populations remained stable or constant, evolution certainly could not occur. The Hardy-Weinberg principle states only the statics of Mendelian populations but it must consider the dynamics of populations. Existence of genetic forces disrupt the genetic equilibrium and modify the gene frequencies in populations. The forces are:

(a) Mutation,
(b) Environmental change,
(c) Natural selection,
(d) Random genetic drift in small populations,
(e) Differential migration.

As for example, a change in the environment causes alteration of the selection pressure, favouring a particular gene (or genes) whose frequency increases and this is the genetic basis of progressive selection.

- **Genetic Drift**

This is one of the factors which can upset the genetic equilibrium, which brings about evolutionary change.

A gene may be lost altogether from a population as a result of pure chance resulting a total loss of that particular gene. But it may be comparatively rare but is likely that a sufficient proportion may be lost in a small population resulting significant drop in the gene frequency. These variations in gene frequency are known as *genetic drift* and it happens in smaller rate in larger populations and produces significant changes in small populations.

## Factors Affecting the Hardy-Weinberg Equilibrium

In genetic disorders many factors can disturb the Hardy-Weinberg equilibrium by influencing either:

I. The distribution of genes in the population, or
II. By altering the gene frequencies.

The Hardy-Weinberg equilibrium holds exactly only for large populations in which there is random mating and in which there is no selection, mutation, or migration. Deviation from any of these conditions can alter allele frequency in a population and lead to an increase or decrease in allele frequencies from one generation to another.

- **Factors**

I. Nonrandom mating
II. Small population
III. Selection
IV. Mutation
V. Migration and gene flow

### I. Nonrandom mating:

In human populations, mating is seldom random. It is usually assortative in that members of a particular population are more likely to mate with other members of that same subpopulation, whether it is defined by racial, ethnic, religious, or other criteria. As for example congenitally deaf individuals often marry congenitally deaf partners. A special form of nonrandom mating in the human population is *consanguinity* or mating among close relatives. Consanguinity (or inbreeding) does not change allele frequencies, but it does increase the proportion of homozygotes in the next generation at the expense of heterozygotes, thereby exposing disadvantageous recessive phenotypes to selection.

### II. Small population:

For any reasons, e.g. political, religious or geographic reasons, a small subgroup of the population may become physically and/or socially isolated from the rest of the population, forming a genetic isolate. The founder members of such an isolate may, by chance, be carriers of mutant alleles for certain recessive traits, so that the frequency of these alleles will be higher within the isolate than within the population at large. Furthermore, the actual frequency in such a small population will vary widely from one generation to the next, a phenomenon called *genetic drift*. By chance, one allele may fail to be passed on to the next generation and thus disappear from that line of descent, leaving only the alternative allele, the mechanism is known as *founder effect,* may account for the high frequency of certain rare diseases among genetic isolates.

### III. Selection:

Selection represents the action of environmental factors on a particular phenotype, and hence its genotype, and may be positive or negative. It is the consequence of differences in the *biological fitness* ($f$) of individual phenotype. Biological fitness is a measure of fertility and therefore of the contribution to the gene pool of the succeeding generation.

**IV. Mutation:**

Change in the genetic material is mutation and the *mutation rate* ($\mu$) is the frequency of such change and is expressed as the number of mutations *per locus per gamete per generation.* Human mutation rates comes from studies of rare autosomal dominant traits for which it is much easier to estimate mutation rates than for recessive traits. For rare autosomal dominant traits, the mutation rate is

$$\mu = n/2N$$

where,

$n$ = Number of affected patients with unaffected parents,

$N$ = Total number of births.

The denominator is $2N$ rather than $N$, because mutation of either allele at an autosomal locus could result in the mutant phenotype.

**V. Migration and gene flow:**

Migration of populations into new regions and intermarriage with indigenous populations can result in a change in allele frequencies in both populations, a phenomenon known as *gene flow*.

## Applications of the Hardy-Weinberg equilibrium

I. **Estimation of carrier frequencies:** If the incidence of an autosomal recessive disorder is known then it is possible to calculate the carrier frequency using some relatively simple algebra.

II. **Estimation of mutation rates:**

(a) **Direct method:** If an autosomal dominant disorder shows full penetrance and is therefore always expressed in heterozygotes, an estimate of its mutation rate can be calculated by counting the number of new cases in a defined number of births.

(b) **Indirect method:** For an autosomal dominant disorder with reproductive fitness ($f$) equal to zero, all cases must result from new mutations. If the incidence of a disorder is denoted as $I$ and the mutation rate as $\mu$, then as each child inherits two alleles, either of which can mutate to cause the disorder, the incidence equals twice the mutation rate, i.e. $I = 2\mu$. When fitness is greater than zero and the disorder is in Hardy-Weinberg equilibrium, then genes lost through reduced fitness must be counterbalanced by new mutations. Therefore,

$$2\mu = (I - f) \text{ or } \mu = [I(I - f)]/2.$$

## Reasons for Usefulness of Knowing Mutation Rate

Knowledge of these formulae which link mutation rates with disease incidence and fitness is useful for the following reasons:

I. **Estimation of gene size:** If a disorder has a high mutation rate this could yield information about the structure of the gene which would help expedite its isolation and characterization.

II. **Determination of mutagenic potential:** If valuable information is to be gained about the potential mutagenic effects of nuclear accidents, it will be necessary to have accurate methods for determining mutation rates and how these are related to observed changes in disease incidence.

III. **Consequences of treatment of genetic diseases:** Knowledge of the relationship between disease incidence, fitness and mutation rate is necessary to determine the possible effects of improved treatment for serious genetic disorders.

## Hardy-Weinberg Law

Population genetics is based on the principle demonstrated independently by Hardy (English mathematician) and Weinberg (German physician). The *principle is concerned* with:

1. Determination of frequencies of genes in a population.
2. Homozygous and heterozygous genotypes in Mendelian populations.

   Hardy and Weinberg showed that in the absence genetic forces (e.g. mutation, environmental change and natural selection) which change the gene frequencies, populations may have any proportions of dominant and recessive traits, the relative frequencies of each gene allele tend to remain constant from generation to generation.

   The mathematical expression is known as the *Hardy-Weinberg formula.*

It was derived to explain:

1. Why dominant traits do not replace automatically replace recessive traits in the population.
2. Why in a large population with random mating, allele frequencies do not change from generation to generation under medical genetics.
3. How, for any genetic locus, the genotype frequencies are determined by the relative frequencies of the alleles at that locus.

Suppose that alleles $A$ and $a$ are present in a population with the frequencies of $p$ and $q$. Then $p + q = 1.0$, where 1.0 represents 100 per cent of the population. In the population, the individuals may be either homozygous for the dominant allele (*AA*), double recessive (*aa*) or heterozygous (*Aa*). The gene frequencies of these different genotypes in a population of any given generation are determined by the way the gametes fuse. The gametes are expected to fuse in the following order (shown in the table) if the mating is random, and all individuals produce approximately equal numbers of gametes, and the genes $A$ and $a$ do not mutate.

| Gametes | A<br>p | a<br>q |
|---|---|---|
| A<br>p | AA<br>$p^2$ | Aa<br>pq |
| a<br>q | Aa<br>pq | aa<br>$q^2$ |

A and a = Alleles, p and q = Frequencies

Then, the frequency of *AA* individuals $= p^2$,

The frequency of *aa* individual $= q^2$,

The frequency of *Aa* individuals $= 2pq$.

It can be expressed in the following way:

$$p^2 + 2pq + q^2 = 1.0$$

where 1.0 designates 100 per cent of the population and this is known as *Hardy-Weinberg formula* which is *used to determine the frequency of the different genotypes in a population* provided:

1. The population is numerically so large that accidents of sampling may be ignored,
2. The mates are chosen at random,
3. The mutation from $A$ to $a$ or from $a$ to $A$ is infrequent,
4. The carriers of the genotypes *AA, Aa* and *aa* are equal in survival and reproduction rates.

- **Use of Formula**

For example, we can consider tongue-rolling in a human population caused by a dominant allele. In a certain population 84 per cent of people can roll whereas remaining 16 per cent cannot.

With the help of Hardy-Weinberg formula, number of heterozygous and homozygous for tongue-rolling can be determined.

Suppose, $R$ = gene for tongue-rolling and $r$ = gene for non-rolling. These people are usually either homozygous dominant ($RR$) or heterozygous ($Rr$); non-roller are homozygous recessive ($rr$). The distribution of genotypes according to the Hardy-Weinberg formula will be:

$$p^2 + 2pq + q^2 = 1.0$$

where, $p^2$ = Frequency of homozygous dominant individuals ($RR$),

$2pq$ = Frequency of heterozygous individuals ($Rr$),

$q^2$ = Frequency of double recessives ($rr$).

In case of non-rollers which is 16 per cent of the population must be homozygous recessive $rr$ and $q^2$ is the frequency of double recessive $rr$.

Then, $q^2 = 0.16$

Therefore, $q = \sqrt{0.16} = 0.4$

Again, $p + q = 1.0$

Therefore, $p = p + q - q$

$= 1.0 - 0.4 = 0.6$

So, $2pq = 2 \times 0.6 \times 0.4 = 0.48$

Then the *percentage* of the *heterozygous* individuals is expected to be 48 per cent. The percentage of tongue-rollers in the population is 84 per cent, the *percentage* of individuals *homozygous* for the dominant allele is expected to be 84 – 48 = 36 per cent.

**Genotype Frequency:**

The genotype frequency in the population would be like this:

$RR$ = 36 per cent

$Rr$ = 48 per cent

$rr$ = 16 per cent

To determine their *genotype frequencies* in the *next generation,* it is known that:

$$p^2 + 2pq + q^2 = 1.0$$

where, $p^2$ = Frequency of $RR$ individuals,

$2pq$ = Frequency of $Rr$ individuals,

$q^2$ = Frequency of $rr$ individuals.

Now, $p^2 = 0.36$, $2pq = 0.48$ and $q^2 = 0.16$.

Therefore, the frequency of the $R$ allele in the population, i.e.

$$p = \sqrt{0.36} = 0.6$$

The frequency of the $r$ allele in the population, i.e.

$$q = \sqrt{0.16} = 0.4$$

The frequencies of different genotypes in the next generation can be determined by the fusion of gametes, provided that:

(a) Mating is random,

(b) All individuals produce approximately equal numbers of gametes,

(c) Genes $A$ and $a$ do not mutate.

The fusion of gametes is expected to be:

| Gametes | R<br>0.6 | r<br>0.4 |
|---|---|---|
| R<br>0.6 | RR<br>0.36 | Rr<br>0.24 |
| r<br>0.4 | Rr<br>0.24 | rr<br>0.16 |

Therefore, the genotype frequencies in the next generation is expected to be:

RR — 36 per cent

Rr — 24 + 24 = 48 per cent

rr — 16 per cent

So, if we compare the *genotype frequencies* of *previous generation* with the *next generation,* if will be the *same.*

The genotype frequencies change only if:

(a) The population is small,

(b) A large number of individuals fail to produce approximately equal number of gametes,

(c) Mating is nonrandom,

(d) The genes are lost from the population.

- **Disadvantage of the Law**

Evolution is a change in the genetic composition. If the frequencies of all genes in all populations always remained constant, evolution certainly could not occur. The Hardy-Weinberg law states only the statics of Mendelian populations. But from the evolutionary point of view, we should consider the dynamics of populations. For evolution to occur, there should be changes in the genotype frequencies by genetic forces that disturb the genetic equilibrium and modify the gene frequencies in populations. The genetic forces are:

1. Mutation,
2. Random genetic drift in small populations,
3. Selection,
4. Differential migration,
5. Environmental change.

## QUESTION

1. What is population genetics?
2. Describe the effect of natural selection on genes.
3. What is genetic death? Describe it.
4. What is polymorphism? Describe it.
5. Describe the influence of natural selection on population genetics.
6. Define gene frequency, genetic drift, genetic flow
7. Describe the genetic equilibrium.
8. Describe briefly the Hardy-Weinberg Law. What is the disadvantage of the law?
9. Explain—Genetic load, gene pool.
10. State briefly Hardy-Weinberg Law.

# 28

# GENETICS OF HISTOCOMPATIBILITY

1. Introduction, 2. Nature of Allograft Rejection, 3. Mechanism of Graft Rejection, 4. Histocompatibility Antigens, 5. Genetics of Histocompatibility, 6. Other Relation of HLA Gene.

## Introduction

The replacement of diseased organs or tissue by a transplant of healthy organ or tissue is known as *transplantation* or *grafting*. But it has created frustration due to uncooperative attempts of the body to reject graft from their individuals, the phenomena is known as *rejection phenomena*.

## Nature of Allograft Rejection Phenomena

It was pointed out that normal and neoplastic tissues contain *iso-antigenic* factors which are genetically determined. Iso-antigenic factors present in the grafted tissues and absent in the host are capable of eliciting a response which results in destruction of the graft.

Graft rejection is an *immune response* and *cell-mediated*. Though similar in principle, the rejection of grafts involves somewhat *different types* of immune response. In this case the *antibodies* are *cell-bound* and *not free in the blood plasma*.

## Mechanism of Graft Rejection

There are two main types of grafts rejection phenomena:

1. The so-called 'first set' response which occurs about 10 days after a first graft from an unrelated donor.
2. The 'second set' response occurring in about 7 days in an animal which had previously received a graft from the same unrelated donor.

It involves a sequence of *separate phenomenon* developing from initial contact of the foreign antigens with the host immunological system through the rejection itself. The sequence of events includes:

Establishment of lymphatic drainage → Antigen release → Contact with cells of the immune system → Proliferation of antibody-forming cells of the lymphoid tissues → The inflammatory response → Release of pharmacological mediators → Cytotoxic effects of lymphoid cells → And/or antibody on the graft tissues.

## Histocompatibility Antigens

Histocompatibility antigens are tissue cell surface proteins. There are different numbers of antigenic combination on tissue surfaces which creates difficulties in getting a good tissue match for transplantation. Some antigens are strong and some are weak. H-2 antigens in mice HLA-antigen in human have been studied and both have many genetic variants. The H-2 antigen in mice is strongly antigenic and is made of heavy chain and light chain.

The light chain in a protein and it is called β-2-microglobulin. The light and heavy chains interact but not by covalent bonds.

The proteolytic enzyme papain cleaves the heavy chain in such a way that a water-soluble portion called the *Fs fragment* is released and a small piece, the *Fm fragment* is left in the membrane. Partial sequences for the heavy chains of both H-2 and HLA antigens and similarities are seen between the two types of heavy chains.

## Genetics of Histocompatibility

It has been observed that normal and neoplastic tissues contain iso-antigenic factors which are genetically determined. Iso-antigenic factors present in the grafted tissue and absent in the recipient are capable of producing immune response which causes destruction of the graft. This type of histocompatibility is controlled by a region on chromosome-6, which are known as HLA–A, HLA–B, HLA–C and HLA–D can control the major histocompatibility and the sequence of arrangement of these four regions or locus on chromosome-6 are HLA–D, –B, –C, –A. Every region or locus contains a series of alleles. Each locus only contains maximum of two alleles.

The HLA–D, –B, –C, –A loci are situated close to each other on chromosome-6 and therefore, the alleles at each of these four loci on the maternal chromosome will nearly always be inherited together and likewise with the four alleles that happen to be on the paternal chromosome.

The inheritance of tissue type is just as the inheritance of blood groups but with a difference of genetic linkage which exists in between the different loci of HLA. The particular combination of the alleles in HLA–B, –C, –D, –A loci on the same chromosome is known as *haplotype* and the haplotype of one chromosome is not, of course, necessarily the same as the other chromosome. In fact, 90 per cent of the people are heterozygous with two different alleles at each locus. The parental haplotypes are inherited unchanged to offsprings which is the common path of inheritance. Occasionally chromosomal crossing over occurs in between the four loci and the production of recombinant haplotypes may occur. It is not rare for parents to have one haplotype in common.

## Other Relations of HLA Gene

HLA gene is not only important for tissue transplantation but it has also been opined that genes concerned with the immune response are also located in the HLA region and both are in close genetic linkage. It has also been demonstrated that there are association between a certain disease and a particular allele of HLA–A or HLA–B loci.

Recent study indicates that Y-chromosome also contains a histocompatibility gene which is known as H-Y.

# 29

# CANCER GENETICS

1. Introduction, 2. Classification, 3. Characteristics, 4. Cause, 5. Mechanism of action of radiation, 6. Chemical compounds, 7. Mechanism of action, 8. Oncogenes, 9. Conversion of proto-oncogen to cellular oncogenes, 10. Comparison—viral oncogenes, 11. Function of oncogenes.

## Introduction

Cell divisions is a very complex process which is controlled by a regulatory mechanism at both molecular and cellular level.

In non-circulatory tissues, such uncontrolled cell growth produces cell mass called *tumors* or *neoplasm.*

Tumors may be:

I. Benign,

II. Malignant tumor or cancer or carcinoma.

**I. Benign Tumors:** The characteristics of benign tumors (e.g. skin wart) are:

1. Remains confined to its original site.
2. Do not invade the surrounding normal tissues.
3. Do not spread to distant parts of the body.
4. Tumor cells closely resemble normal cells may function like normal cells.
5. Generally harmless but may be harmful when these are located in brain, liver and turn to malignancy.

**II. Malignant tumor or Cancer or Carcinoma**

**What is Cancer?**

Cancer exhibits uncontrolled cell growth and divisions without differentiation.

**Characteristics**

1. Does not remain confined to its original site.
2. Capable of invading the surrounding normal tissues.
3. Spreads to other sites via the circulatory or lymphatic system and this process known as *metastasis.*
4. Harmful and life-threatening.

**Characteristics of Cancer cells**

1. **Uncontrolled cell division:** The process of cell change in which a normal cell loses its ability to control its rate of division and thus becomes a tumor cell known as *cell transformation.* In cancer cells there is uncontrolled division of cells.

2. **Survival:** Human cell in culture dies after about 50 generations. But cancer cells can remain in good condition and survive generation after generation indefinitely.

3. **Invasiveness:** Cancer cells have the ability to invade other tissues for the following reasons:
   (a) Cancer cells generally secrete *proteases* which digest extra-cellular matrix components.
   (b) Secretion of collagenase, digests and penetrate through basal laminal to invade the underlying connective tissue.
   (c) Cancer cells also secrete *growth factors* that promote the formation of new blood vessels known as *angiogenesis* to supply oxygen and nutrition, and provide the opportunity to enter the circulatory system for metastasis.

4. **Loss of contact inhibition:** In tissue culture, normal cells while growing tend to make cell contacts by adhesions to neighbouring cells. At adhesion points some kind of electron dense plaque is formed in both contacting cells. There is slowing down of the amoeboid process which results in contact inhibition of movement.

   Whereas cancer cells are unable to form adhesive junctions and do not show contact inhibition.

5. **Cell layer:** In culture, normal cells are surrounded by other cells, mobility is stopped and form a *monolayer,* inhibition of growth, number of cells remain constant. But cancer cells continue to multiply and form irregular masses *several layers deep.*

6. **Changes in cell membrane:** Cancer cells undergo a change in the property of cell membranes and cell coat, e.g. disappearance of gap junction, loss of coupling changes in glycolipid and glycoprotein and reduction in gangliosides. And these changes enable the cells to dissociate from neighbouring cells and show loss of contact inhibition.

7. **Reduction of cellular adhesion:** Cancer cells are less adhesive than the normal cells due to reduced expression of cell surface adhesive molecules. But when normal cells are transformed into cancer cells there is a change in stickiness of their cell membranes.

8. **Specificity of adhesiveness:** Adhesiveness normal cell shows specificity, e.g. kidney cell tends to stick with another kidney cell but never with liver cell, whereas cancer cell can mix and stick to any type of normal cells, e.g. malignant liver cell can mix and stick to normal kidney cell.

9. **No differentiation:** Cancer cells cannot differentiate. Normal cells are fully differentiated and that time cell division ceases.

10. **Autostimulation of cell division:** Cancer cells produce growth factor that stimulates their own cell division and leads to continuous auto-stimulation of cell division known as *autocrine growth stimulation.* Hence, cancer cells are less dependent on general growth factor produced physiologically due to abnormalities in intracellular signalling system.

11. **Microscopical cellular changes:** Cancer cell can be distinguished from normal cell under microscope. Such as cancer cell shows the following:
    (a) High nucleus to cytoplasm ratio;
    (b) Prominent nucleoli;
    (c) Many mitosis,
    (d) Relatively little specialised structure,
    (e) Cytoskeleton—Depolymerisation,
    (f) Microtubules—Disaggregate.

12. **Density-dependent inhibition:** Normal cells show density-dependent inhibition of cell division in culture, whereas cancer cells continue to proliferate independent of cell density.
13. **Programmed cell death—apoptosis:** Apoptosis means programmed cell death. Cancer cells fail to undergo apoptosis which leads to tumor development. The survival of many normal cells is dependent on signals from growth factors or from the extracellular matrix that inhibits apoptosis, but undergo apoptosis following DNA damage.
14. **Chromosomal damage:** Human normal cells having 46 chromosomes but in cancer cell the chromosomes can undergo both structural and numerical changes. Series of abnormal divisions, the cancer cells contain series of chromosome number and Karyotype. Chromosomes swell up and the number of chromosome sets increase due to the growth of cancer cell and the condition is known as *aneuploidy*.
15. **Cancer cell and interaction with immune system:** Transformation of normal cell to precancer cell occurs everyday in our life due to radiation, chemical, carcinogens and certain viruses. These abnormal cells, have some surface proteins differ from normal cells. These proteins act as antigens and stimulate an immune response which generally destroy these abnormal precancer cells. If the surface proteins are not changed, the immune system may fail to recognize and these cells remain antigenetically similar to normal cells and sometimes to recognize but unable to destroy them.

- **Cause of cancer**

The following are the agents causing cancer:

I. Radiation compounds,

II. Chemical,

III. Viruses.

These have been found to induce cancer in both humans and experimental animals. Agents causing cancers are called *carcinogens.*

### I. Radiation

The ionizing radiations having biological importance and absorbed by the tissues are the following: Alpha particles, beta particles, gamma rays, X-rays, proton, deuterons, heavy nuclei, neutrons, negative pimesons (pions).

These are called ionizing radiations because the main way of energy dissipation is the ejection of electrons from the atoms of interacting materials.

Radiation and chemical carcinogens act by (1) damaging DNA and (2) inducing somatic mutations. These carcinogens are generally called as *initiating agent* because induction of mutations in target genes is supposed to be the initial event leading to cancer development.

Other carcinogens induce the cancer development by stimulating cell proliferation and the compounds are called *promoters.*

- **Mechanism of Action of Radiation**

1. Ultraviolet rays and $\gamma$-rays damage DNA by disruption of linear arrangement of bases and the damaging effects are:
   (a) Inhibition of DNA synthesis
   (b) Synthesis of defective DNA
   (c) Inhibition of mitosis
   (d) Synthesis of defective protein.
2. Radiation (e.g. ultraviolet) may cause formation of pyrimidine dimers.
3. Apurinic or apyrimidinic sites may form by elimination of corresponding bases.
4. Single- and double-strand breaks or cross-linking of strands may occur.
5. Free radicals are formed in the tissues by radiation and the resultant OH, superoxide etc. may interact with DNA

which leads to molecular damage producing carcinogenic effects.

**II. Chemical compounds**

**Source**

1. Environmental chemicals
2. Exposure due to occupation, e.g. benzene, asbestos
3. Diet—Aflatoxin $B_1$ (from mold Aspergillus flavus)
4. Habit—Smoking

**Chemical compounds**

These may be:

1. Organic, 2. Inorganic.

| Chemical compound (Carcinogen) | Site of induced cancer |
|---|---|
| **A. ORGANIC** | |
| 1. Acrylonitrile | Lung, colon |
| 2. 4-Aminodiphenyl | Urinary bladder |
| 3. Aniline derivatives | Do |
| 4. Benzene | Leukemia |
| 5. Diethylstilbesterol (DES) | Uterus, Vagina |
| 6. O-Naphthylamine | Urinary bladder |
| 7. Vinyl chloride | Liver, lung, brain |
| **B. INORGANIC** | |
| 1. Arsenic compounds | Lung, skin |
| 2. Astestos | Lung |
| 3. Cadmium salts | Prostate, lung |
| 4. Chromium and chromates | Lung, nasal sinuses |
| 5. Lead | Kidney |
| 6. Nickel | Lung, nose |

- **Mechanism of Action**

Some carcinogens induce some normal cells to become irreversibly altered to a preneoplastic state, the process is known as *initiation* and the carcinogens are known as *initiating agents.*

Methylcholanthrene rarely cause the development of tumor but subsequent application of castor oil induces the formation of multiple tumors. Here methylcholanthrene is an *initiator* and castor oil is *promoter*.

Initiation is a quick, irreversible process that causes a permanent changes in the DNA. The carcinogenic chemicals acting as initiating agents, are capable to bind with DNA and induce somatic mutation and consequently produces a stable, inheritable changes in the cell's properties.

On the basis of action of chemical carcinogens on DNA, they are grouped into two:

1. **Direct carcinogens:** These compounds are highly *electrophilic* (molecule deficient of electron) and react with DNA.
2. **Indirect carcinogens:** These are converted metabollically (procarcinogens) to become carcinogenic by introduction of *electrophilic centres.* This readily attacks *neucleophilic (electron-rich)* group in DNA, RNA and proteins. This is enzyme-catalyzed reaction for conversion is called *metabolic activation.* The intermediate product is called *proximate carcinogen* and final product is called *ultimate carcinogen.*

   Procarcinogen → Proximate carcinogen A → Proximate carcinogen B → Ultimate carcinogen.

   Most of the *chemical carcinogens are mutagens.*

**III. Virus:** There are many viruses which are capable of causing tumor in animals, human and plants. These virues are called *tumor viruses* or *on covirus.* Tumor viruses are grouped into two:

1. *DNA tumor virus*—Viruses having RNA genome,
2. *Retrovirus*—Viruses having DNA genome.

## Oncogenes

- **Definition**

Oncogenes is a type of *specific viral gene* that is capable of *inducing cancer* or *cell*

**Table 29.1:** Tumor viruses

| Class | Example | Tumor induced |
|---|---|---|
| **DNA Viruses** | | |
| 1. Herpes viruses | Lucke virus | Kidney adeno-carcinoma |
| 2. Papoviruses | Marek's disease virus | Lymphoma |
| 3. Hepatitis B virus | | Liver cancer |
| 4. Adenovirus | Human adenoviruses | Subcutaneous, intraperitoneal, intracranial |
| **RNA Viruses** | | |
| 1. B-type viruses | Bittner mammary tumor virus | Mammary carcinoma |
| 2. C-type viruses | Rous sarcoma virus | Sarcoma |

*transformation* either in the body of host or in the culture (viral oncogenes) and the virus is called *oncogenic virus*.

## Cell Transformation

The cells in culture undergoes malignant transformation when infected with certain oncogenic viruses. The changes are in cell shape, motility, growth and a number of biochemical processes which reflect the conversion from the normal cell to the malignant cell and the process is known as *transformation*. This does not indicate the same process will occur *in vivo*.

### • Occurrence of oncogenes

After identification of *src* oncogenes in RSV, more than 40 different oncogenic retroviruses have been isolated from rat, cat, chickens, mice, monkeys, etc. These viruses contain at least one and in some cases two oncogenes. These oncogenes are not required for viral replication but is needed for cell transformation. In some cases, different viruses contain same oncogenes. Many of these genes encodes protein which in turn acts as the main components of signalling mechanism that induces cell transformation.

## Proto-oncogenes

Using nucleic acid hybridization techniques it has been observed that DNA sequence which are *homologous* but *not identical* with the Rous *src* gene can be detected in the *genome of normal cells* of different organisms including humans. The term *proto-oncogenes* has been introduced to refer these normal cellular genes that closely resemble oncogenes.

Two different studies such as:

1. Observations on cellular proto-oncogenes homologous to the oncogenes of animal viruses.
2. Studies on cancer-causing genes in the genome of cancer cells by transfection experiment. DNA of tumor cells is isolated and added to normal tissue culture cells and converting them to cancerous cells. These observations have revealed the involvement of 15 different normal cellular genes—the proto-oncogenes producing certain types of cancer in animals.

### • Conversion of proto-oncogenes to cellular oncogenes

The normal cellular proto-oncogenes can be converted into tumor-causing cellular oncogenes by at least 5 mechanisms are given below:

1. **Point Mutation:** This involves a single base pair substitution or point mutation.

2. **Local DNA Rearrangement:** DNA rearrangements cause either deletion or base sequence exchange between proto-oncogene and surrounding genes.
3. **Insertional Mutagenesis:** Some cancer-causing retroviruses lack oncogenes. These viruses cause cancer by integrating DNA copy of their genetic information into a host chromosome in a region where a proto-oncogene is located. Thus disrupting the structure of the host proto-oncogenes and thereby converting it into an oncogene.
4. **Gene Amplification:** Gene amplification requires to increase the number of copies of a particular protogene. The excess production of copies of a particular proto-oncogene leads to malignant transformation.
5. **Chromosomal Translocation:** By this process, a portion of one chromosome is broken and joined to another chromosome. Then the broken segment containing proto-oncogene is transferred from its normal location to a new location where it is converted to oncogene.
6. Other oncogenes encode products that are identical to growth hormone as well as hormone receptors.
7. Protein tyrosine kinase (oncogene encoded) is a transmembrane receptor transmits a perfect signal for cell division. Alteration in the structure of this enzyme transmits wrong signal for cell division when it normally should not divide and leads to tumor formation.
8. Some oncogenes *code for proteins* that function within the nucleus particularly in the *regulation of gene transcription.*

- **Causes of oncogenicity of oncogene**

1. Oncogenicity was found as a result from a *single base-pair substitution.*

**Table 29.2:** Difference between virus oncogenes and cellular proto-oncogenes

| Viral oncogenes | Cellular proto-oncogenes |
|---|---|
| 1. **Source**<br>Some of the retroviruses contain a gene called on-oncogenes. | 1. **Source**<br>Set of homologous genes in the genomes of normal cells (of normal animals and humans) called proto-oncogenes. |
| 2. **No. of oncogenes**<br>At least 15 different viral oncogenes found in different retroviruses that induce similar type of cancer often carry the same oncogenes. | 2. **No. of oncogenes**<br>In human chromosomes there are at least 9 oncogenes are present and one in each chromosome No. 6, 8, 9, 11, 12, 15, 20 and 22 and 15 oncogenes in animals. |
| 3. **Conversion**<br>Viruses store their genetic information in single-stranded RNA genome and then convert it into an homologous double-stranded DNA form after infecting a host cell. | 3. **Conversion**<br>The normal cellular proto-oncogenes can be converted intotum or-causing cellular oncogenes by 5 mechanisms:<br>a. Point mutation<br>b. Local DNA rearrangement<br>c. Insertional mutagenesis<br>d. Gene amplification<br>e. Chromosomal translocation |
| 4. **Homologous sequence**<br>When viral oncogenes are cloned by recombinant DNA techniques and are used as hybridization probes to search for homologous sequences in normal host cells, such sequences are almost always found. | 4. **Homologous sequence**<br>Such sequences are almost always found. |
| 5. **Coding sequence**<br>Integrated | 5. **Coding sequence**<br>Interrupted coding sequences like eukaryotic genes. |

*contd.*

| Viral oncogenes | Cellular proto-oncogenes |
|---|---|
| 6. **Exons and introns** Single exons | 6. **Exons and introns:** Have multiple exons separated by introns. |
| 7. **Coded protein** Protein coded (Protein kinase) has similar size and structure like cellular oncogenes. | 7. **Coded protein:** Same |

This was evident when oncogene from human bladder carcinoma cells was sequenced, and its sequence was compared with the sequence of the homologous proto-oncogene.

2. Production of oncogene from its proto-oncogene by *GC-TA transversion* and this mutation results in the *substitution of valine for the glycine* present as 12th amino acid in the normal protein kinase.
3. Oncogene causes a predisposition, rather than an intermediate changes to the malignancy state. It has been explained that chromosome breaks and rearrangements may cause alteration of expression of proto-oncogenes or cellular oncogenes.
4. Regarding induction of cancers by oncogenes, it can be stated that
   (a) Retroviral oncogenes have evolved from normal cellular genes.
   (b) Cancerous state is the end-product of multi-step process.

- **Retroviruses or RNA tumor viruses**

The name retroviruses has been derived from the fact these viruses store their genetic information in a single-stranded RNA genome and then convert it into an homologous double-stranded DNA form after infecting a host cell. Thus, there is a backward flow (retro) of genetic information by a special enzyme reverse transcriptase.

Retrovirus replicate via DNA provirus in the infected cells.

- **Functions of oncogenes**

1. Oncogenes that code for proteins are either abnormal in structure or inappropriate amounts which drives the *uncontrolled proliferation* of the cancer cells, whereas normal genes encoded proteins which regulate *normal cell proliferation.*
2. Some oncogenes products are involved for defective differentiation and failure to undergo programmed death in case of cancer cells.
3. Majority of oncogene proteins function as elements of signalling pathways which regulate cell proliferation in response to growth factor stimulation. These oncogene proteins are:
   (a) Polypeptide growth factors,
   (b) Growth factor receptors,
   (c) Elements of intra-cellular signalling pathway,
   (d) Transcriptional factors.
4. Product of some oncogenes is closely related to a polypeptide growth hormone called *platelet-derived* growth factor (PDGF) produced by platelets promote wound healing by stimulating cell growth and transform fibroblasts to tumor cells in culture.

# PART-VI

# IMMUNOGENETICS

# 30

# IMMUNOGENETICS

1. Immunoglobulin-definitions, Source, Chemistry, Molecular structure, Comparison of different immunoglobulins, 2. Structure of IgG, 3. Immunoglobulins allotypes, 4. Antibody diversity, 5. Specificity of antibody genetically, 6. Agammaglobulinemia, 7. Genetics and immunity, Genetic control of immuno response, 8. Regulation of immune responses, 9. Immunodeficiency, 10. Inherited immunodeficiency disorders, 11. Rejection of grafts.

## Immunoglobulins

### • Definition

Electrophoretic techniques, ion-exchange chromatography and ultracentrifugal analysis have shown that *antibodies* are *located* in the *serum γ-globulins* which are therefore called *immunoglobulins* (the globulins related to the immunity processes). It is formed in the body due to entrance of foreign proteins. Specific antibodies are formed and reacts against specific antigen.

### • Source

Apart from serum, immunoglobulins are found (in varying proportions) in extravascular fluids, exocrine secretions (saliva, intestinal juices, colostrum) and the surface of some lymphocytes.

### • Chemistry

The immunoglobulins are *heterogeneous group of proteins* accounting for about 20% of the plasma proteins. These are *glycoproteins* composed of 82–96% polypeptides and 18–5% of carbohydrates. The *polypeptide component possesses almost all* the biological *activities possessed by the antibody molecules.* They are the *slowest moving* serum proteins *on electrophoresis* and majority of the immunoglobulins migrate to the zone designated for *γ-globulins* but significant amounts are also found in the *β-globulin* zone. The molecular weight of the immunoglobulins vary between 150000 to 900000 and their isoelectric pH vary between 6.3 to 7.3.

### • Molecular structure

Enormous advances have been made regarding knowledge of chemical structure of immunoglobulins and the relationship of the *structure* to the *biological activities.*

Immunoglobulins or serum antibodies contains *one basic unit* (domain) or monomer comprising of *4 polypeptide chains covalently linked* by *disulphide* (—S—S) bonds mostly within the chains and few between chains. Within this framework the total immunoglobulin population shows heterogenecity related to diversity of function of cellular origin.

The immunoglobulins consists *of two types of polypeptide chains*, i.e. *light chain* (L) and *heavy chain* (H), and most of them are having *two light* and *two heavy chains.* Some of the immunoglobulins, though have combinations of greater than two heavy and two light chains which give rise to the much larger molecular weight. Yet, in all of these, *each heavy chain is paralleled by a light chain at one of its ends,* thus forming a *heavy-light chain pair* and there are *always at least two such pairs* in each immunoglobulin molecule. The *chains can be separated from each other* by *chemical reduction of S—S bonds.* In IgG the heavy chain has a molecular weight of 50000 and the light chain a molecular weight of 20000. The IgG molecule with 2 heavy +2 light chains, therefore has a molecular weight of 140000. High molecular weight immunoglobulins (e.g. IgM) consists of polymers of the four chain units. Apart from H and L chains, there *are other two chains:*

(a) *T-chain* (found in IgA and IgM): It is a small glycopeptide with a high content of aspartic and glutamic acid, and having mol. wt. 25000. It helps to join together the monomeric units of the molecule.

(b) *Secretory component*: (found in IgA): It is present in dimeric immunoglobulin of tear or saliva. It is a single polypeptide chain, glycoprote is of mol. wt. 58000.

**Variable Region or Portion** (Amino terminal portion): Figure shows the designated end of each of the light and each of the heavy chains (amino terminal region) called the *variable region* or portion. This part is different for each specificity of antibody and it is *this portion* that *allows the antibody to attach specifically to a particular type of antigen.*

**Invariable or Constant Region or Portion:** The remainder (carbohydrate terminal portion) of each chain is called constant region. It—

(a) determines the gross physical and chemical properties of the antibody,

(b) establishes such factors as mobility of the antibody in the tissues,

(c) adheres of the antibody to specific structures within the tissues,

(d) attachment to complement,

(e) the ease with which the antibodies pass through membrane (e.g. placental transfer),

(f) other biological properties of the antibody..

The terms of variable and constant region indicate the comparative *variability of the amino acid sequence.* Each antibody that is specific for a particular antigen has a *different organization of amino acid residue in the variable portions of both the light and heavy chains.*

**Switch Region:** The zone where the variable and constant region meet is called the switch region.

**Hinge Region:** In all except IgM, the area of the H chain of a particular type has several regions of constant amino acid sequence ($CH_1$, $CH_2$, $CH_3$) or between Fd and Fc portion and $CH_2$ a stretch of 15 amino acids which is more flexible region and is exposed to enzymes and chemicals.

All L chains have a mol. wt. of 23000 but can be classified into two types, "κ" (kappa) and "λ" (lambda) on the basis of multiple structural differences in the constant region which are reflected in antigenic differences. A given immunoglobulin molecule always contains either "κ" or "λ", never a mixture of the two. The ratio of 'κ' : 'λ' in immunoglobulin molecules varies from species to species and is 2 : 1 in humans (*may be designated* K and L).

Five classes of H chains have been found in the humans—they are designated as γ (in IgG), α(in IgA), μ (in IgM), δ (in IgD) and Σ (in IgE). The different classes of H chains vary in mol. wt between 50–70000. The γ type is divided into 4 subclasses ($\gamma_1$, $\gamma_2$, $\gamma_3$, $\gamma_4$), a into 2 ($\alpha_1$, $\alpha_2$) and μ into 2($\mu_1$, $\mu_2$)..

Each L chain of a particular type has a region of constant sequence of amino acids ($C_L$) and a region of variable sequence of amino acids ($V_L$) in its peptide chain, all human immunoglobulin molecules have identical $C_L$ in the L chains. Each H chain of a particular type has several regions of constant amino acid sequences ($CH_1$, $CH_2$, $CH_3$, etc.) and a region of variable amino acid sequence ($V_H$), the latter being similar to $V_L$ in length.

Papain (a proteolytic enzyme) digestion of Ig molecule causes splitting at the hinge region producing two Fab fragments (ab = antibody, F = fragment) which includes Fd region of H chain plus L chain. These are the sites for antigen-antibody combination and FC fragment (C = crystallizable, F= fragment) which is concerned with:

(a) Skin fixation
(b) Placental transfer
(c) Complement binding
(d) Reactivity with rheumatoid factor
(e) Macrophage fixation
(f) Regulation of catabolism

Examination of *crystallized immunoglobulins* or their parts of X-ray diffraction analysis have shown that all domain have a characteristic pattern of folding, regardless of their origin thus V and C region domains from L or H chains have a very similar appearance. In addition there is a close approximation between corresponding domains, i.e., $V_H$ and $V_L$, $C_L$ and $C_H$ and the identical H chain domains. The hypothetical model of IgG immunoglobulin illustrated in Fig. 30.2.

- **Classification**

Different classes and subclasses of immunoglobulin differ in structure though they are composed of the basic 4-chain unit. All classes have similar L chain associated with properties common to all immunoglobulins. H chains of different classes have distinct physicochemical properties. Retaining the biological properties of different fragments after chemical cleavage by papain or pepsin digestion, they are classified into five classes, i.e. IgG, IgM, IgA, IgE and IgD based on marked antigenic differences in the heavy chain. Immunoglobulin is designated by the symbol Ig and followed by the capital letter signifies the class to which it belongs according to the antigenic differences in the H chains. Two of these classes of antibodies are of particular importance; IgG which comprises about 75% of the antibodies of the normal person and IgE which constitutes only a small per cent of the antibodies but which is especially involved in allergy.

Subclass: $IgG = IgG_1, IgG_2, IgG_3, IgG_4$
$IgA = IgA_1, IgA_2$
$IgM = IgM_1, IgM_2$.

- **Antigenic Classification**

1. Individuals can be subdivided on the basis of certain characteristics of their Ig molecules which is comparable to the blood groups. It depends on the antigenic differences in the L and H chains of Ig molecule. The genetically determined sub-groups are known as *allotypes*. Three main groups *Gm, Km* and *Am* have been found and each can be subdivided.
2. Fab fragments having having different amino acid sequences in the variable portion of L and H chains show different specificity have different antigenic determinants. These are called idiotypes.
3. Isotypes also exist and are due to antigenic determinants on all molecules of a certain class or subclass of H chains or of certain L chains.
4. **Allotypes:** This depends upon the existence of allelic forms (encoded by alleles or alternative genes at a single locus) like RBC which is genetically differ by blood groups.
5. **Idiotypes:** Idiotypic determinants unique to a given immunoglobulin are recognizable by anti-idiotypic antibodies and are associated with the hypervariable regions forming the antigen binding site.

**Functions:**

IgG = 1. Can cross the placental barrier and responsible for protection of the newborn.
2. Fixes serum complement.
3. Arming the macrophages.
4. Serves as precipitins, antitoxin and late antibodies.

IgA = Primary defence mechanism against local infection.

IgM = 1. Prominent in early immune response.
2. Predominates in responses of blood group antibodies.
3. Most efficient complement fixing immunoglobulin.
4. Adheres on the surface of β-cell

IgD = It is affective against antigens like insulin, penicillin, milk protein, diphtheria toxoid, nuclear antigens and thyroid antigens.

IgE = 1. Triggers the release of mast cells mediators responsible for wheal and flare skin reactions.
2. Binds the allergin.

IgT = 1. It acts as lymphokinase stimulating B-lymphocytes maturation and 2. antibody production.

Papain inhibits at the hinge region into two Fab and one Fc. Fab is concerned with antigen + antibody combining sites and Fc is concerned with placental transfer and complement binding. It is Fab composed of two heavy polypeptide chains and two light polypeptide chains. The antigen binds at two different sites on the "variable portions" of the chains (i.e. designated end of each of the light and each of the heavy chains).

## Immunoglobulins Allotypes

The five classes of immunoglobulins present in all normal persons but genetically determined variants or *immunoglobulin allotypes* of these have also been identified. These are *Gm* system associated with the heavy chain of IgG, the *Am* system associated with the IgA heavy chain, the *Km* and *Inv* systems associated with the κ light chain, *OZ* system for the λ light chain and the *Em* allotype for the IgE heavy chain. The *Gm* and *Km* systems are independent of each other and are polymorphic, the frequencies of the different alleles varying in different ethnic groups.

**Table 30.1:** Human immunoglobulins

| Immuno-globulins | Serum concen-tration (mg/100 ml) | Molecular weight | Half life (days) | Light chain | Heavy chain | Characteristic properties |
|---|---|---|---|---|---|---|
| IgG | 900-1800 | 160,000 | 18-23 | K and L | γ | Precipitin antitoxin complement fixation Late antibody |
| IgA | 156-294 | 170,000 and polymers | 5-6.5 | K and L | α | Surface protection |
| IgM | 67-145 | 960,000 | 5 | K and L | μ | Agglutinins opsonins dysins, complement fixation Early antibody |
| IgD | 0.3-40 | 184,000 | 2.8 | K and L | δ | Not known |
| IgE | 10-130 (μg) | 188,000 | 2.3 | K and L | ε | Reaginic antibody |

Male—Lower level of IgG and IgM, higher IgA

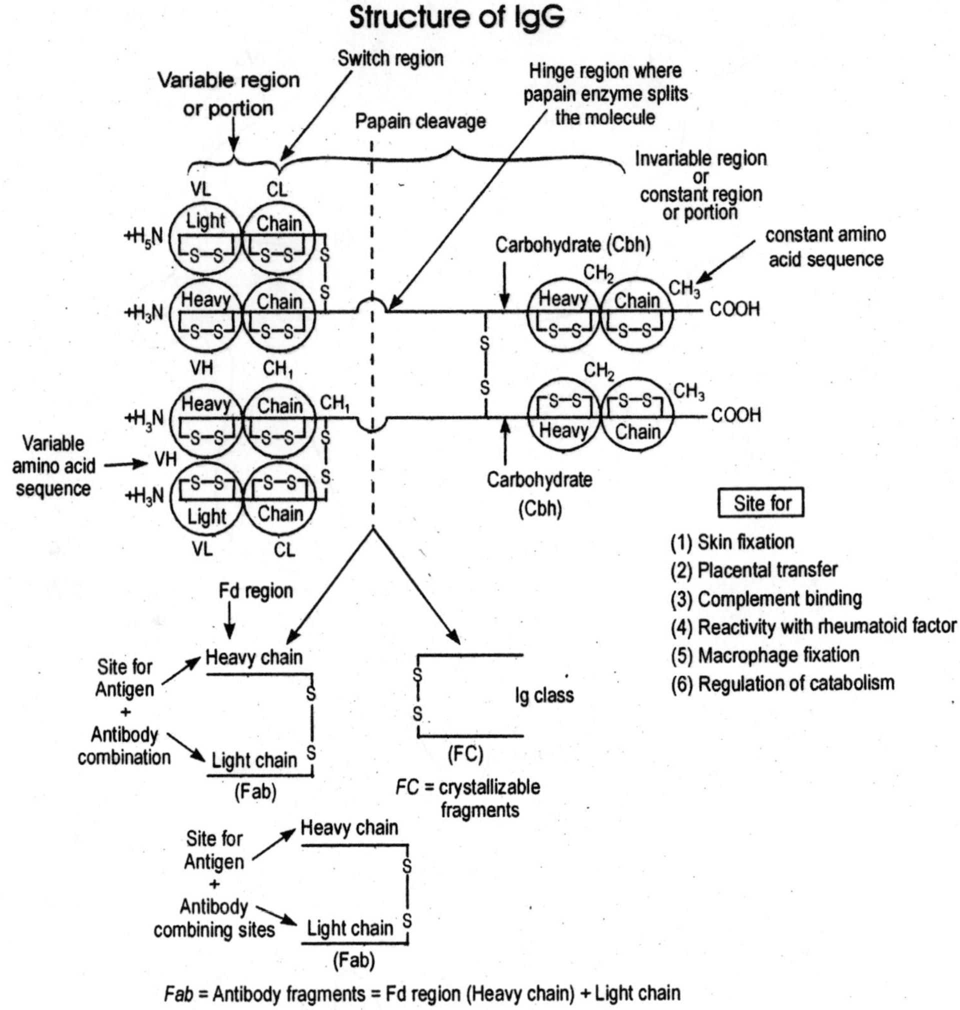


**Fig. 30.1:** Structure of IgG antibody or immunoglobulins (papain at the hinge region into two Fab and one Fc. Fab is concerned with antigen + antibody combing sites and Fc is concerned with placental transfer and complement binding. It is Fab composed of two heavy polypeptide chain and two light polypeptide chain. The antigen binds at two different sites on the "variable portions" of the chains (i.e., designated end of each of the light and each of the heavy chains).

## Antibody Diversity

A single protein molecule can exhibit structural heterogeneity to have specificity for a wide variety of antigens. Different combinations of heavy and light chains could, to some extent, account for this diversity. It would require, however, thousands of structural genes for each chain type to account for the large number of antibodies produced in response to the equally large number of antigens to which individuals can be exposed.

The patients with *multiple myeloma,* a cancer of plasma cells, produce antibody which is a *single antibody species* in large quantities. These patients excrete abundant quantities of this *single or monoclonal* protein in the urine called *Bence Jones Protein* consists of an *immunoglobulin light chain.* This protein has been compared with other different myeloma patients and revealed that *amino terminal ends* of the molecule to be quite *variable* in their amino acid sequence whilst the *carboxyl terminal*

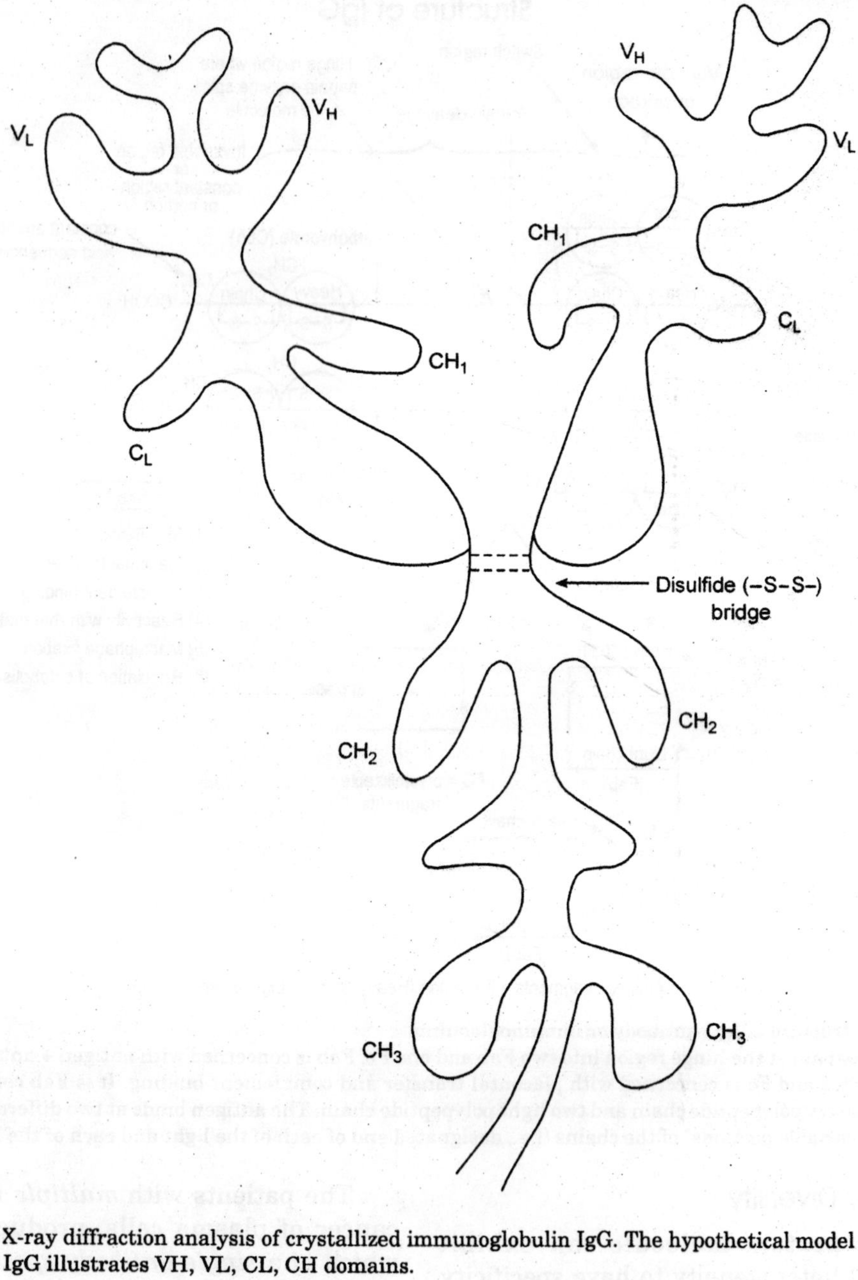


**Fig. 30.2:** X-ray diffraction analysis of crystallized immunoglobulin IgG. The hypothetical model of IgG illustrates VH, VL, CL, CH domains.

*ends* were relatively *constant*. These are known as:

(a) *Constant or C*, (b) *Variable or V*.

Further analysis of the amino acid sequence of the V regions of different myeloma proteins showed *four* regions:

1. **Framework regions:** Vary little from one antibody to another.
2. **Hypervariable regions:** Three markedly variable regions interspersed between the framework regions (I, II, III of Fig. 30.3).

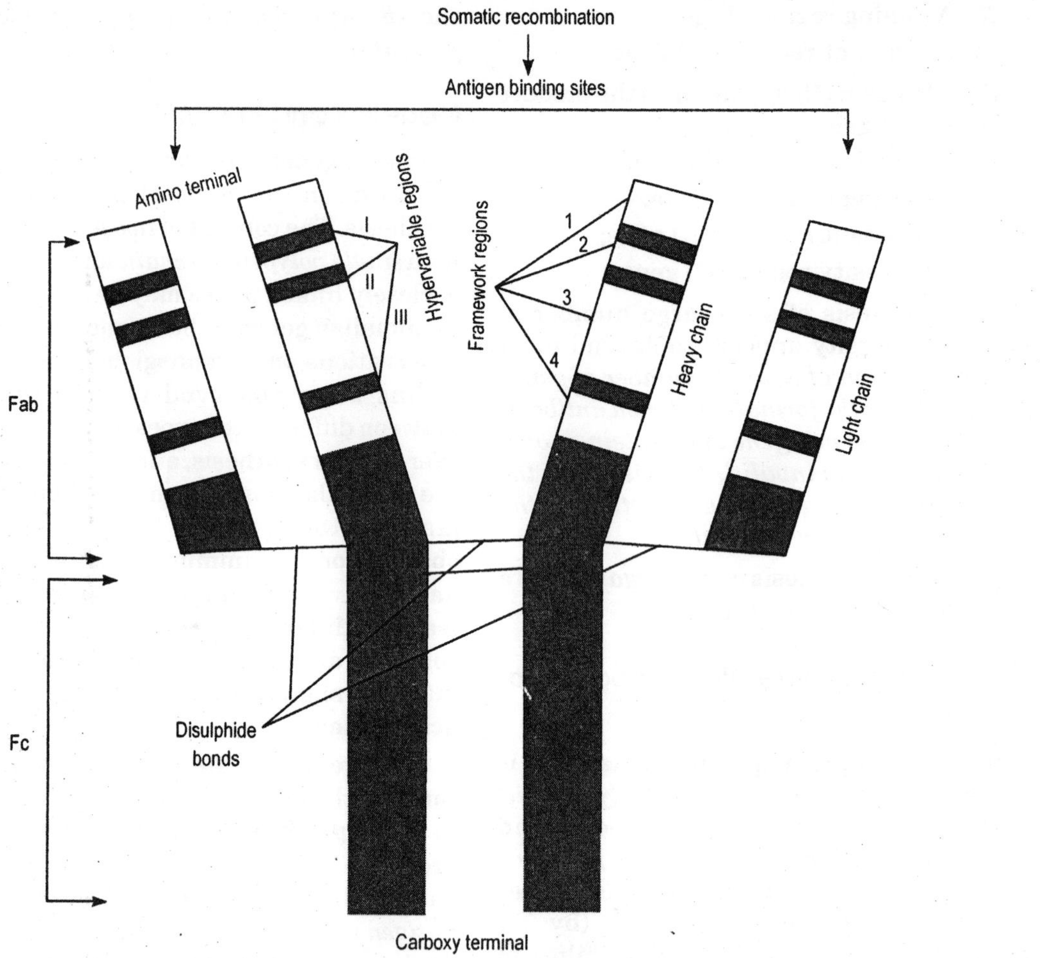


**Fig. 30.3:** Structure of antibody molecule of multiple myeloma patients.

- **Antibody Diversity and DNA**

Comparison of the restriction maps of the DNA segments coding for the C and V regions of the immunoglobulins λ light chain in embryonic and antibody-producing cells showed that they were far apart in the former whilst they were close together in the latter.

The immunoglobulin chains are produced from combinations of separate groups of DNA segments. These consist of one from a variable number of DNA segments coding for the constant (C). Variable (V) and joining (j) regions between the V and C regions for the κ and λ light chains and the various types of heavy chains. The heavy chains also contains a diversity (D) region located between the V and j regions. The total number of possible antibodies combinations of these DNA segments accounts for the antibody diversity seen in man.

## Specificity of Antibody Genetically

There are millions of antigens, hence millions of antibodies are required. How can such large number of antibody molecules be genetically specified? Does each individual possesses all the genes necessary to specify the millions of antibodies?

The L-chain (Light chain) is the product of at least 3 separate structural genes:

1. A variable region ($V_L$) gene;

2. A joining region (J) gene;
3. A constant region ($C_L$) gene.

Each H-chain (Heavy chain) is the product of 4 different genes:

1. A variable region ($V_H$) gene;
2. A joining region (J) gene;
3. A constant region ($C_H$) gene;
4. A diversity region (D) gene.

The synthesis of such large number of antibody diversity appears to depend upon *combined action of a small number of genes contributing to the formation of each antibody chain.* Thus the *one-gene one-protein or one-enzyme concept is invalid.* The *cistron is the genetic unit coding for the structure of the subunit of a protein molecule.*

Thus the hypothesis *can be regarded* as a *one-cistron one-subunit concept.*

## Agammaglobulinemia (Hypogammaglobulinemia)

In *congenital* agammaglobulinemia there is (1) *deficiency of gut-associated lymphoid tissue* (2) *absence of plasma cells* and (3) *marked reduction in serum gamma-globulin levels.* After the maternal antibodies have disappeared from the blood (by 3-4 months) there is a great susceptibility to bacterial infections (e.g. pneumonia, meningitis). On the other hand, the common virus infections of childhood, such as measles, mumps and chickenpox pursue a normal course leading to subsequent immunity and vaccination against smallpox is successful.

Burnet (1968) suggests that there are *two sites* in the body where *stem cells* from the bone marrow are *differentiated into immunocytes.* (1) These sites are the lymphoid tissue of the *thymus* where immunocytes responsible for delayed hypersensitivity and homograft immunity are produced and (2) the *lymphoid tissue of the gut* which forms immunocytes that give rise to plasma cells which in turn synthesize IgM, IgG and IgA antibodies. Thus in agammaglobulinemia the *gut-dependent immunocytes are missing,* but the *thymus-dependent immunocytes are present.*

## Genetics and Immunity

Studies of genetic control of protein synthesis has made in the study of immunoglobulin synthesis. The concept of *two genes* may code for a *single polypeptide chain* and that groups of closely linked genes may be basic units of mammalian genomes have emerged from the observations on immunoglobulin synthesis. It has been observed that interaction between different cell types may exert control over protein synthesis, affecting for example the particular class of immunoglobulin being manufactured by a lymphocyte. The observations on immunoglobulin structure have shown that in the human there are at least 10 different types and that each type contains an enormous number of variants differing slightly from each other in amino acid sequence.

The gene complex of the mouse that controls the cell surface antigens important in transplantation (graft) rejection has proved to be an invaluable general model for the study of the *major transplantation antigen systems* of higher vertebrates. Recent studies have shown that small segment of chromosome 17 contains this gene complex has created considerable interest to geneticist and immunologist.

In addition to controlling cell surface antigens, there are closely linked genes that determine:

(a) Differences in level of antibody responses to many antigens;
(b) Differences in susceptibility to tumour viruses;
(c) A number of immunological traits.

This gene complex is now the most important segment of chromosome and it constitutes a very valuable model for:

(a) Sudies of gene evolution;
(b) Gene action;
(c) Gene organization.

**1. Genetic Control of Immune Response:** It has been shown that the—

(a) *Ability of an individual to respond to a particular antigen is under genetic control.* There appear to be two forms of this control:

1. Control of *general capacity* to produce an antibody response.
   Experimental evidences show a uniformly good or a uniformly poor antibody response to several antigens. The poor response seems to be due to the slower mitotic division time of the responding cells.
2. Other genes are able to control the immune response to a *particular antigen*, e.g. a synthetic polypeptide in mice.

(b) The genes controlling the *ability to produce an immune response* are called *immune response* or *Ir genes* and is closely related in their position on a chromosome (number 17 in the mouse) to the genes controlling certain surface antigens of cells. These surface antigens are the major antigenic determinants in graft rejection and are called histocompatibility antigens.

After the discovery of immune response genes and their close linkage with the *major histocompatibility antigen system* encouraged to a search for an association between susceptibility to a variety of:

(a) Neoplastic diseases;
(b) Autoimmune diseases;
(c) Infectious diseases

and the activity of genes.

**2. Cell Cooperation and Genetic Control of the Immune Response:**

A. Alloantisera obtained by immunization between different mouse strains have identified two major regions, each defined by a major genetic locus. These are the K and D regions, which are *important antigens ingraft rejection.* The K and D antigens are glycoprotein having MW 43,000 associated with $L_2$ microglobulin.
   The genes making up a given histocompatibility complex are termed as a *haplotype.*

B. The I region appears to determine certain surface antigens on T and B lymphocytes. Some of these antigens can be found on both T

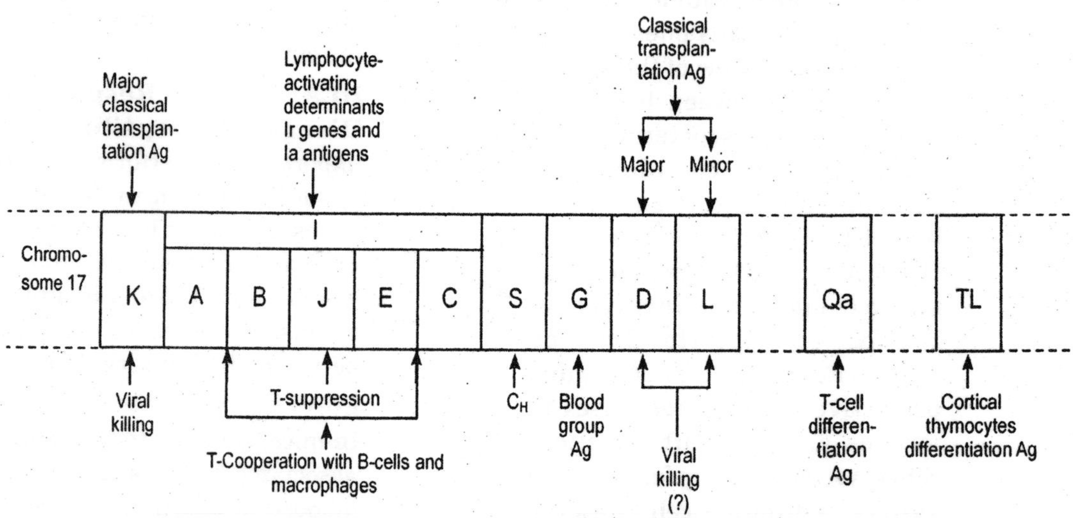


**Fig. 30.4:** The major histocompatibility gene complex in the mouse (H-2) chromosome 17. The sub regions of I are designated IA, IB, etc.

and B cells whilst other appear to be restricted to one cell type.

C. The I region contains genes which function as lymphocyte-activating determinants and code for membrane-associated antigens, the Ia antigens. Various roles have been found for the different I regions in the immune response. These antigens appear to play a crucial role in determining *antigen presentation* and *T cell-B cell interactions.*

D. **Susceptibility to virus infections and the role of the major histocompatibility antigens in cellular response:** There is genetic control of immune response by a specific group of genes called *Ir genes* and these have been shown to be closely related to those *genes controlling the major antigenic determinants of graft rejection (MHC genes)*. The possibility that a gene product closely associated with the MHC genes is involved in cooperation between T and B lymphocytes leading to an immune response has been seen. This phenomenon is of interest in view of the known relationship between histocompatibility antigens (determined by the MHC gene complex) and susceptibility to certain virus—induced leukaemias in mice. The H-2K and H-2D regions control surface molecules involved in cytotoxic T lymphocytes (Tc) (type 1) activities and thus determine protective immune responses to viruses as well graft rejection. Since viruses can infect a large variety of different cell types the ubiquitious distribution of H-2K and H-2D antigens enable Tc cells to deal with virus infections of all cell types.

E. **Transplantation Antigen System:** It has been observed that just as in the red cell antigenic system, some transplantation antigens were very 'strong' and others very weak. The strong transplantation or 'histocompatibility' antigens of the mouse are controlled by the H-2 genetic locus and *in man* by the *HLA locus.*

In man, two strong and several week histocompatibility systems have been found. HLA system (human lymphocyte antigen system), consisting of several hundred genes on the short arm of chromosome 6. HLA is the strong system with the ABH group system next in importance. The major loci are very complex, the antigens being determined by a chromosomal region rather than by a single genetic locus. HLA-A and HL-A-B of chromosome 6 are the human counterpart of the mouse H-2K and H-2D loci, and are the antigens most often used in immune typing.

F. **HLA, Immune Response and Disease Susceptibility:** The important role performed by HLA-associated immune response genes in relation to the expression of acquired immunity has been worked out. There has also been extensive studies for associations between the expression of these genes and the diseases.

In mice, some histocompatibility-associated Ir genes appear to predispose to autoimmune thyroiditis and susceptibility to viral meningitis infection. This suggests that

**Table 30.2:** Some diseases associated with HLA types

| Disease | HLA types |
|---|---|
| Ankylosing spondylitis<br>Reiter's syndrome<br>Anterior uveitis<br>Juvenile arthritis | B27 |
| Coeliac disease<br>Addison's disease<br>Chronic active hepatitis<br>Insulin-dependent diabetes | B8 |
| Thyrotoxicosis | B7 |
| Multiple sclerosis | DW2 |
| Rheumatoid arthritis | DW4<br>DRW4 |
| Ragweed allergy<br>Early onset type-1 diabetes | HLA identity<br>in siblings |

these genes, possibly expressed as receptors on lymphocytes, may be important in genetic susceptibility or resistance to autoimmune diseases, infection and neoplasia.

## Regulation of Immune Responses

The immune response is regulated by different control mechanisms. These help the immune system to restore to a resting stage when reaction to a given antigen is no longer necessary. The factors controlling immune response are as follows:

1. Regulation by antigen.
2. Regulation by antibody.
3. Regulation by immune-complexes.
4. Regulation by lymphocytes.
5. Idiotypic modulation of responses.
6. Neuroendocrine modulation of immune responses.
7. Genetic control of immune responses.
8. MHC (major histocompatibility complex)-linked immune response genes.
9. Non-MHC (major histocompatibility complex)-linked immune response genes.

1. **Regulation by Antigen:** The *main components* of the immune system are *T-cells* and *B-cells*. In *absence of antigen,* these cells remain *inactive* and immune response does not occur. The effective immune response starts when these cells are triggered by antigens. When the antigen activity is removed or destroyed then the cells return again to the initial stage. In this way the immune response is regulated.
2. **Regulation by Antibody:** Regulation by antibody is a form of feedback control on immune response. Passive administration of IgM antibody together with antigen enhances the immune response to antigen but IgG antibody suppresses the response and the mechanism of regulation by antibody is not known.
3. **Regulation by Immune-complexes:** Immune-complexes either inhibit or augment the immune-response. When FC receptor of B-cell is cross-linked to its antigen receptor by an antigen-antibody complex, a signal is delivered to the B-cell inhibiting it from entering

the antibody production phase. Antibody encourages presentation of antigen to B-cells when it is present on an antigen-presenting cell (APC) bound via FC receptors.

4. **Regulation by Lymphocytes:** T-lymphocytes in immune system stimulates either humoral or cell-mediated immunity. It has been observed that $CD_4^+$ T-cells are generated following administration of high doses of autoantigen and prevent further induction of autoimmunity. The exact mechanism of negative response of T-lymphocytes is not known. But the recent view is that T-lymphocytes produce cytokines which either partially or totally suppress an immune response.
5. **Idiotypic Modulation of Immune Response:** Idiotype means the antigenic characteristics of the antigen-binding site of an antibody. Antibodies formed against these antigen-binding sites are known as anti-idiotypic antibodies. They have the capacity to influence the outcome of an immune response. When an antibody response is triggered by antigen this antibody will in turn invoke an anti-idiotypic response to itself. There is good evidence that anti-idiotypic response can affect the representation of recognised idiotypes in an immune system.
6. **Neuroendocrine Modulation of Immune Response:** Nervous and endocrine system can regulate the immune response. Stressful condition may suppress the immune response due to corticoids. The nervous system directly or indirectly controls the output of various hormones (e.g. epinephrine) which in turn also modulate immune response.
7. **Genetic Control of Immune Responses (at the Cellular Level):** There are many ways in which genetic factors may play some role in influencing the immune response. Pedigree analysis and/or recording a family history of susceptibility to any infection will suggest that resistance or susceptibility may be an inherited characteristic. Again variation among individuals in respect to immune response is mainly due to combination of different gene or genes or due to lack of combination of gene or genes controlling the production of mediators of immunity.
8. **MHC-linked Immune Response Gene:** MHC-linked immune response *Ir genes* or *I regions* control the immune response. The activation of T-lymphocytes is induced by MHC molecules. During development T-lymphocytes are subjected to two selection processes in the thymus.
   (a) *Positive selection*: Based on an interaction of TCR (T-cell receptor) with MHC on *thymic cortical epithelium.*
   (b) *Negative selection:* As a result of a high effinity interaction between TCR (T-cell receptor) and MHC peptide presented on *bone marrow derived cells in the thymus medulla.*

   Thus both positive and negative selection of T-Cell affect the immune response. *MHC-linked genes* play a role in the *immune response to infections agents* and also *to self antigens* and to auto-immune disease (Fig. 30.5).
9. **Non-MHC linked Immune Response Genes:** The immune response is modulated by some *genes outside the MHC region genes* regulating the macrophages activity determine the outcome of many immune responses. Most autoimmune diseases have been shown to be influenced not only by MHC-linked gene, but also by *genes not linked to this complex.*

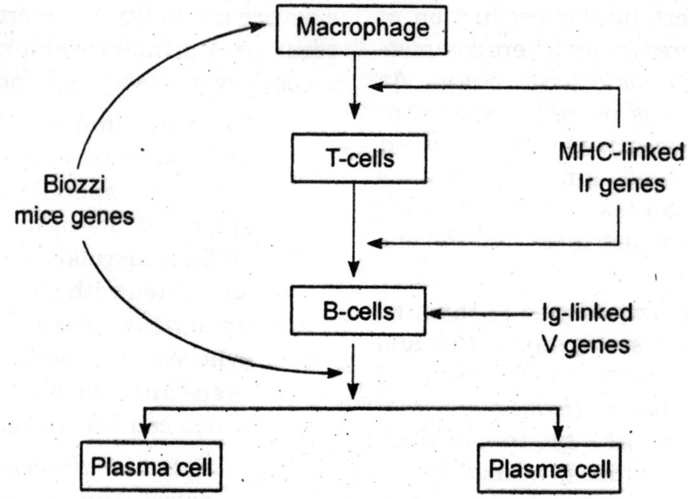


**Fig. 30.5:** Genetic control of antibody response.

[*MHC*—The cell surface antigens are known as *histocompatibility* antigens and the group of genes that determines the expression of these antigens as the *major histocompatibility complex (MHC)*. In the mouse these genes are present in chromosome number 17 and the genes control the immune response—referred to as the immune response (*Ir*) *genes* or *I region*.]

## Immunodeficiency

The condition is due to deficiency of lymphocyte-mediated immunity. There may be failure at any stage in the development and function of the lympho-reticular system and that these defects are associated with a diminution of resistance to microbial infection.

- **Immunodeficiency**

I. Defects in phagocytic function

A. *Quantitative deficiency* of granulocytes

(a) *Congenital:* Defects in granulocyte production, e.g. infantile agranulocytosis.

(b) *Acquired:* Replacement of bone marrow by tumour tissue or toxic chemicals.

B. *Qualitative deficiency* of some substance in granulocytes such as:

(a) Enzyme defect (NADPH) of neutrophils myeloperoxidase.

(b) Deficiency of glucose-6-phosphate dehydrogenase in granulocytes.

(c) Failure to respond chemotactic stimuli, e.g. lazy leucocyte syndrome and defect in lysosome, e.g., *Chediak-Higashi syndrome.* Normal phagocytosis present but fail to digest ingested microbes.

(d) In *chronic granulomatous disease,* there is defective production of $H_2O_2$.

II. Congenital defects of any:

(a) Affecting immunoglobulin synthesis.

(b) Cell-mediated immune mechanism.

(c) Both.

**The defect and the disease:**

1. Failure of stem cell production—*reticular dysgenesis.*
2. *Agammaglobulinemia,* in which there is deficient production of immunoglobulins. This may affect
   (a) All the immunoglobulin classes, or
   (b) May be restricted to a single class.

In severe *combined immuno deficiency,*

there is defect lies in production of *lymphocyte precursors* where *complete absence of lymphoid tissue* occurs. As a result there is no cell-mediated or humoral immunity, e.g., *X-linked* severe combined immunodeficiency known as *Swiss type agammaglobulinemia* which is treated by bone marrow graft.

3. *Deficiency of lymphocytes* in thymus-dependent areas of lymph nodes and spleen and there is *no cell-mediated immunity* occurs in *thymus hypoplasia*, e.g. *Di George's syndrome* which is treated by thymus grafting.
4. *X-linked hypogammaglobulinemia*. There is *absence of germinal centres* in lymph nodes and spleen and no humoral immunity, e.g. *Bruton type* which is treated by passive immunoglobulin.

## Inherited Immunodedeficiency Disorders

These are uncommon but are associated with a severe morbidity and mortality. They may occur as a primary isolated disorder or as secondary or associated finding.

A. **Primary Immunodeficiency Disorders:** These disorders occur due to defect at specific developmental stages of the immune system (Fig. 30.6). Abnormal cellular immunity is associated with increased susceptibility to virus infections. Deficiency of Ig synthesis is associated with reduced resistance to bacterial infections which can lead to death in infancy.

1. *Reticular dysgenesis:* The site of defect between stem cells of bone marrow, fetal liver or spleen and stem cells of leukocytes, i.e. No. 1 of Fig. 30.7. Absence of thymus gland, reduced lymphocyte count, reduced cell-mediated immunity, reduced plasma cells, reduced Ig synthesis are seen. Genetically autosomal recessive (?) Treatment is bone marrow transplantation.

## SUMMARY

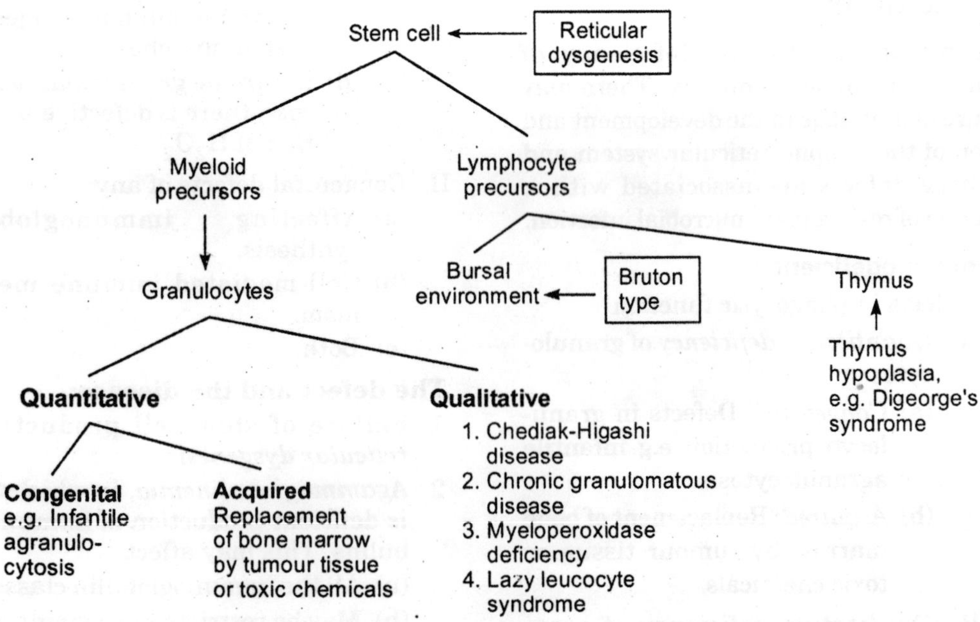


**Fig. 30.6:** Cellular basis of primary immunodeficiency

2. *Severe combined immuno-deficiency:* Defect lies between the stem cells of leucocyte and stem cells of lymphoid tissue (No. 2, Fig. 30.7). There is vestigial thymus gland, reduction of lymphocytes, reduction of cell-mediated immunity, reduction of plasma cells, reduction of Ig synthesis. Genetically autosomal recessive and X-linked recessive. Treatments are bone marrow transplant, enzyme replacement for adenosine deaminase deficiency.

3. *Bruton-type agammaglobulinemia:* Defect lies between stem cells of lymphoid tissue and Peyer's patches and related lymphoid tissue (No. 3, Fig. 30.7). It shows presence of thymus gland, lymphocytes, normal cell-mediated immunity, reduction of plasma cells and Ig synthesis. Genetically X-linked recessive. Treatment is Ig injections and antibiotics.

B. **Secondary (or Associated) immunodeficiency:**

1. *Di-George syndrome*: This occur in children and with the syndrome

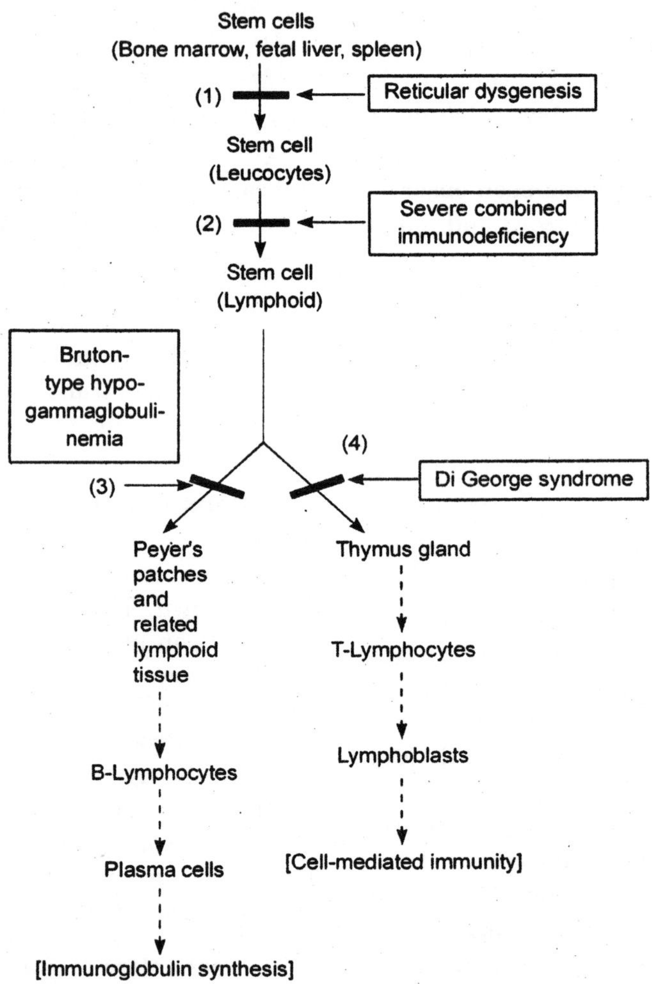


**Fig. 30.7:** Sites of blocks in certain immunological deficiency diseases.

show recurrent viral illness and abnormal cellular immunity. The defect lies between stem cells of lymphoid tissue and thymus gland (No. 4, Fig. 30.7).

This is characterized by absence of thymus gland (parathyroid absent as well), increased in lymphocytes, abnormal cellular immunity due to severely reduced or absent T-lymphocytes, reduction of cell-mediated immunity, increase in plasma cells, normal Ig synthesis. This has been found to be associated with a number of characteristics, congenital abnormalities such as congenital heart disease. There is tetany in the newborn due to low calcium for absence of parathyroid.

The syndrome has been observed to be part of the third and fourth pharyngeal pouches as a consequence of a deletion of a particular region of the long arm of chromosome 22.

2. *Ataxia telangiectasis:* This is an autosomal recessive disorder present in early childhood and characterized with difficulty in control of movement and balance (cerebellar ataxia), dilated blood vessels of the conjunctiva, ears and face (oculocutaneous telangiectasia) and susceptible to sinus and pulmonary infections, low serum IgA levels and hypoplastic thymus. Diagnosis is done by low or absent serum IgA and IgG, and characteristic chromosome abnormalities on culture of peripheral blood lymphocytes, so-called chromo-some instability. Person shows increased risk of developing leukemia or lymphoid malignancies.

3. *Wiskott-Aldrich syndrome:* This is an X-linked recessive disorder in which the affected boys have eczema, diarrhoea, recurrent infections, low platelet count, low serum IgM levels, possibility of death due to haemorrhage or B-cell malignancies.

## Rejection of Grafts

- **Definition**

The rejection phenomenon is a by-product of immune response. If living tissue is transplanted from one person to another, the *recipient* recognizes the *foreign antigens* in the *allograft-transplanted* tissues and destroy it. Destruction of a graft in this way is described as *rejection of grafts.*

- **Tissues for allograft rejection**

Avascular tissues, e.g. cornea and cartilage do not arouse the allograft reaction but grafts of tissues which possess a fibrous stroma and a blood supply are all liable to rejection. Some organs are better tolerated than others; a transplanted kidney arouses less reaction than an allograft skin. Successful organ transplantation can only be achieved by reducing the allograft reaction to a minimum.

- **Nature of allograft rejection mechanism**

Graft rejection is an *immune response* and *cell-mediated.* Though similar in principle, the rejection of grafts involves somewhat *different types* of immune response. In this case the *antibodies are cell-bound* and *not free in the blood plasma.*

Transplanted organs such as kidney and heart are rejected unless the *two individuals share exactly the same complement of antigens.* This will only be the case if they are *genetically identical*, i.e. *identical twins.*

- **Mechanism of graft rejection**

There are two main types of grafts rejection processes:

1. The so-called 'first set' response which occurs about 10 days after a first graft from an unrelated donor.
2. The 'second set' response occurring in about 7 days in an animal which had previously received a graft from the same unrelated donor.

It involves a *sequence of separate phenomena* developing from initial contact of the foreign antigens with the host immunological system through to rejection itself (Fig. 30.8).

The sequence of events includes:

Establishment of lymphatic drainage → Antigen release → Contact with cells of the immune system → proliferation of antibody-forming cells of the lymphoid tissues → The inflammatory response → Release of pharmacological mediators → Cytotoxic effects of lymphoid cells → and/or Antibody on the graft tissues.

- **Prevention of rejection** (Recipient's acceptance of foreign tissue) (Fig. 30.8).

Under the following circumstances the recipients can accept the donor tissue.

1. **Immunological tolerance** (*Introducing foreign tissue into the body*): The principle is the formation of antibody against the foreign tissue. One way of achieving this is to introduce some of the foreign tissue into the recipient before or very soon after birth. Medwar injected cells obtained from an adult mouse into a fetus. He found that after birth the recipient would accept, permanently, grafts of skin and other tissues from the donors. The recipient has been made *immunologically tolerant* to the donor's tissue.

   The maintenance of tolerance state requires the persistence of the antigen in the graft tissues but it does not occur. So tolerance persists indefinitely when the animal continues to be exposed indefinitely to the tolerizing antigen.

   The immunological tolerance is the result of a change in the lymphoid cell population.
2. **Re-equipping the recipient with lymphocytes:** Such induced tolerance was found to be abolished by re-equipping the recipient with lymphocytes from another individual of the same strain.
3. **Antilymphocytic globulin:** An antilymphocytic globulin (ALG) or serum prepared by injection of lymphocytes of one species into an animal of another species, prevents lymphocytes from participating in immune reaction. ALG has been used to suppress reactions to human allografts, e.g. kidney transplant.
4. **Histocompatibility antigens and tests:** Attempts have been made to detect histocompatibility antigens (HLA) so that the most suitable donor can be selected.

   The cell surface antigens are known as *histocompatibility antigens* and the *genetic region determines* the expression of these antigens termed the *major histocompatibility complex* (MHC) exerts a predominant influence on the survival of grafts.
5. **Identical (monozygotic) twins:** A transplanted organ can survive indefinitely if the donor and recipient are identical (monozygotic) twins.
6. **Immunosuppressive agents:**
   (a) X-ray irradiation prolonged the survival of allograft.
   (b) Drugs
      i. Steroids—Cortisone and related steroids;
      ii. Drugs inhibiting the growth of malignant cells.

## Summary of Rejection Mechanism and Their Prevention

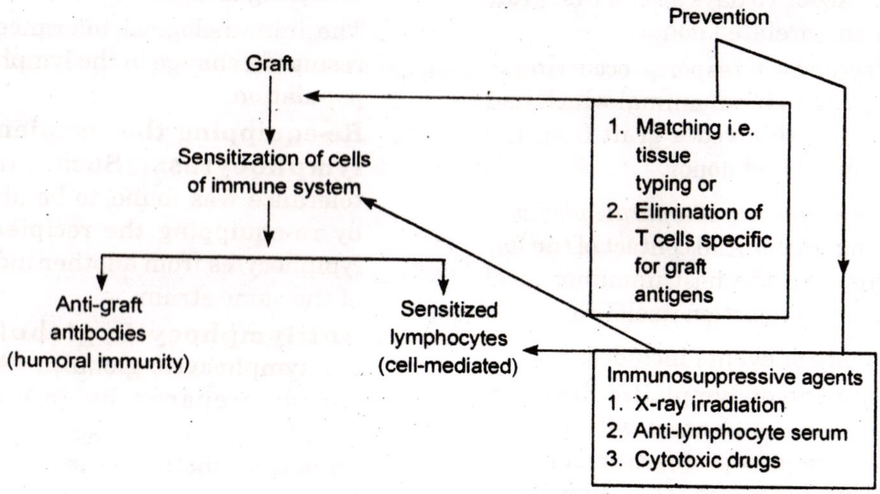


**Fig. 30.8:** Schematic representation of rejection mechanisms and their prevention.

# 31

# PHARMACOGENETICS

1. What is pharmacogenetics, 2. Drug metabolism and genetics, 3. Modification of drugs-biochemically, 4. Kinetics, 5. Evaluation of genetic variation by the effects of drug, 6. Hereditary disorders drugs with altered drug response.

## What is Pharmacogenetics?

Pharmacogenetics is defined as the study of genetically determined variations revealed by the action of drugs. Hereditary disorders, where symptoms occur spontaneous can be increased or precipitated by drugs are included under the purview of pharmacogenetics.

## Drug Metabolism and Genetics

I. **Metabolism of Drugs:** Drug is metabolized usually by a common path. The drug is absorbed by intestine or mouth or skin or injected to the blood stream and distributed throughout the body tissue and tissue fluids. A small amount shows pharmacological effect and rest amount is broken down or excreted in unchanged form (Fig. 31.1).

II. **Modification of Drug Biochemically:** Usually the breakdown process takes place in the liver but with variations with different drugs. Such as:

1. Some are completely oxidized to $CO_2$ and is exhaled by the lungs.
2. Excreted in modified form through the kidney.
3. Excreted in the bile and then by stool.

Some drugs are biochemically modified for:

1. To increase their solubility
2. More readily excreted
   (a) Among the biochemical modifications is conjugation in the liver with glucuronic acid, e.g. elimination of morphine and codeine.
   (b) Isoniazid, sulphonamides are modified by introduction with acetyl group, i.e. acetylation.

## Kinetics

Metabolic study and the effect of a particular drug after giving standard dose is usually done to observe the response after certain time interval. This can be done either by:

1. Measuring the amount of the drug circulating in the blood;
2. Determining the rate at which it is metabolised.

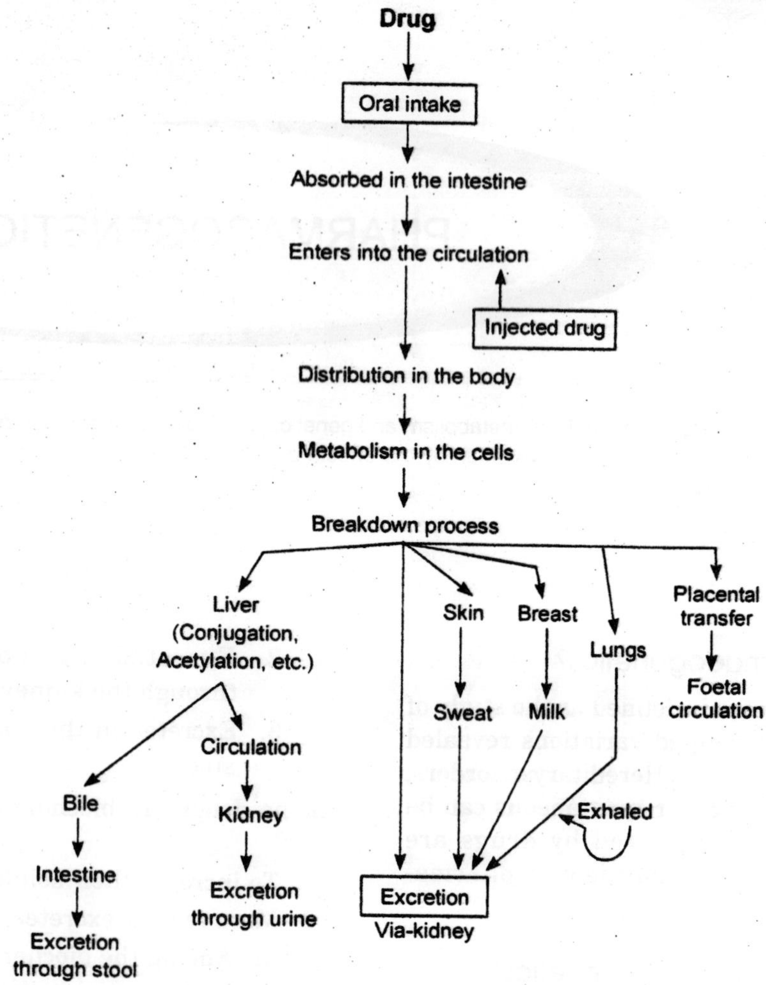


**Fig. 31.1:** Schematic representation of drug metabolism.

These observations show that there is significant variation of responding to certain drugs in individuals. This variation may be:

A. Continuous;

B. Discontinuous.

If the graph is plotted for responses in a large number of subjects then the result will be:

1. **Continuous variation:** The results will show a belt-shaped or *unimodal distribution.* A unimodal distribution. reveals that the metabolism of the drug given is under the control of many genes, i.e. *polygenic* (Fig. 31.2).

2. **Discontinuous variaton:** A discontinuous response reveals that the metabolism of the drug is under *monogenic control.* Such as when the normal metabolism of a drug is controlled by a *dominant gene,* R, and if the subjects are *unable to metabolise the drug* because they are *homozygous* for a *recessive gene,* r, then there will be three classes of subjects:

   (a) RR

   (b) Rr

   (c) rr

1. If the responses of RR and Rr are not distinguishable, a bimodal

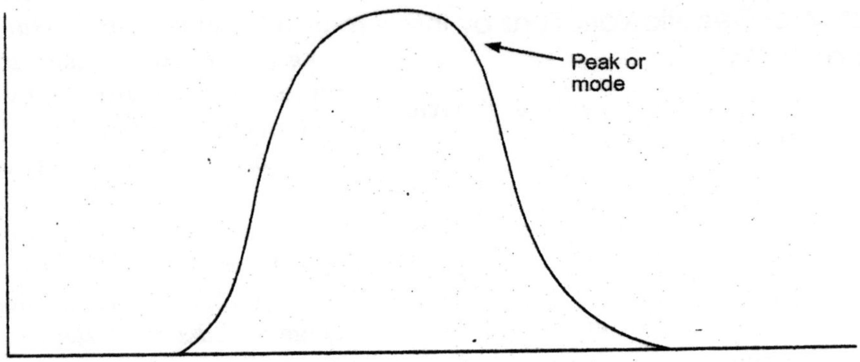


**Fig. 31.2:** Various types of response to different drugs. Continuous variation, multifactional control of drug metabolism.

distribution is observed (Fig. 31.3) and is under *monogenic control*, (B) and (C) discontinuous variation, monogenic control of drug metabolism.

2. If RR and Rr are distinguished, a trimodal distribution is observed, *each peak or mode representing a different phenotype* and is under *monogenic* control of drug metabolism.

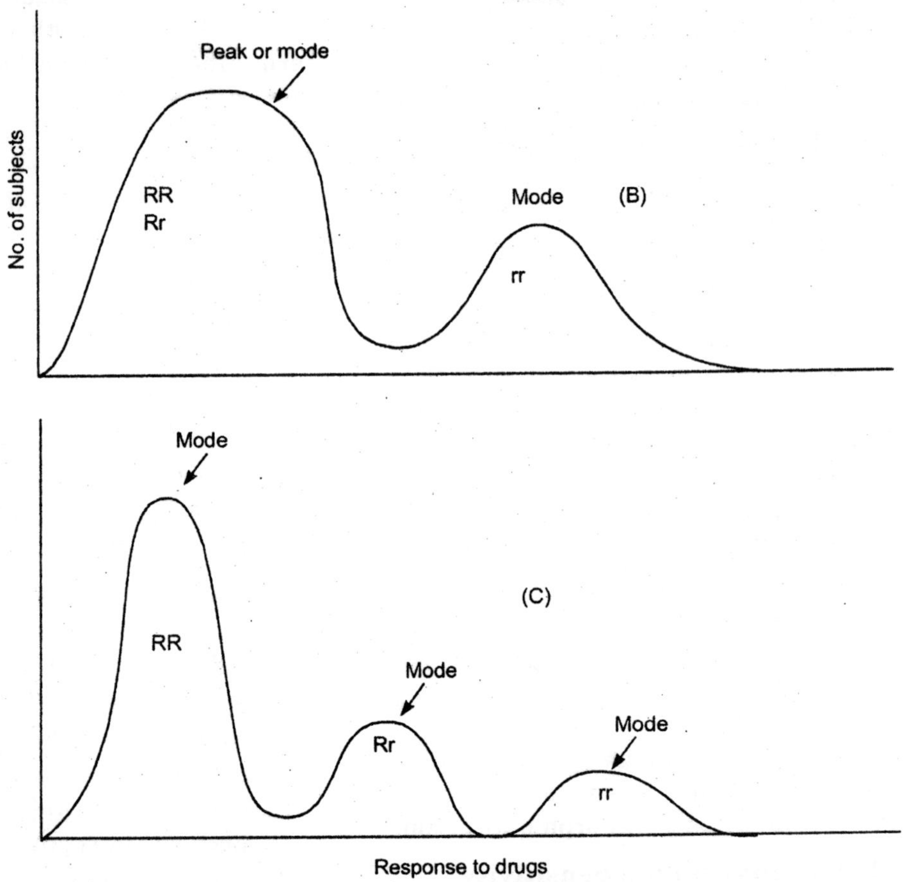


**Fig. 31.3:** Various types of response to different drugs.

## Evaluation of Genetic Variations by the Effects of Drugs

Drugs responsible for revealing genetic variations are:

I. Isoniazid
II. Succinylcholine
III. Primaquine
IV. Certain anticoagulants
V. Certain anaesthetic agents
VI. Thiopurines
VII. Phenylbutazone
VIII. Debrisoquine
IX. Alcohol

### I. Isoniazid

Isoniazid is anti-tuberculous drug which is rapidly absorbed resulting initial high level in blood and then slowly reduced by excretion after inactivation. Isoniazid metabolism revealed two groups of subjects:

1. **Rapid inactivators:** Blood levels of the drug fall rapidly after an oral dose.
2. **Slow inactivators:** Blood levels remain high for sometime. Family studies have revealed that these groups are homozygous for an autosomal recessive allele of the liver enzyme N-acetyltransferase with lower activity levels and its activity varies in different population. DNA studies have shown that *cloning of the genes is responsible* for *N-acetyltransferase* activity in human. These studies also revealed that there are *three genes:*
   (a) One is not expressed and represents a pseudogene (NATP);
   (b) One does not exhibit differences in activity between individuals ($NAT_1$);
   (c) One shows mutations ($NAT_2$) which are responsible for the inherited polymorphic variation.

### II. Succinylcholine Sensitivity

*Curare* a plant extract used in surgery for muscle relaxation. *Succinylcholine* (Suxamethonium) also causes muscle relaxation like curare by a different mechanism, i.e. relaxation of skeletal and respiratory muscles and apnoea for a short time. But very few subjects, apnoea lasts for an hour or more which is corrected by blood or plasma transfusion from a normal person. Succinylcholine is normally destroyed by the plasma *Pseudocholinesterase*.

(a) In subjects of highly sensitive to succinylcholine, the plasma pseudocholinesterase, destroys the drug at a significant slower rate than normal.
(b) In rare cases, the enzyme is absent. Family studies have revealed that succinylcholine sensitivity is *inherited as an autosomal recessive trait.*

Studying plasma pseudocholinesterase activity with local anaesthetic *dibucaine* and determining the percentage inhibition is termed *dibucaine number.* The frequency distribution of dibucaine number values in families with succinylcholine-sensitive individuals gives a *trimodal curve.* Three modes represent:

(a) Normal homozygotes;
(b) Heterozygotes;
(c) Affected homozygotes.

The experimental evidence indicates that there are at least two different forms of plasma pseudocholinesterase:

(a) one in the normal homozygote
(b) one in the affected homozygote

The two enzymes differ in the following ways:

1. The way in which they are inhibited by dibucaine;
2. The enzyme in the affected homozygote being less inhibited.
3. In enzyme kinetics, the normal enzyme being more efficient than the abnormal enzyme in destroying acetylcholine and other choline esters.

### III. Glucose-6-phosphate Dehydrogenase (G6PD)

Antimalarial drug primaquine which is better than quinine in preventing relapse. Some patients were found to be sensitive to the drug. Few days there is no adverse reaction but then suddenly the subject passes very dark, often black urine, jaundice, gradual fall of RBC count and haemoglobin concentration as a consquence of haemolysis. Though the affected person recovers from the haemolytic episode but occasionally the haemolysis is so extensive that it becomes fatal. The cause of sensivity is due to deficiency of enzyme in the RBC known as glucose-6-phosphate dehydrogenase (G6PD).

Family studies revealed that G6PD deficiency is *inherited as an X-linked recessive trait.*

Persons with G6PD deficiency are sensitive not only to primaquine but also to phenacetin, nitrofurantoin and certain sulphonamides.

### IV. Coumarin Metabolism

Coumarin anticoagulant drugs, e.g. warfarin which is used for preventing clotting. There is discontinuous variation in the response of some patients taking these drugs which involves an increased resistance to the effects of the anticoagulant which appears to be transmitted as an *autosomal dominant trait.*

### V. Malignant Hyperthermia

Malignant hyperthermia is a rare complication of anaesthesia, usually occurs by Halothane and particularly with succinylcholine. Susceptible person develops muscle rigidity, increased temperature (hyperthermia), often 42.3°C (108°F).

Susceptibility appears to be inherited as an autosomal dominant trait affecting 1 in 10000 persons. These persons often show elevation of *serum creatine kinase.* It is genetically heterogeneous with several loci identified as being responsible; only minor number of families show linkage to or mutations in the ryanodine receptor gene on the long arm of chromosome 19.

### VI. Thiopurine Methyltransferase

A group of toxic substances collectively known as thiopurines (e.g. 6-mercaptopurine, 6-thioguanine, azathioprine) are used in leukaemia (1) to suppress the immune response with autoimmune disorders, e.g. systemic lupus erythematosus and (2) to prevent rejection in recipients of organ transplants. It produces leucopenia and severe liver damage and not possible to predict who is likely to produce side-effects or therapeutic effect because the response of the drugs varies widely. The main step of the metabolism of these drugs is methylation catalyzed by the enzyme thiopurine methyltransferase (TPMT). It seems that RBC TPMT activity is polymorphic. In case of high TPMT activity such as in childhood leukaemia, these drugs give better therapeutic effect.

### VII. Phenylbutazone

Phenylbutazone is used in severe arthritis. The metabolism of the drug appears to be *polygenic control.* The patients having slow metabolism of this drug, show side-effects, e.g. hypoplastic anemia.

### VIII. Debrisoquine Metabolism

Debrisoquine is often used for treatment of hypertension. There is a bimodal distribution in the response to the drug in the general population but 5-10 percent of Europeans show poor metabolism and are homozygotes for an autosomal recessive gene with reduced hydroxylation. Such people are more susceptible to overreact to the drug and develop hypotension during treatment. They are also more prone to side-effects to other drugs, e.g. beta blockers propranolol, metoprolol, antidepressants, e.g. amitryptiline, imipramine and the antipsychotics, e.g. thioridazine and haloperidol.

Molecular studies have observed that the gene involved in metabolism is one of the P450 family of genes on chromosome 22 known as CYP2D6. The mutations responsible for the poor metabolizer phenotype are heterogeneous, either a splice site mutation resulting in a truncated gene product or a deletion of the gene.

### IX. Alcohol Metabolism

Genetic factors can be involved in alcohol metabolism. The evidence is based on twin studies and have shown high incidence rates and family studies has shown a high prevalence of alcohol-related problem among relatives of alcoholics. Possible role of genetic factors revealed from the study of enzymes involved in alcohol metabolism. Alcohol is metabolized in the liver by alcohol dehydrogenase (ADH) to acetaldehyde and then further degraded by acetaldehyde dehydrogenase (ALDH). Human ADH consists of dimers of various combinations of subunits of three different polypeptide units coded for by three loci. $ADH_1$ coding for the α-subunit, $ADH_2$ for the β-subunit and $ADH_3$ for the γ-subunits. $ADH_1$ is primarily expressed in early foetal life, while $ADH_2$ is expressed in adult life. The flushing reaction to alcohol in one group of people than the other group suggest that this sensitivity could be due to difference in the rate of acetaldehyde metabolism. There are two major acetaldehyde dehydrogenase variants or isozymes. $ALDH_1$ which is present in the cytosol and $ALDH_2$ which is present in the mitochondria. The acute flushing reaction has been observed to be due to absence of $ALDH_2$ activity.

## II. Hereditary Disorders with Altered Drug Response

### A. Porphyria Variegata

Porphyria variegata is one type of porphyria is inherited as an autosomal dominant trait. Some affected persons develop skin lesions particularly on sun-exposed surfaces. Others develop severe abdominal pain, muscular paralysis, mental disturbance. An acute attack may be fatal to the person.

### B. Haemoglobinopathies

Sulphonamide drugs can cause severe haemolysis in some persons with certain haemoglobinopathies, e.g. haemoglobin H, a haemoglobin which consists of β-globin chain and haemoglobin Zurich which was found in a swiss family. These haemoglobinopathies are rare.

### C. Gout and Chlorothiazide

Diuretics are used in congestive heart failure which increases excretion of water to reduce edema. The widely used one is chlorothiazide which can produce gout as side-affect. In genetically predisposed persons attacks often follow dietary excesses which lead to an elevation in serum uric acid. This drug has a similar effect by raising the serum level of uric acid in genetically predisposed persons.

### D. Crigler-Najjar Syndrome

Crigler-Najjar syndrome is manifested by a severe non-haemolytic jaundice, appearing on the first or second day after birth which is associated deposition of unconjugated bilirubin in the CNS leading to neurological sequelae. Affected baby often dies in infancy. The manifestation is inherited as an autosomal recessive trait, heterozygotes being perfectly healthy. The basic defect lies in the liver which shows inability to conjugate bilirubin with glucuronide due to deficiency of the enzyme glucuronyl transferase. Affected persons are unable to conjugate certain drugs as salicylates.

### E. Non-insulin-Dependent Diabetes

Non-insulin-dependent diabetes is inherited as an autosomal dominant trait. It has been opined that those at risk of developing the disease, they develop facial flushing after alcohol if they are first given the oral hypoglyaemic drug chlorpropamide. It may prove a useful test for familial susceptibility to diabetes and more recently to identify those at risk of developing complication of diabetes, there have been no consistent findings.

## PART-VII

# DIAGNOSIS OF PRENATAL GENETICAL DISEASES

# 32

# DIAGNOSIS OF PRENATAL GENETICAL DISEASES

1. Importance of diagnosis of prenatal genetic diseases, 2. Amniocentesis, 3. Analysis of human Genome, 4. Blotting procedure, 5. Polymerase chain reaction, 6. Application, 7. Merits of the method, 8. Diagnosis by DNA analysis, 9. Single-gene disorders by DNA analysis, 10. Methods of prenatal diagnosis.

## Importance of Diagnosis of Prenatal Genetical Diseases

Because the birth of a child with congenital defects or mental deficiency or metabolic disorder causes mental upset to the parents and family. Before birth there is no indication of the anomalies which would have alerted the parents. In case of genetic disorders, the parents and family members are eager to know the reason and method to avert the repetition of the shocking events and to know how it can be diagnosed, etc. A genetic counsellor must have a confirmed diagnosis of the patient, an accurate family history and the mode of inheritance of the disorder so that a risk of occurrence can be averted. Considering other reproductive options such as long-term contraception, sterilization, termination of all pregnancies or other alternative, like adoption, long-term fostering and artificial insemination by donor sperm.

- **Criteria for Prenatal Diagnosis**

1. High genetic risk
2. Severe disorder
3. Treatment not available
4. Reliable prenatal test
5. Termination of pregnancy acceptable

## Methods of Diagnosis in Prenatal Life

I. **Early Period** (12-20 weeks of pregnancy)

A. **Amniocentesis:** Examination of amniotic cells and fluid for alpha-fetoprotein acetylcholine esterase, chromosomal analysis.

B. **Chorionic Villous Sampling (CVS):** DNA analysis chromosomal analysis, biochemical analysis.

C. **Ultrasonography:** Structural abnormalities.

D. Fetoscopy

E. Foetal and Maternal Blood and Urine Analysis

(a) Flow cytometric test for foetal cells

(b) Serum alpha-fetoprotein, oestriol, human chorionic gonadotropin

(c) urine of pregnant mother.

**F. Molecular Biology**

(a) *Analysis of Human Genome*

1. Molecular cloning and DNA sequence
2. Nucleic acid hybridizatiion

(b) Restriction endonucleases

(c) Blotting procedure

i. Southern blotting

ii. Northern blotting

iii. Western blotting.

(d) DNA amplification using the polymerase chain reaction (PCR)

(e) Detection of mutant genes

(f) Fluorescence in situ hybridization (FISH) technique.

**II. Late Period**

(1) X-ray, (2) Ultrasonogram, etc. (3) Cordocentesis—performed in 2nd trimester or (4) Foetal tissue biopsy—performed in 2nd trimester—skin, liver, lung, kidney muscle.

**III. Preimplantation Embryo Biopsy:**

*In very early stage* which has limited application and availability. It offers an alternative to current prenatal diagnostic techniques. In this method *in vitro* fertilisation and embryo culture is followed by biopsy of one or two outer embryonal cells at the 6-10 cell stage of development. Affected embryos are identified by using molecular genetic techniques such as (1) DNA analysis of a single cell and (2) possibly chromosomal analysis by in situ hybridisation and only embryos free of a particular genetic defect would be reimplanted. This method may occasionally be more acceptable than other forms of prenatal diagnosis but has a limited application and availability.

**Indications**

1. Couples at risk of passing an X-linked disorder for which there is as yet, no specific molecular diagnosis. Recently in such cases all male embryos are aborted, as they are hemizygous for an X-linked trait and display full impact of disease if affected.
2. Those couples who had to suffer repeated terminations of pregnancy.
3. Subfertile couples who may resort to *in vitro* fertilisation treatment and desire to increase the chances of a normal zygote.
4. Those with moral or religious objections to pregnancy termination, this group would prefer selection to

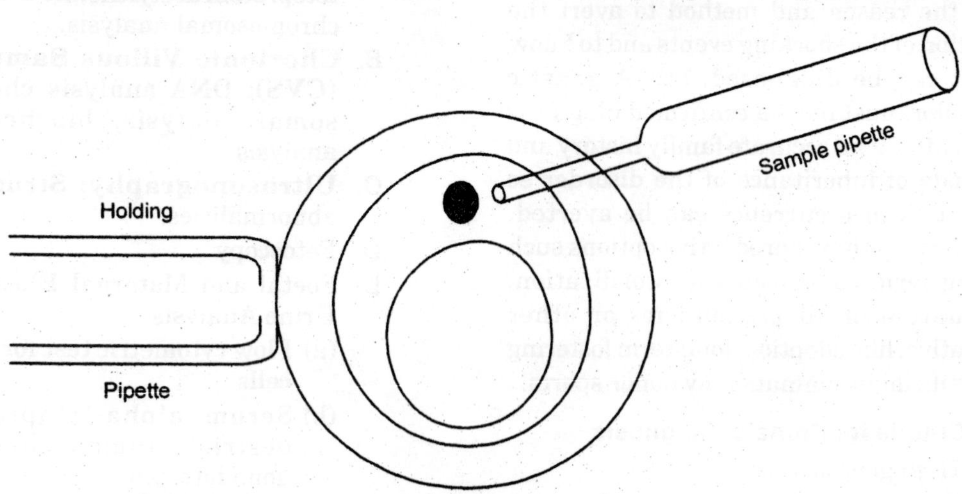


**Fig. 32.1:** Polar body biopsy.

occur at the preimplantation stage giving them a reasonable chance of starting with a normal pregnancy.

- **Procedure for the preimplantation diagnosis of genetic defects**

1. **Polar body biopsy:** The completion of first meitotic division causes extrusion of the first polar body. It has a diploid complement of chromatin and is considered to be metabolically inactive. This chromatin is considered as an image of the chromatin of the oocyte. Since the first polar body does not contribute to the development of the embryo and can be removed with minimal damage on the oocyte (Fig. 32.1).
2. **Multicell biopsy:** Before the late eight-cell stage, the blastomeres of the preembryo are dissociated with pipetting after drilling a small hole in the zona pellucida, that heals rapidly within short period. 1 to 3 cells are removed by using micropipettes (Fig. 32.2).
3. **Blastocyst biopsy:** A human blastocyst containing 60 or more cells programmed for differentiation is used. From trophoblast placenta a number of cells can be safely removed for analysis without adversely affecting the fetus (Fig. 32.3).

The diagnosis of genetic diseases are usually done in the prenatal period.

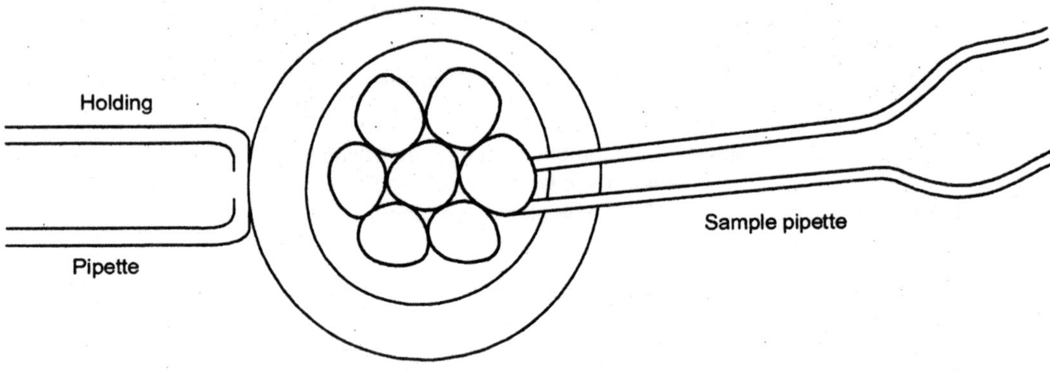


**Fig. 32.2:** Multicell biopsy.

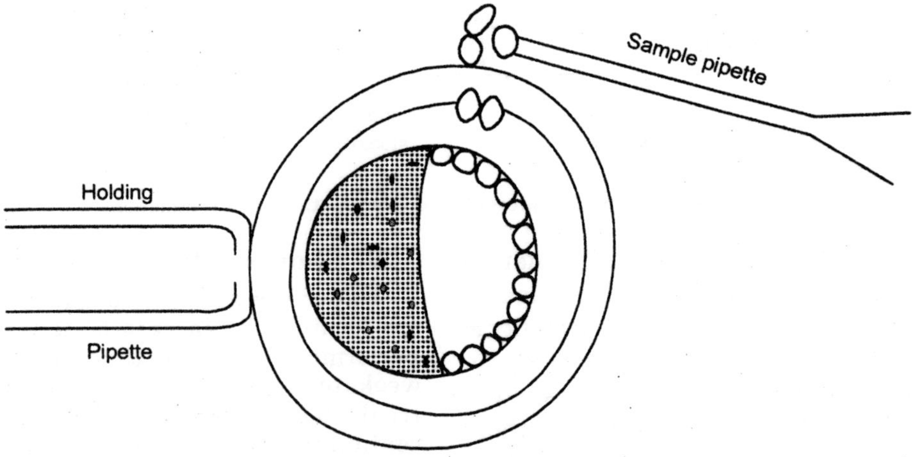


**Fig. 32.3:** Blastocyst biopsy.

## Method of diagnosis in Prenatal Life

### I. Early Period

#### A. Amniocentesis

For prenatal diagnosis, examination of amniotic cells and fluids is most successful method. It is done under the guidance of ultrasonography by suprapubic amniocentesis, i.e. penetration of amnion sac above the symphysis pubis with a hollow needle. It is done between 12-20 weeks of pregnancy. The amount of fluid is drawn between 5-20 ml. The procedure is not painful and does not create any danger. Risk of miscarriage seems to be less than one percent. After collection, the fluid is centrifuged and the precipitated cells are cultured which takes few weeks for chromosomal anomalies. The supernatant amniotic fluid is biochemically investigated to determine the biochemical abnormality of the embryo such as determination of *alpha-fetoprotein* which is present in traces in the amniotic fluid if the fetus is normal but there is massive increase in case of anencephaly or *spina bifida*.

- **Indications**

This *method* is useful to detect defects due to chromosomal abnormalities, neural e.g. Down's syndrome, fragile X-syndromes, tube defect and biochemical disorders, sex determination (99% accurate) (Fig. 32.4).

- **Indications for amniocentesis**

1. Advanced maternal age
2. Abnormal maternal serum triple marker screening test
3. Family history of chromosomal abnormality
4. Neural tube defect/abnormal maternal serum α-fetoprotein.
5. Mendelian disease detectable by biochemical or DNA analysis.

**Procedure:**

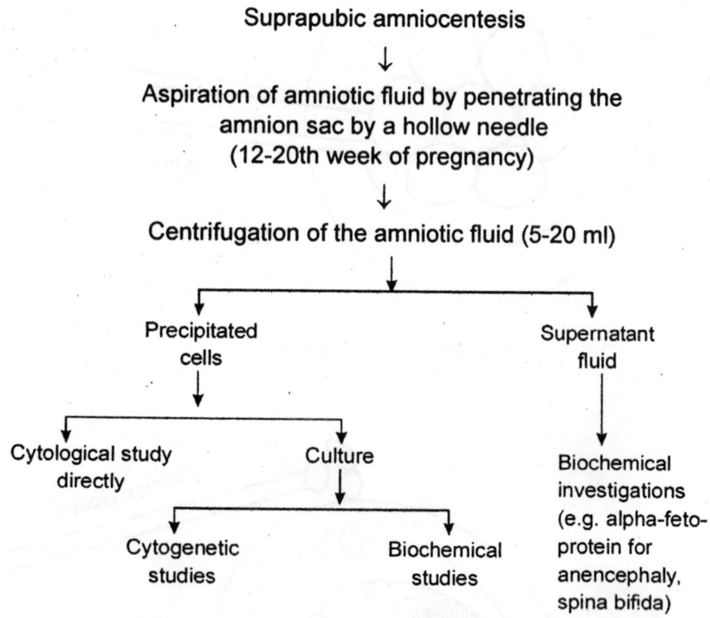


**Fig. 32.4:** Schematic representation of amniotic fluid investigation.

#### B. Chorionic Villous Sampling (CVS)

Under the guidance of ultrasonography the location of placenta is detected before performing CVS. The time for CVS is 9-12 weeks of pregnancy when chorionic villi float freely within the intervillous space and contain mitotically active villous cells. CVS may be done:

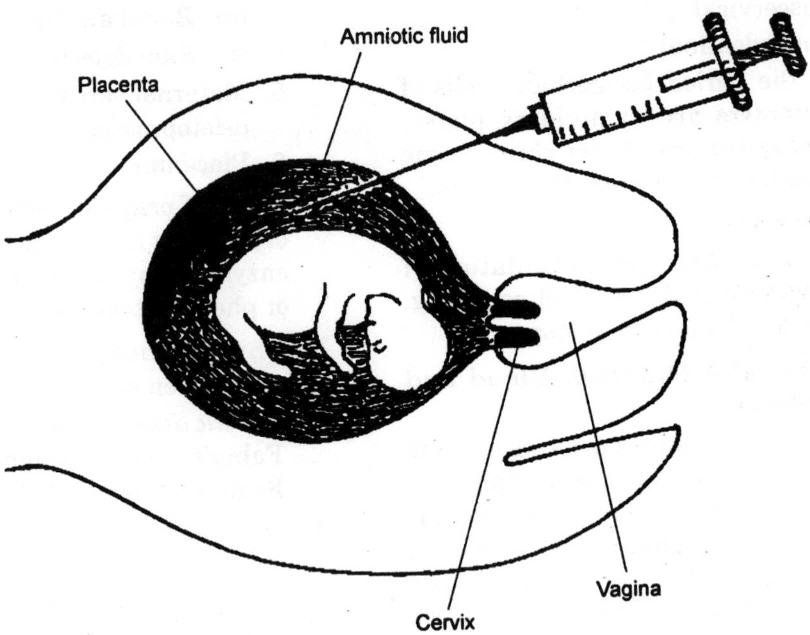


**Fig. 32.5:** The technique of amniocentesis.

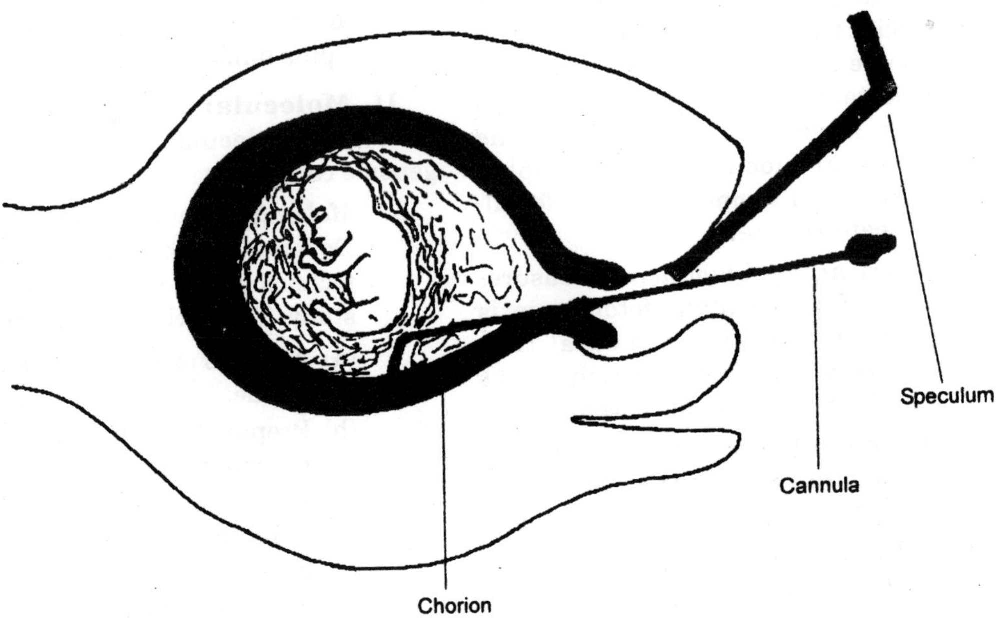


**Fig. 32.6:** Transvaginal chorionic villus sampling technique.

1. Transcervical
2. Transabdominal.

During the period Langerhan's cells of cytotrophoblasts are in dividing phase, without doing culture, direct chromosome analysis can be made within 24-48 hours.

**C. Fetoscopy**

Fetoscopy is done for cannulation of umbilical vessels and for blood sampling, transfusion and foetal tissue biopsy.

**D. Foetal and Maternal Blood and Urine Analysis**

1. *Foetal cells*: Maternal blood is being used recently for isolation and analysis of circulating foetal cells because it is a safest and non-invasive procedure for prenatal diagnosis. It is known that foetal cells appear in maternal blood as early as 15 weeks of pregnancy.

   *Flow cytometric test* of maternal blood with anti-gamma-globulin MAb (monoclonal antibody to gamma chain haemoglobin molecule) is highly specific for examining fetal cells. 20 ml sample of maternal blood contains 0-20 foetal cells and three types of foetal cells have been observed in the maternal blood—syncytio-trophoblast, foetal lymphocytes and foetal-nucleated red cells.
2. *Serum alpha-fetoprotein*: Measurement of serum alpha-fetoprotein is done for screening of neural tube defects of the foetus which shows significant increase in concentration.

**Increased maternal serum α-fetoprotein concentration**

1. Under-estimated gestational age
2. Threatened abortion
3. Multiple pregnancy
4. Foetal abnormality
   (a) Anencephaly
   (b) Open Neural tube defect
   (c) Anterior abdominal wall defect
   (d) Turner's syndrome
   (e) Bowel atresias
   (f) Skin defects
5. Maternal hereditary persistence of α-fetoprotein
6. Placental haemangioma

2. *Urine of pregnant mother*: The urine of pregnant mother can be analysed for enzyme assay of foetal homocystinuria or phenylketonuria.

By this method, diagnosis of some diseases has been achieved such as:

(a) *Genetic disorders of lipid metabolism*: Fabry's disease, Gaucher's disease, Krabbe's disease, Wolman's disease.
(b) *Genetic disorders of carbohydrate metabolism*: Galactosemia, Glycogen storage disease, G-6-P-D deficiency.
(c) *Genetic disorders of amino acid metabolism and other related disorders*: Cystinosis, Histidinemia, Hunter's disease, Adrenogenital syndrome, Xeroderma pigmentosum, Sickle-cell anemia, Thalassemia, Anencephaly, Spina bifida.

**II. Molecular Analysis of Human Genome:** Molecular analysis of human genome comprises:

A. *Molecular cloning and DNA sequence*: Development of recombinant DNA technology has enhanced this type of study. This technology has provided:
   (a) Analysis of one DNA frgments at a time.
   (b) Preparation an "index" for the human genome.
   (c) Location and isolation of individual genes or portions of genes.
   (d) Making of an unlimited number of copies of such DNA fragments.
   (e) Determination of the nucleotide sequence of the DNA.
   (f) To transcribe and translate the isolated genes.

The genes and their products are then studied for gene structure and function in normal and disease states.

The cloning of DNA involves isolation of DNA fragments and insertion of a sequence into the nucleic acid from another biologic source, i.e., vector for manipulation and propagation (for cloning—see recombinant DNA technology).

B. *Nucleic Acid Hybridization:* Linear pieces of double-stranded DNA (native) after treatment with acid or alkali dissociates into single—stranded i.e. denatured DNA. This denatured DNA after incubation is investigated for nucleic acid hybridization, i.e. recognition of two complementary strands and reformation of double-stranded molecules by base-pairing. The specificity of nucleic acid hybridization, often in combination with fractionation or amplification procedures, gives opportunity for detection of a single gene among thousands or of a viral sequence in the midst of other nucleic acid sequences.

## III. Restriction Endonucleases

Recombinant DNA manipulation has been facilitated by the discovery of restriction endonucleases or restriction enzymes in microorganisms.

The enzymes recognize a specific oligonucleotide sequence in double-stranded DNA and cleave the DNA at this site. The sequence specificity of restriction enzymes is the best means in dissection of large genome.

(For details—see recombinant DNA technology)

## IV. Blotting Procedure

A. *Southern Blotting:* Human genome analysis involves a specific application of DNA-DNA hybridization, the blotting procedure developed by E.M. Southern. Steps of the method are given below:

Isolation of high molecular weight genomic DNA (cells of blood leucocytes and fetal cells)

↓ Digested with restriction enzyme

Series of reproducible fragments

↓

Separated by electrophoresis in agarose gel

↓ Transferred

To a membrane that binds DNA

↓ Treated

Denatured DNA

↓ Soaked

In a solution of radion of radioactive single-stranded nucleic acid probe

↓

Probe forms double-stranded nucleic acid complex at the sites on the membrane where homologous DNA present

↓

The membrane is washed to remove unbound radioactivity and regions on the membrane where homologous DNA sequences were bound are detected by autoradiogram.

By this method analysis can be done with a tiny portion of the primary structure of human genome DNA taken from an individual.

B. *Northern Blotting:* Another analogous procedure is northern blotting which starts with RNA for analysis. In this method, the presence or absence of a particular mRNA as well as its approximate size can be determined.

C. *Western Blotting (immunoblotting):* This method is designed for analysis of protein antigens. Proteins are separated by electrophoresis and transferred to a solid membrane through a blotting procedure. The membrane is analyzed by incubation with antibodies followed by a second step for enzymatic or radioactive detection of bound antibody.

Thus, these blotting procedures each combines a fractionation and a detection method to offer a sensitive technique for the analysis of DNA, RNA and protein antigens.

The reaction is usually carried by using heat-resistant *Tag polymerase* which remains active during the temperature cycle (50-90°C). After several cycles, thousands of copies of the original target sequence are synthesized. The bulk product is a double-stranded DNA fragment of a specific length.

- **Application of the Technique**

1. Preparation of recombinant DNA constructs.

**Table 32.1:** Different Blotting Procedures for Analysis

| Blotting Method | Sample for Analysis | Method of Fractionaction | Detection |
|---|---|---|---|
| Sourthern | DNA | Electrophoresis | Nucleic acid hybridization |
| Northern | RNA | Electrophoresis | Nucleic acid hybridization |
| Western or Immunoblast | Protein antigens | Electrophoresis Electrophoresis | Immunologic |

### V. DNA amplification using the polymerase chain reaction (PCR)

The method is based on knowing the nucleic acid sequence for a region for diagnosis.

This is analyzed repeatedly for preparation of oligonucleotide primers. These are complementary to opposite strands of the DNA and are separated by up to a few hundred base pair. The oligonucleotide primers are incubated with the target DNA to be amplified and with a DNA polymerase that synthesizes a complementary strand in a $5' \rightarrow 3'$ direction. The reaction is subjected to a series of temperature variations including a denaturing temperature. At this denaturing temperature, double-stranded DNA is dissociated to single-stranded DNA. At an annealing temperature, oligonucleotide primers hybridize to target DNA. A polymerization temperature for the synthetic step is applied.

2. Mutagenesis of cloned DNA.
3. Detection of rare nucleotide sequences.
4. Detection of nucleotide sequences of infectious agents.

- **Merits of the Method**

1. Rapid analysis in 24 hours
2. Case of automation
3. Relative economy
4. Very much specific

## New Techniques for Tissue Aspiration under Investigation

1. Transcervical cell collection of foetal cells by intrauterine lavage.
2. Cervical mucus aspiration.
3. Cytobrush—cytobrush sampling and mucus aspiration both at the level of external os are least invasive.
4. Intracervical and intrauterine lavage provide fetal nucleated cells but these are more invasive and warrants further investigation to establish as safe procedure for ongoing pregnancies.

**Table 32.2:** Prenatal diagnosis by DNA analysis

1. Detection of known point mutations, splicing mutations, or mal-insertions or deletions by:
   i. Restriction endonuclease analysis to detect mutations that alter restriction sites.
   ii. Allele-specific oligonucleotide hybridization.
   iii. PCR (polymerase chain reaction) amplification and direct automated sequencing.
2. Detection of unknown mutations by PCR amplification followed by:
   i. Single-strand conformation polymorphism analysis.
   ii. Heteroduplex analysis.
   iii. Mismatch cleavage
   iv. DNA sequencing.
3. Detection of large insertions, expansions, deletions or other major structural rearrangements by:
   i. Southern blot analysis
   ii. PCR amplification
4. Linkage-based analysis using intragenic DNA polymorphisms to diagnose diseases in which there are many pathogenic mutations, and the specific mutations, and the specific mutation in a family is unknown.

**Table 32.3:** Some single-gene disorders and prenatal or presymptomatic diagnosis by DNA analysis

**I. Autosomal dominant**

1. Myotonic dystrophy
2. Adult polycystic kidney disease
3. Huntington disease
4. Neurofibromatosis-I
5. Familial breast cancer

**II. Autosomal recessive**

1. Sickle-cell anemia
2. α-Thalassemia, β-Thalassemia
3. Cystic fibrosis
4. Phenylketonuria
5. $\alpha_1$-Antitrypsin deficiency
6. Tay-Sacss disease

**III. X-linked recessive**

1. Hemophilia A and B
2. Duchenne and Becker muscular dystrophy
3. Fragile X syndrome
4. Ornithine transcarbamylase deficiency

- **In Summary**

**Effect of prenatal diagnosis on gene frequency and disease frequency**

Rapid advances in molecular genetics and in the mapping and cloning of genes and identification of disease causing mutations, have dramatically increased the scope of prenatal diagnosis. Combined with sensitive and supportive genetic counselling, the availability of prenatal diagnosis allows couples of risk for having a child with a serious genetic disease to choose among reproductive option and to avoid, if they so choose, the birth of an affected child.

**Table 32.4:** Comparison of procedures of prenatal diagnosis

| Procedure | Gestation (weeks) | Culture time (hours, days, weeks) | Sample aspirated (mg,ml) | Risk (fetal loss) |
|---|---|---|---|---|
| 1. Ultrasonography | 6-40 | Nil | Nil | Negligible |
| 2. CV sampling | 9-12 | 2 hours-2 days | 5-10 mg | 1-2% |
| 3. Amniocentesis | 15-18 | 2-3 weeks | 15-20 ml | 1-1.5% |
| 4. Maternal blood sampling | 15-18 | 2-3 days | 20 ml | Negligible |
| 5. Fetal blood and tissue sampling | 18-20<br>18-20 | 2-3 days<br>2-3 days | 0.2-0.5 ml<br>0.2-0.5 ml<br>10-25 mg | 2-5%<br>2-5% |

**Table 32.5:** Methods used in prenatal diagnosis in gestation period

| Method | Time (in weeks) | Disorders diagnosed |
|---|---|---|
| I. **Non-invasive maternal blood screening** | | |
| 1. Alpha-fetoprotein | 16 | Neural tube defects |
| 2. Triple test | 16 | Down's syndrome |
| 3. Ultrasonography | 18 | Structural deformities, e.g. CNS, heart, kidney and limbs |
| II. **Invasive** | | |
| 1. Amniocentesis<br>(a) Fluid<br>(b) Cells | 16 | Neural tube defects, chromosome abnormalities, metabolic disorders, molecular defects |
| 2. Chorionic villous sampling | 10-12 | Chromosome abnormalities, metabolic disorders, molecular defects |
| 3. Fetoscopy | | |
| (a) Blood | | Chromosome abnormalities, haematological disorders, congenital infection |
| (b) Liver | | Metabolic disorders, e.g. ornithine transcarbamylase deficiency |
| (c) Skin | | Hereditary skin disorders, e.g. epidermolysis bullosa |

## Summary

1. Most of the available procedures are invasive.
2. New approaches of noninvasive character to investigate the chromosomal defect in fetus during gestation period.
3. Prenatal diagnosis for inherited diseases can be done by studying fetal cell obtained by aminocentesis and chorionic villus sampling between 9-20 weeks of pregnancy. If the fetus having serious genetic disorder, abortion is indicated.

4. But invasive techniques may cause an inherent risk for fetal damage and loss for which application of these procedure is limited to the patients having chromosomal abnormalities for risk.
5. Non-invasive procedures like screening biochemical markers in maternal blood is used for identification of patients at risk.
6. Screening of serum analyte and ultrasonography are being used for trisomies but not giving maximum result because sensitivity of the serum analyte is 60-90 percent depending upon age of material.

   Ultrasound requires expert opinion in selected centres.
7. Lesser invasive and lesser risky technique is the study of fetal cells. Fetal cells in maternal circulation may give information prenatal genetic disorders. Three types of fetal cells have been identified in the maternal circulation such as:
   (a) Syncitotrophoblant
   (b) Fetal lymphocytes
   (c) Fetal nucleated RBC

   Culturing of each of them requires separate method.

   Cord blood is better than maternal blood because it has more fetal nucleated RBC. But the procedure is difficult and risky for fetus.
8. But excellent method is preimplantation diagnosis of genetic abnormalities. In this method one or more cells are removed from the embryo for genetic analysis and unaffected embryos are identified and transferred to the uterus to establish a pregnancy. The cells are obtained at any stage between the two cells and blastocyst stage. However, the decreased success rate in establishing pregnancy and with low proportion of embryos reaching this stage is likely to decrease the number unaffected embryos available for transfer to the uterus. Perhaps this method will be the earliest diagnosis of fetal genetic defects.

## QUESTION

1. What is the importance of prenatal diagnosis for genetical diseases?

   What are the criteria for prenatal diagnosis?
2. What are the methods for prenatal diagnosis?
3. What is amniocentesis? State briefly the procedure and indications.
4. How fetal and maternal blood and urine analysis help for diagnosis of genetic diseases?
5. Describe briefly the molecular analysis for prenatal diagnosis.
6. Enumerate the prenatal diagnosis by DNA analysis.
7. Enumerate some single-gene disorders and prenatal or presymptomatic diagnosis by DNA analysis.
8. State the following:
   (a) Effect of prenatal diagnosis on gene frequency and disease frequency.
   (b) Comparison of procedures of prenatal diagnosis.

# 33

# HUMAN GENETICAL DISEASES

1. Human genetical diseases, 2. Cytogenetic chromosome disorders, 3. Indication for chromosome analysis, 4. Genetic factors in common diseases, 5. Procedure to show genetic susceptibility to common diseases, 6. Single gene disorders, 7. Haemophilia, 8. Mutational analysis of single gene disorders, 9. Congenital abnormalities or Anomalies and genetics (Dysmorphology), 10. Causes of malformation, 11. Teratogens, 12. Polygenic or multifactorial disorders, 12. Identification of genes which are responsible for multifactorial disorders.

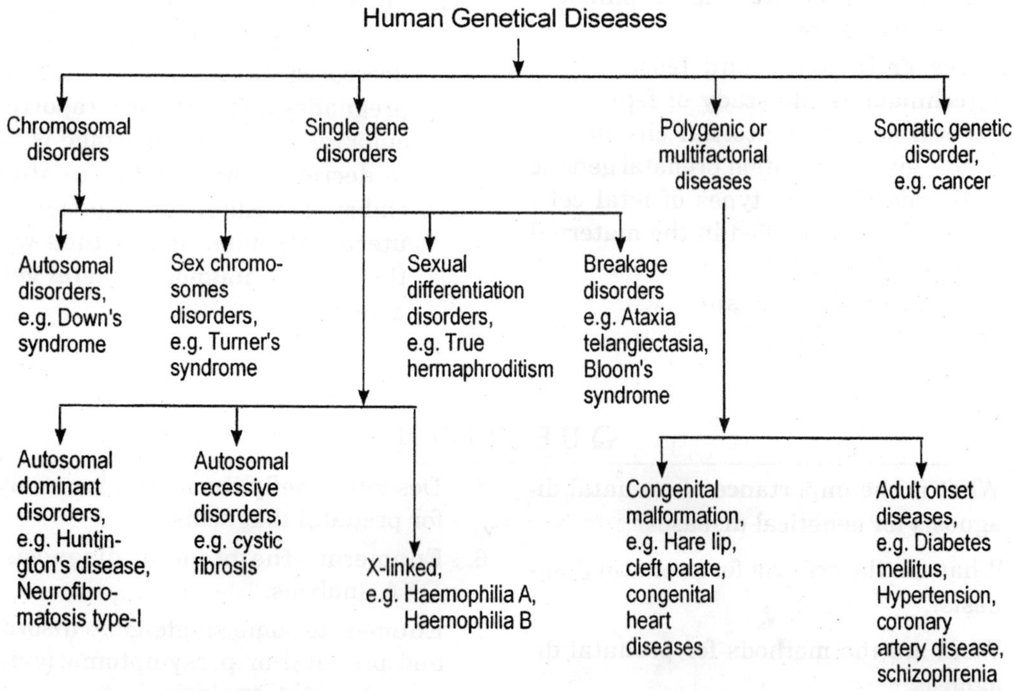


## CYTOGENETICS

### Chromosome Disorders

Chromosome disorders account for 50 per cent of all spontaneous miscarriages and are present in 0.5-10 per cent of all newborn infants.

Chromosome abnormalities are present in at least 10 per cent of all spermatozoa and 25 per cent of all recognized pregnancies. Many more zygotes and embryos are so abnormal that survival beyond the first few days or weeks after fertilization is not possible. These diseases are the result of the addition or deletion of entire chromosomes or parts of chromosomes. Because each chromosome contains tens of thousands of genes, physical manifestations of chromosome disorders are characterized by growth retardation, mental retardation and a variety of somatic abnormalities. Clinically significant chromosome abnormalities occur in nearly 1% of liveborn babies and account for about 1% of pediatric hospital admission and 2.5% of childhood. Three major disorders caused by altered numbers of chromosomes are Down syndrome, Turner syndrome, and Klinefelter syndrome.

- **Chromosome Disorders**

**I. Autosome disorders**

1. Down's syndrome (Trisomy 21)
2. Patau's syndrome (Trisomy 13)
   Edward's syndrome (Trisomy 18)
3. Chromosome deletion syndrome
   (a) Wolf-Hirschhorn syndrome
   (b) Cri-du-chat syndrome
4. Microdeletion syndrome.

**II. Sex chromosomes disorders**

1. Klinefelter's syndrome (47, XXY)
2. Turner's syndrome (45, X)
3. XXX Females
4. XXY Males
5. The Fragile X syndrome.

**III. Sexual differentiation disorders**

1. True hermaphroditism.
2. Male Pseudohermaphroditism.
3. Female Pseudohermaphroditism.

**IV. Chromosome breakage syndromes**

1. Ataxia telangiectasia
2. Bloom's syndrome
3. Fanconi's anemia
4. Xeroderma pigmentosa
5. Chromosome breakage and sister chromatid exchange.

## I. AUTOSOME DISORDERS

**Down's Syndrome (Trisomy 21):** This syndrome caused by trisomy 21, or three copies of chromosome 21 instead of the usual two copies. This abnormality occurs in approximately 1 in 800 liveborn infants and increases in frequency with advancing maternal age.

- **Clinical findings**

It is characterized by growth retardation, variable but often severe mental retardation, and characteristic physical abnormalities including the upward slanting eyes, congenital heart defects with are the major cause of death in children.

**Table 33.1:** Findings in Down's syndrome

1. **Newborn period**
   Hypotonia, excess nuchal skin, sleepy.
2. **Craniofacial**
   Brachycephaly, protruding tongue, epicanthic folds, small ears, upward slanting eyes.
3. **Cardiac**
   Atrial and ventricular septal defect, patent ductus arteriosus, common atrioventricular canal.
4. **Limbs**
   Single palmar crease, small middle phalanx of 5th finger, wide gap between first and second toes.
5. **Other**
   Duodenal atresia, anal atresia, short stature, strabismus.

### • Patau's Syndrome (Trisomy 13) Edward's Syndrome (Trisomy 18)

The incidence of these syndromes is approximately 1 : 5000. The prognosis is grave with most affected infants dying during the first few weeks of life. In the unusual event of long-term survival there is severe mental retardation. Cardiac abnormalities occur in at least 90 per cent of all cases.

Chromosome analysis reveals straightforward trisomy. Both of these disorders show an association between increasing incidence and advanced maternal age, and in both the additional chromosome is usually of maternal origin. Approximately 10 per cent of cases are due to mosaicism or unbalanced rearrangements, particularly. Robertsonian translocations in cases of Patau's syndrome.

### • Chromosome Deletion Syndromes

(a) Wolf-Hirschhorn syndrome

(b) Cri-du-Chat syndrome

Microscopically visible deletions of the terminal portions of chromosomes 4 and 5 cause the above syndromes. In both conditions there is severe mental retardation, often in association with failure to thrive. The Cri-du-Chat syndrome derives its name from the characteristic cat-like cry of affected neonates which results from underdevelopment of the larynx. Both conditions are very rare with estimated incidences of approximately 1 : 50000 births. In some children with clinical features of one of these syndromes but apparently normal chromosomes, the presence of a very subtle deletion can be demonstrated by FISH using locus-specific probes for 4p or 5p.

### • Microdeletion Syndromes

By using a combination of high resolution prometaphase banding and FISH, an increasing number of previously unexplained syndromes are being shown to be due to submicroscopic or 'micro' deletions. Some of these *microdeletions* involve loss of a few genes at closely adjacent loci, resulting in what are known as *contiguous gene syndromes*. In several other microdeletion syndromes it is likely that more than a few loci are involved. The microdeletion syndrome are given below in Table 33.2 with the involvement of chromosome.

**Table 33.2:** Microdeletion syndromes

| Syndromes | Chromosome |
|---|---|
| 1. Williams | 7 |
| 2. Langer-Giedion | 8 |
| 3. Wilms' tumour, Aniridia, Genitourinary malformations and Retardation of growth and development (WAGR) | 11 |
| 4. Angelman | 15 |
| 5. Prader-Willi | 15 |
| 6. Rubinstein-Taybi | 16 |
| 7. Miller-Dicker | 17 |
| 8. Smith-Magennis | 17 |
| 9. DiGeorge | 22 |
| 10. Shprintzen | 22 |

## II. SEX CHROMOSOMES DISORDERS

**Klinefelter's Syndrome (47, XXY):** In childhood the diagnosis can be suspected in a male presenting with clumsiness or mild learning difficulties, particularly in relation to verbal skills. The verbal IQ is reduced by 10 to 20 points below that of unaffected siblings. It occurs approximately 1 per 1000 male live-births, is characterized by postpubertal testicular failure. The patient is a phenotype male with small testes, hyalinized testicular tubules, and azoospermia (low sperm count), resulting in infertility and variable signs of hypogonadism. These individuals are normal in appearance before puberty, but after puberty are recognized by their small testes and occasionally by enlargement of the breast (gynecomastia) and abnormal body proportions. The karyotype shows additional X chromosome i.e. 47, XXY chromosome constitution. A small proportion of cases show mosaicism, i.e. 46, XY/47, XXY. Rarely a male with more than two X chromosomes can be encountered such as 48, XXXY or 49, XXXXY. These are usually severely retarded.

- **Turner's Syndrome (45, X)**

It is characterized by a phenotypic female with gonadal dysgenesis and sexual immaturity. These women have primary amenorrhea (absence of menstruation) and infertility. It is associated with short stature and a host of somatic abnormalities including webbing of the neck, swelling of the hands and feet in infancy, increased carrying angle of the elbow, and cardiovascular and renal abnormalities.

Presentation can be at any time from pregnancy to adult life. Increasingly this syndrome is being detected during the second trimester as a result of routine detailed ultrasonography which can reveal either generalized edema (hydrops) or swelling localized to the neck (nuchal cyst or thickened nuchal pad). At birth some babies with the syndrome look entirely normal.

The most common finding is 45, X, sometimes erroneously referred to as 45, XO. It arises in 80 per cent of instances through a sex chromosome (X or Y) being lost in paternal meiosis. The frequency of this syndrome in liveborn girls is about 1/5000 but the frequency of monosomy X in consequence may be as high as 4 per cent. It has been observed that more than 99 per cent of 45, X conceptuses are spontaneously aborted. The chromosome findings are given in Table 33.3.

**Table 33.3:** Chromosome findings

| Karyotype | Frequency in per cent |
|---|---|
| 1. Monosomy X-45, X | 55 |
| 2. Mosaicism, e.g. 45, X/46, XX | 10 |
| 3. Isochromosome — 46, X, i (Xq) | 20 |
| 4. Ring — 46, X, r (X) | 5 |
| 5. Deletion — 46, X, del (Xp) | 5 |
| 6. Other | 5 |

- **XXX Females**

Approximately 0.1 per cent of all females have a 47, XXX karyotype. These women usually have no physical abnormalities but can show a mild reduction of between 10 and 20 points in intellectual skills. Observations revealed that the additional X chromosome is of maternal origin in 95 per cent of cases and usually arises from an error in meiosis I. Women with a 47, XXX karyotype usually show normal fertility and have children with normal karyotypes.

As with males who have more than two X chromosomes, women with more than three X chromosomes show a high incidence of mental retardation, the severity of which is directly related to the number of X chromosomes present.

- **XYY Male or XYY Syndrome**

One survey led to the discovery of individuals with aneuploidy of Y chromosomes such as 47, XYY and 48, XXYY. 47 XYY occurs in approximately 1 : 1000 male livebirths but was found in 4-20 per 1000 in the survey group. XYY males have normal sexual development and are fertile. The only phenotype uniformly associated with the XYY karyotype is tall stature. The additional Y chromosome must arise as a result of non-disjunction in paternal meiosis II or as a post zygotic event.

- **The Fragile X Syndrome**

It is better to classify this syndrome equally as a single gene disorder rather than a chromosomal abnormality. It affects approximately 1 in 2000 males and accounts for 4 to 8 per cent of all males with mental retardation. In older boys an adult males usually are characterized by facial abnormalities, including a narrow face with prominent forehead, prominent jaw and large ears. After puberty most affected males have large testes (macroorchidism, which is about 90 per cent).

There is mild connective tissue abnormalities and moderate to severe mental retardation. X chromosome which shows a *fragile site* close to the telomere at the end of the long arm at X q 27.3. A fragile site is a non-staining gap usually involving both chromatids at a point at which the chromosome is liable to break.

In this condition, detection of the fragile site involves the use of special culture techniques such as folate or thymidine depletion, which can result in the fragile site being detectable in up to 50 per cent of cells from affected males. Demonstration of the fragile site in female carriers is much more difficult and cytogenetic studies alone are not a reliable means of carrier detection but absence of the fragile site does not exclude a woman from being a carrier.

## III. DISORDERS OF SEXUAL DIFFERENTIATION

- **True Hermaphroditism**

This is rare condition and an individual has both testicular and ovarian tissue with ambiguous genitalia. In these cases, an ovary can be seen on one side and a testis on the other. There can be a mixture of ovarian and testicular tissue in the gonad known *ovotestis.* Most subjects having true hermaphroditism have a 46, XX karyotype and in many of these subjects the paternally derived X chromosome carries Y chromosome—specific DNA sequences as a result of illegitimate crossing-over between the X and Y chromosomes during meiosis I in spermatogenesis. A small proportion of subjects with true hermaphroditism are found to be Chimaers with both 46, XX and 46, XY cell lines, a situation analogous to free martins in cattle.

- **Male Pseudohermaphroditism**

In this condition there is gonadal tissue of one sex. The external genitalia may be ambiguous or of the sex opposite to that of the chromosomes. So in case of male pseudohermaphroditism there is a 46, XY karyotype with ambiguous or female genitalia.

The *cause* of this condition is *androgen insensitivity*. This condition is also known as testicular-feminization syndrome, the karyotype is normal male and the external phenotype is essentially that of a normal female. Internally the vagina ends blindly and the uterus and Fallopian tube are absent. Testes are located in the abdomen or in the inguinal canal which may be mistaken for inguinal hernia. This condition is caused by the absence of androgen receptors in the target organs, so that although testosterone is formed normally, its peripheral masculinizing effects are blocked. These androgen receptors are coded for by a gene on the X chromosome in which both deletions and point mutations have been identified. Other causes are:

(a) An incomplete form of androgen insensitivity known as *Reifenstein syndrome* in male having hypospadias, small testis and gynaecomastia.
(b) 5α-reductase deficiency, the enzyme responsible for androgen synthesis.
(c) Chromosome mosaicism (45X/46, XY) in which most individuals are normal males but a small proportion have ambiguous or female external genitalia.
(d) Campomelic dysplasia which is caused by mutations in the S0 X9 gene on chromosome 17, S0 X9 is believed to be an important gene in the regulatory pathway by which SRY causes masculinization of the undifferentiated fetal gonads.

- **Female Pseudohermaphroditism**

In this condition, the karyotype is female and the external genitalia are virilized so that they either resemble those of a normal male or are ambiguous.

The important cause is congenital hyperplasia. This can be caused by several different enzyme defects in the adrenal cortex, all of which show autosomal recessive inheritance. Reduced cortical production leads to an increase in ACTH secretion from anterior pituitary which in turn cause hyperplasia of the adrenal glands. In common form of congenital adrenal hyperplasia due to 21-hydroxylase deficiency, hormone synthesis switches from the manufacture of cortisol and aldosterone to the androgen pathway, leading virilization of a female fetus. The lack of cortisol and aldosterone usually leads to rapid collapse shortly after birth which can prove fatal unless appropriate hormone and electrolyte supplementation are initiated.

Rare causes may be an androgen secreting tumour and maternal androgen ingestion during pregnancy.

## Chromosome Breakage Syndromes

### • Chromosome Breakage and Sister Chromatid Exchange

Increased *chromosome instability* is demonstrated by an increased number of *sister chromatid exchanges* in cultured cells. A sister chromatid exchange, i.e. crossing-over of genetic material between the two chromatids of a chromosomes in mitosis, in contrast to recombination in meiosis I which is between homologous chromatids. Sister chromatid exchange can be demonstrated by differences in the uptake of certain stains by the two chromatids of each metaphase chromosome after two rounds of cell division in the presence of the thymidine analogue 5-bromodeoxyuridine (Bud R), which becomes incorporated in the newly synthesized DNA. There are normally about 10 sister chromatid exchange per cell from subjects with this syndrome and xeroderma pigmentosa. In the latter condition this is only apparent after the cells have been exposed to ultraviolet light.

It has been recognized that a small number of hereditary disorders are characterized by an excess of chromosome breaks and gaps as well as an increased susceptibility to neoplasia.

### • Ataxia Telangiectasia

It is present in early childhood with ataxia, oculocutaneous telangiectasia, radiation sensitivity and susceptibility to sinus and pulmonary infection. There is risk of leukaemia or lymphoma (10-20%). Cells of subject show chromosomal abnormalities such as chromatid gaps and breaks, which are enhanced by radiation. The gene for this condition is known as ATM (M = mutation) and maps to chromosome 11q 23. Its precise role is nuclear but when functioning normally it is probably responsible for alerting the cell to radiation damage thereby initiating DNA repair mechanisms and arresting cell division so that DNA repair can be completed before the cell enters mitosis. This is an autosomal recessive disorder.

### • Bloom's Syndrome

This occurs in small children with autosomal recessive disorder. They show a light-sensitive facial rash and reduced immunoglobulin levels (IgA and IgM). The there is risk of lymphoreticular malignancy approximately 20 per cent. Cultured cells show an increased frequency of chromosome breaks, particularly if they are exposed *in vitro* to ultraviolet light. The gene for this condition maps to chromosome 15q 26 where it encodes one member of a group of enzymes called DNA heli-cases. These are responsible for unwinding double-stranded DNA prior to replication, repair and recombination. Normally the Bloom's syndrome gene plays a major role in maintaining genome stability. When defective in the homozygous state, DNA repair is impaired and the rate of recombination between sister chromatids is increased. This can be demonstrated by looking for sister chromatid exchanges.

- **Fanconi's Anemia**

This is an autosomal recessive disorder associated with upper limb abnormalities involving the radius and thumb, increased pigmentation and a failure of the bone marrow leading to deficiency of all types of blood cells, i.e. pancytopenia. There is also an increased risk of neoplasia, particularly leukaemia, lymphoma and hepatic carcinoma. Multiple chromosomal breaks are observed in cultured cells and the basic defect lies in the repair of DNA strand cross-links. There are five known subtypes of Fanconi's anemia, each caused by recessive mutations at different autosomal loci. The commonest type A, maps to chromosome 16q 24. It is not known how any of the Fanconi's anemia genes are involved in maintaining the integrity of DNA cross-links.

- **Xeroderma Pigmentosa**

This condition exists in at least seven different types, all of which show autosomal recessive inheritance. Patients present with a light sensitive pigmented rash and usually die of skin malignancy in sun-exposed areas before the age of 20 years. Cells cultured from these patients only show chromosome abnormalities after exposure to ultraviolet light. These disorders are due to defects in the nucleotide excision repair pathway. This involves endonuclease cleavage 5′ and 3′ to each damaged nucleotide, excision of the damaged nucleotide(s) and finally restoration of the damaged strand using opposite strand as a template.

**Table 33.4:** Disorder of sexual differentiation

1. Seminiferous tubule dysgenesis (Klinefelter's syndrome)
   47, XXY, 48, XXYY, 49, XXXXY
2. Ovarian dysgenesis (Turner's syndrome)
   45, X, 46, X, i(Xq), 46, X, del (Xp), 46, X, r(X)
3. True hermaphroditism
   46, XX with Y-derived sequences
   46, XX/46, XY chimerism
4. Male pseudohermaphroditism androgen insensitivity Complete—testicular feminization Incomplete—Reifenstein syndrome. Inborn errors of testosterone biosynthesis. e.g. 5α-reductase deficiency, 45, X/46, XY mosaicism.
5. Female pseudohermaphroditism, congenital adrenal hyperplasia, Maternal androgen ingestion or androgen secreting tumour.

## Indications for Chromosome Analysis

Chromosome abnormalities can present in many different ways hence it is necessary to consider the indications for chromosome analysis under a number of different conditions. The indications are:

1. Known or Suspected Chromosome Abnormality
2. Multiple Congenital Anomalies
3. Unexplained or Undiagnosed Mental Retardation
4. Disorders of Sexual Differentiation
5. Multiple or Recurrent Miscarriages
6. Infertility
7. Unexplained Stillbirth
8. Malignancies and Chromosome Breakage Syndrome

1. **Known or Suspected Chromosome Abnormality:** Subject with a known or suspected chromosome abnormality have the diagnosis made or confirmed by a karyotypic analysis. But sometimes it is missed to diagnose, e.g. Down syndrome in a newborn. Hence it is necessary to follow the most accurate diagnostic procedure such as in case of Down syndrome it is necessary to determine whether the individual has trisomy 21 or an unbalanced translocation that might be inherited from a balanced carrier parent.
2. **Multiple Congenital Anomalies:** For the characteristic phenotype of major chromosome abnormalities is the combination of growth retardation, mental retardation, and somatic abnormalities always suggest the

chromosome abnormalities and require prometaphase analysis or FISH and when available, for microdeletion syndromes.

3. **Unexplained or Undiagnosed Mental Retardation:** Chromosome abnormalities cause at least one-third of the 50 per cent of mental retardation which is attributed to genetic factors. If the fragile X syndrome is a possibility it is important that cytogenetics laboratory should conduct correct culture. In many laboratories the Fragile X syndrome is now diagnosed by molecular methods rather than chromosome analysis.

4. **Disorders of Sexual Differentiation:** The birth of a baby with doubtful genitalia is a serious problem because not only for the emotional state of the family but also for ruling out the potentially life-threatening diagnosis of salt-losing congenital adrenal hyperplasia. A chromosome analysis should be done first beside other investigation.

   Disorders in later life with problems, e.g. delayed puberty, primary amenorrhoea or male gynecomastia are also strong indications for chromosome analysis which should be first investigation. Thus we can get a diagnosis like Turner's syndrome (45, X) or Klinefelter's syndrome (47, XXY). If the karyotype is normal then it leads to find out for endocrine abnormality.

5. **Multiple or Recurrent Miscarraiges:** 15 per cent of all recognized pregnancies terminate in spontaneous miscarraige and 50 per cent of these cases because of a chromosome abnormality. Sometimes some couples experience recurrent pregnancies loss, which is usually defined as three or more spontaneous miscarriages. Often no explanation can be given and many couples go on to have successful pregnancies. In such condition, 3-6 per cent of these couples, one partner is found to carry a chromosome rearrangement which predisposes to severe imbalance through malsegregation at meiosis. For which it is the usual procedure to do chromosome analysis for all couples who have experienced three or more spontaneous pregnancy losses.

6. **Infertility:** Infertility should undergo chromosome analysis particularly in case of azoospermia in the male partner. About 5 per cent of such men are found to have Klinefelter's syndrome. Rarely a complex chromosome rearrangment such as a translocation can cause such severe mechanical disruption in meiosis that complete failure of gametogenesis ensures.

7. **Unexplained Stillbirth:** One of the indications for chromosome analysis based on analysis of blood or skin collected from the baby before or as soon after death when there is growth retardation and at least one congenital abnormality in a stillbirth or neonatal death. Skin fibroblasts continue to be viable for several days after death. Chromosome abnormalities account for 5 per cent of all stillbirths and neonatal deaths and not all of these babies have multiple abnormalities which would immediately suggest a chromosomal cause.

8. **Malignancies and Chromosome Breakage Syndrome:** The frequent finding of specific chromosome abnormalities in leukaemia and solid tumours like retinoblastoma and Wilms' tumour has led to consider the diagnostic and prognostic value. Therefore, it has established the value of chromosome analysis in diagnosis and in prognosis of these disorders. In addition there are certain genetic diseases associated with chromosome

instability and a high risk of leukemia and other malignancies.

Clinical features suggestive of a chromosome breakage syndrome, such as a combination of photosensitivity and short stature, should also lead to appropriate chromosome fragility studies, such as analysis of sister chromatid exchange.

## Genetic Factors in Common Diseases

In common diseases it is not expected that either genetic or environmental factors to be solely responsible. In most cases both genetic and environmental factors are usually contributory, although sometimes one can appear more important than the other (Table 33.5).

**Table 33.5:** Showing the diseases and the factor responsible for it

| Genetic | Between the two | Environmental |
|---|---|---|
| 1. Duchenne muscular dystrophy | Diabetes mellitus | Tuberculosis |
| 2. Haemophilia | Hypertension | Scurvy |
| 3. Osteogenesis imperfecta | Cerebro-vascular diseases | |
| 4. Phenylketonuria | Coronary artery disease | |
| 5. Galactosaemia | Schizophrenia common | |
| 6. Club foot | cancers | |
| 7. Pyloric stenosis | Peptic ulcer | |
| 8. Dislocation of hip | | |

### A. Polygenic Inheritance

Some common diseases in small proportion are due to single gene causes but major proportion of the genetic basis can be considered to be due to an inherited predisposition or genetic susceptibility. These diseases result from a complex interaction of the effects of multiple different genes known as *polygenic inheritance.*

- **Evidence for genetic factor in common diseases and malformations**

1. *From epidemiologic study:* Observations from epidemiologic studies comparing the frequency of disease among genetically related individuals with that in the general population.
2. *From twin study:* More convincing line of evidence supporting the role of genetic factors in the causation of common diseases and malformations come from twin studies. It has been shown that between 45 per cent and 96 per cent of identical twins are concordant for diabetes whereas between only 3 per cent and 37 per cent of non-identical twins are concordant.
3. *Family history:* About 25 per cent to 50 per cent of diabetics have a family history of diabetics as opposed to less than 15 per cent in the general population.

### Procedure to show genetic susceptibility to common diseases

1. Family studies
2. Twin studies
3. Adoption studies
4. Population studies
5. Immigration studies
6. Polymorphic association
7. Biochemical studies
8. Animal models

- **Genetic susceptibility—Types and Mechanism**

1. Single gene inheritance of an abnormal gene product involved in a

particular metabolic pathway, e.g. early coronary artery disease due to familial hypercholesterolaemia (FH).

2. In some case, with a mutation in FH gene, the genetic susceptibility is the main determinant of the development of coronary artery but this can be modified by environmental alteration, e.g. reduction in dietary cholesterol and avoidance of other risk factors such as obesity, lack of exercise and smoking.
3. Inheritance of a single gene susceptibility does not necessarily lead to development of disease. In some diseases exposure to specific environmental factors will be the main determinant in the development of the disease, e.g. smoking or occupational dust exposures in the development of pulmonary emphysema in persons with $\alpha_1$-antitrypsin deficiency.

## I. Diabetes Mellitus

1. Insulin-dependent diabetes or type I or IDDM—accounts for 5-10 percent.
2. Non-insulin-dependent diabetes mellitus or type II or NIDDM.
3. Maturity onset diabetes of the young or MOD—accounts for 1-2 per cent.
4. Gestation diabetes—occurs during pregnancy and accounts for 1-3 per cent.

Diabetes can occur secondary to a variety of other rare genetic syndromes and non-genetic disorders, e.g. autosomal dominant muscle disorders myotonic dystrophy have an increased risk of developing diabetes. Evidence with animal model has shown that genetic factors are important.

## II. Hypertension

Hypertension are of two types:

A. **Secondary hypertension:** It is a consequence of another disorders, e.g. kidney disease or disorders of certain endocrine glands.

B. **Essential hypertension:** More common hypertension is prevalent in middle age group and its aetiology is known.

Family and twin studies suggest the importance of genetic factors in the aetiology of hypertension.

## III. Coronary Artery Disease

Twin studies of concordance for coronary artery disease vary from 15 per cent to 25 per cent for dizygotic twins and from 39 per cent to 48 per cent for monozygotic twins. Although these figures support the genetic factors, the low concordance rate for monozygotic twins clearly supports the importance of environmental factors.

## IV. Depressive Illness

Affective disorders such as:

1. Purely depressive or unipolar illness.
2. Manic-depressive or bipolar illness.

Adoption studies show that adoptees with bipolar affective disorders have a genetic basis. The possible mode of transmission has been suggested to be autosomal dominant, autosomal recessive and X-linked. Analysis of the age of onset and severity in familiar bipolar illness suggests the possibility of unstable triplet repeat mutations being involved in the aetiology of the illness.

## V. Alzheimer's Disease (AD)

The characteristic feature is dementia which is manifested by an irreversible and progressive global impairment of intellect, memory, social skills and control of emotional reactions in the presence of normal consciousness. Dementia is aetiologically heterogenous occurring secondary to both a variety of non-genetic causes, such as vascular disease and infections such as AIDS and genetic causes. The absence of linkage to chromosome 21 in many families with familial Alzheimer's disease led to find for other loci. Evidence of linkage was found for another locus mapping the long arm of chromosome 14 and mutations were identified.

## Single Gene Disorders

- **Introduction**

There are about 6000 single gene traits and disorders. Most of these are individually rare but together they affect between 1-2 per cent of the general population.

These disorders are caused by single mutant genes with a large effect on the patient's health. Single gene disorders are inherited in a simple *Mendelian fashion* and, are referred to as *Mendelian diseases*. These are inherited in *autosomal dominant, autosomal recessive,* or *X-linked fashion.*

Single gene disorders are about 5-10 per cent of pediatric hospital patients and childhood mortality. The major impact of single gene disorders occurs in the newborn period and early childhood, although their importance in adult life is increasingly being appreciated. Although many single gene disorders are rare, others are common and pose major problems such as:

1. Familial hypercholesterolemia – with high risk of premature coronary artery disease occurs in 1 in 500 persons.
2. Familial breast cancer and hereditary colon cancer—occurs in 1 in 300.
3. Sickle-cell anemia—affects in 1 in 400.
4. Cystic fibrosis—affects in 1 in 2000.

**Table 33.6:** Single gene disorders in human with molecular defect

| Disease | Molecular defect |
|---|---|
| I. **Autosomal dominant** | |
| 1. Antithrombin III deficiency | Deletions, point mutations, insertions |
| 2. Familial hypercholesterolaemia | Deletions, point mutations, duplication, insertions |
| 3. Huntington's disease | CAG triplet repeat expansion |
| 4. Myotonic dystrophy | CTG triplet repeat expansion |
| 5. Neurofibromatosis type I (von Recklinghausen's disease) | Deletion (Alu repeat), point mutations, insertions |
| 6. Osteogenesis imperfecta | Deletions, point mutations |
| II. **Autosomal recessive** | |
| 1. $\alpha_1$-antitrypsin deficiency | Point mutations, deletions, |
| 2. Congenital adrenal hyperplasia due to 21-hydroxylase deficiency | Point mutations, deletions |
| 3. Cystic fibrosis | Deletions, point mutations, insertions. |
| 4. Growth hormone-deficiency (rare familial form) | Deletions |
| 5. Phenylketonuria | Point mutations, deletions |
| 6. Tay-Sachs disease | Deletions, point mutations. |
| III. **X-linked** | |
| 1. Chronic granulomatous disease | Deletions, point mutations |
| 2. Duchenne/Becker muscular dystrophy | Deletions, duplication, point mutations |

*Contd.*

| Disease | Molecular defect |
|---|---|
| 3. Fragile X mental retardation | CGG triplet repeat expansion |
| 4. Haemophilia A (factor VIII) deficiency) | Point mutations, deletions, insertions, duplications, inversion. |
| 5. Haemophilia B (factor IX deficiency) | Point mutations, deletions, insertions |
| 6. Lesch-Nyhan syndrome *(HGPRT deficiency) | Deletions, point mutations, insertions, duplications |
| 7. Ornithine transcarbamylase deficiency | Deletions, point mutations |
| 8. Emery-Dreifuss muscular dystrophy | Deletion, insertion, point mutations |

*HGPRT = hypoxanthine-guanine phosphoribosyl transferase.

## I. Autosomal Dominant

1. **Huntington's Disease (Chorea) (HD):** It is characterized by slowly progressive selective cell death in the central nervous system. The prevalence is approximately 1 in 15000. Disorder occurs in middle to late adult life but a rare juvenile form may be present in younger age with different clinical features. The clinical features in the adult (around 40 years) are the onset of the disease is characterized by a slowly progressive movement disorder and an insidious impairment of intellectual function with psychiatric disturbance and eventual dementia and duration of the illness is about 15 years. Chorea is the most common movement abnormality manifested by subtle involuntary movements such as facial grimacing, twitching of the face and limbs, folding of the arms and crossing of the legs. As the disease progresses, the gait becomes very unsteady and speech becomes unclear. Memory impairment, poor concentration span, anxiety, aggressive behaviour, paranoia and alcohol abuse can also occur. Gradual deterioration in intellectual function leads to total incapacitation and dementia.

In juvenile Huntington's disease, disorders are seen before 20 years in 10 per cent cases. Instead of chorea there is rigidity with slow voluntary movement and clumsiness. There is decline in school performance heralds the onset of a severe progressive dementia and often in association with epileptic seizures. The average duration of illness is 10 to 15 years.

The genetics show autosomal dominant inheritance with a variable age of onset, close to complete penetrance and a very low mutation rate. Recent observation on identification of HD gene has provided such information. All HD possess an expansion of CAG polyglutamate repeat sequence located in the 5′ region of HD gene. Thus HD is caused by expansion of a triple repeat, a mutational mechanism which was first identified in man.

2. **Myotonia Dystrophy (MD):** It is most common form of muscular dystrophy in adults with an incidence of 1 in 8000. This illness shares many features in common with HD. In particular both illness show autosomal dominant inheritance. It is characterized by slow

**Table 33.7:** Genetical comparison of Huntington's disease and myotonic dystrophy

| | Huntington's disease | Myotonic dystrophy |
|---|---|---|
| I. **Mode of inheritance** | Autosomal dominant | Autosomal dominant |
| 2. **Chromosome locus** | 4p 16.3 | 19q 13.3 |
| 3. **Trinucleotide repeat** | CAG in 5′ translated region | CTG in 3′ untranslated region |
| 4. **Normal repeat size** | 6-35 | 5-35 |
| 5. **Expanded repeat size** | Intermediate 30-38<br>Full mutation 35-120 | Premutation 50-60<br>Full mutation 50-2000 |
| 6. **Protein product** | Huntington | MD protein kinase |
| 7. **Early onset form** | 'Juvenile'—paternally transmitted as most expansion in male meiosis | 'Congenital'—maternally transmitted due to probable size limit of expansion in male meiosis. |

progressive weakness and myotonia, i.e. tonic muscle spasm with prolonged relaxation, cataracts, cardiac conduction defects, disturbed gastrointestinal peristalsis, frontal balding and testicular atrophy. The diagnosis is based on myotonic discharges in EMG but it has been replaced by specific DNA mutational analysis.

Genetically it shows autosomal dominant inheritance with increasing severity in succeeding generations. In unaffected persons the CTG sequence lying 3′ to the DMPK gene consists of up to 30-35 repeats. Affected persons have an expansion of at least 50 copies of the CTG sequence.

Regarding MD protein kinase is not main cause of MD but that instead the RNA produced by the mutant DMPK gene somehow interferes with the cellular processing of RNA produced by other genes.

3. **Hereditary Motor and Sensory Neuropathy (HMSN):** It comprises a group of clinically and genetically heterogenous disorders manifested by slowly progressive distal muscle weakness and wasting. These orders are also known as Charcot-Marie-Tooth disease and peroneal muscular atrophy. The incidence is 1 in 2500.

The autosomal dominant HMSNI is the most common form of HMSM. The onset is of slowly progressive distal muscle weakness and wasting in the lower limbs between ages of 10-30 years. Similar changes occur in the upper limbs often with ataxia and tremor. With age locomotion becomes more difficult and feet tend to show exaggeration of their normal arch but most patients retain reasonable muscle strength and are not too seriously disabled. No impairment of vision, hearing and intellect and palpable thickening of peripheral nerves are detected in one-third cases. In other type of HMSN, the clinical features are same but differ in age of onset, rate of progression and presence of other neurological involvement.

Genetically it shows autosomal dominant, autosomal recessive or X-linked inheritance, although, autosomal dominant forms are by far the most common. In 90% of HMSNI cases, the disorder arises due to duplication of 1.5 megabases of genomic DNA on the short arm of chromosome 17. Located within this 1.5 megabases region on 17p is the gene which codes for a 22kDa glycoprotein known as peripheral myelin protein 22 (PMP-22). This is present in the myelin membranes of perinerves where it plays a major role in arresting Schwann cell division. Changes in this protein cause a peripheral neuropathy in a mutant strain of mice. Recently it has been observed that the HMSNI duplication is generated by misalignment and subsequent recombination between homologous sequences which flank the PMP-22 gene and usually occurs in male gametogenesis rather than female such as in Duchenne muscular dystrophy.

- **Neurofibromatosis (NF)**

Neurofibromatoses are of two types, NF-I and NF-II. NF-I is most common with an incidence of 1 in 3000 and NF-II has an incidence 1 in 35,000 and a prevalence of 1 in 200,000.

The clinical features include small pigmented skin lesions (cafe-au-bit spots), small soft fleshy growths known as neurofibromata which are benign tumours which are commonly in the skin. These appear in late childhood and adult life and show an increase in number with age. Other features include axillary freckling, macrocephaly, mild developmental delay and Lisch nodules which are small harmless raised pigmented hamartomata of the iris.

The characteristic feature of NF-II is the development of tumours in adult life involving eighth cranial nerves called acoustic neuromas and are now known as vestibular schwannomas.

NF-I shows autosomal dominant inheritance. About 50 per cent of NF-I are due to new mutations with the estimated mutation rate being 1 per 10,000 gametes. About 100 different mutations have been identified in the NF-I gene which include deletions, insertions, duplications and point mutations.

- **Cystic Fibrosis (CF)**

Cystic fibrosis (CF) is one of the most common autosomal recessive diseases of white populations, affecting about 1 in 2500 newborns. From Hardy-Weinberg considerations, it can be calculated the frequency of heterozygous carriers is about 1 in 25 individuals.

The disease is characterized clinically by sticky viscous secretions of the pancreas and lungs. If may be recognizable at birth by a condition called "meconium ileus" which is an intestinal obstruction. The diagnosis is commonly made by demonstrating high chloride in sweat.

There is pancreatic insufficiency with reduced enzyme secretion due to blockage of pancreatic duct by secretions. Eventually there is malabsorption and with an increase in fat in stools, which is amenable to treatment with pancreatic enzymes. Other problems are nasal polyp, rectal prolapse, cirrhosis and diabetes mellitus. Almost all males are sterile due to congenital bilateral absence of the vas deferens.

CF is the most serious autosomal recessive disorder. The explanations for high incidence include multiple CF loci, a high mutation rate, meiotic drive and heterozygote advantage.

The CF gene was eventually cloned by careful study of the approximately 500,000-

bp candidate interval defined by genetic analysis. Northern blotting with the fourth candidate gene as probe detected a transcript in lung, pancreas, sweat-glands, intestine and liver affected by CF. This gene is large, containing 26 exons and stretching across 250,000 bp of DNA. The mRNA transcript is 6129 - bp long and encodes a 1480-amino acid protein. The sequence of the protein, termed CFTR (cystic fibrosis transmembrane regulator) suggested that it is anchored in the cell membrane and binds ATP.

- **Spinal Muscular Atrophy (SMA)**

Spinal muscular atrophy (SMA) is a clinically and genetically heterogeneous group of disorders which are common genetic causes of death in childhood. This is characterized by degeneration of the anterior horn cells of the spinal cord leading to progressive muscle weakness and ultimately death. There are three common childhood types of SMA which have been observed with collective incidence of 1 in 10,000.

- **Types**

1. **Type I SMA (Werdnig-Hoffman disease):** Children at birth show severe hypotonia and lack of spontaneous movement, normal intellectual activity, difficulty in swallowing and respiration due to muscle weakness. Death occurs within the first two years of life. Diagnosis is established by nerve conduction.
2. **Type II SMA:** Age of onset is between 6 and 18 months and less severe than type-I SMA. Muscle weakness and hypotonia are the main clinical features. Children can sit without any help but unable for independent locomotion. The rate of progress is slow and affected children survive into early adult life.
3. **Type-III SMA (Kugelberg-Welander disease):** The age of onset is after 18 months, mild form, able to walk without support, slow progress of muscle weakness, scoliosis due to weakness of the spinal muscles.

Genetically all three types show autosomal recessive inheritance. Two distinct genes have been isolated which show a high incidence of deletion in SMA patients.

- **Duchenne Muscular Dystrophy**

Duchenne muscular dystrophy (DMD) is the most common and severe form of muscular dystrophy. Similar but wider form known as Becker muscular dystrophy (BMD) is caused by mutations in the same gene. The incidence of DMD and BMD form are 1 in 3500 males and 1 in 20,000 males.

This occurs in male between the ages of 3 and 5 years with slowly progressive muscle weakness resulting in an awkward gait, inability to run, difficulty to rise from the floor, the patient uses wheel chair at the age of 11 years. Deterioration leads to a lumbar lordosis, joint contractures and cardiorespiratory failure leading to death at the age of 18 years. There is apparent increase in the size of calf muscles due replacement of muscle fibres by fat and connective tissue and this is known as "pseudohypertrophy" for which it also known as pseudohypertrophic muscular dystrophy.

In BMD the clinical feature is very similar with much less aggressive course. Genetically DMD and BMD show X-linked recessive inheritance. The DMD locus has been mapped to chromosome $Xp^{21}$ and gene was found to be the largest in humans. It consists of 2.3 megabase of genomic DNA, of which only 14kb are transcribed into mRNA. The gene contains 79 axons and is transcribed in muscle and also in brain.

The DMD gene encodes a 427-kDa protein known as *dystrophin*. Absence of dystrophin as in DMD results in gradual muscle cell degeneration which is assessed by immunofluorescence. It links intracellular action with extracellular laminin.

- **Haemophilia**

This disease is X-linked recessive inheritance invariably transmitted by females without affecting them but males manifest sign of disease, i.e. *mother carries the disease* and *son is affected and daughter carries the disease only* (Fig. 33.1).

The condition is characterized by *a marked increase in the coagulation time without affecting bleeding time* because minute breaches in the skin are sealed by contraction of the capillaries. Haemophilia blood may take 1-12 hours for coagulation. The abnormality is responsible for deficiency of antihaemophilic globulin (AHG, factor VIII) and the closely related Christmas disease is due to lack of the factor IX.

- **Types**

There are three types of haemophilia:

I. **Haemophilia A:** Deficiency of AHG of factor VIII. It is the most common severe inherited disease with incidence of 1 in 5000 males.
II. **Haemophilia B:** Deficiency of Christmas factor or factor IX, affects approximately 1 in 40,000 males.
III. **Haemophilia C:** Deficiency of plasma thromboplastin antecedent (PTA).

Haemophilic father will have normal son but daughters are carrier. (Many Royal families of Europe are suffering from haemophilia.) If a haemophilic man marries a haemophilic carrier, it is possible to have a female haemophilia. This occurrence is very rare. The disease is also found in domestic animals, e.g. dogs and these are maintained for experimental purpose.

The clinical features, are similar in both haemophilia A and B. It varies from mild bleeding following major injury or surgery to spontaneous haemorrhage into muscles and joints. Levels of factor VIII or IX below 1% are usually associated with a severe haemorrhagic tendency. Haemorrhage into joints causes severe pain and swelling leading to arthropathy and disability.

- **Genetics**

Genetically both forms show X-linked recessive inheritance. The loci lie close together near the distal end of the long arm of the X chromosome.

**Haemophilia A:** The factor VIII gene is located at the tip of the long arm of the X chromosome and is large spanning 186 kb with 26 exons and a 9 kb mRNA transcript. Deletions account for 5% of all cases and usually cause complete absence of factor VIII expression.

Carrier detection is done by

1. Linkage analysis using polymorphic intragenic marker
2. Specific mutation analysis
3. Prenatal diagnosis by mutation or linkage analysis.

**Haemophilia B:** The gene for factor IX is 34 kb long and contains 8 exons. Over 100 different point mutations, deletions and insertions have been reported.

Carrier detection is done by:

1. Linkage analysis
2. Direct mutation analysis.

**Test**

1. Thromboplastin generation test
2. Partial thromboplastin time
3. Linkage analysis
4. Specific mutation analysis.

**Treatment**

1. Blood transfusion improves temporarily.
2. Successful replacement treatment using plasma-derived factor VIII or IX.
3. Gene therapy in animal models offer hope that this will soon be possible in humans.

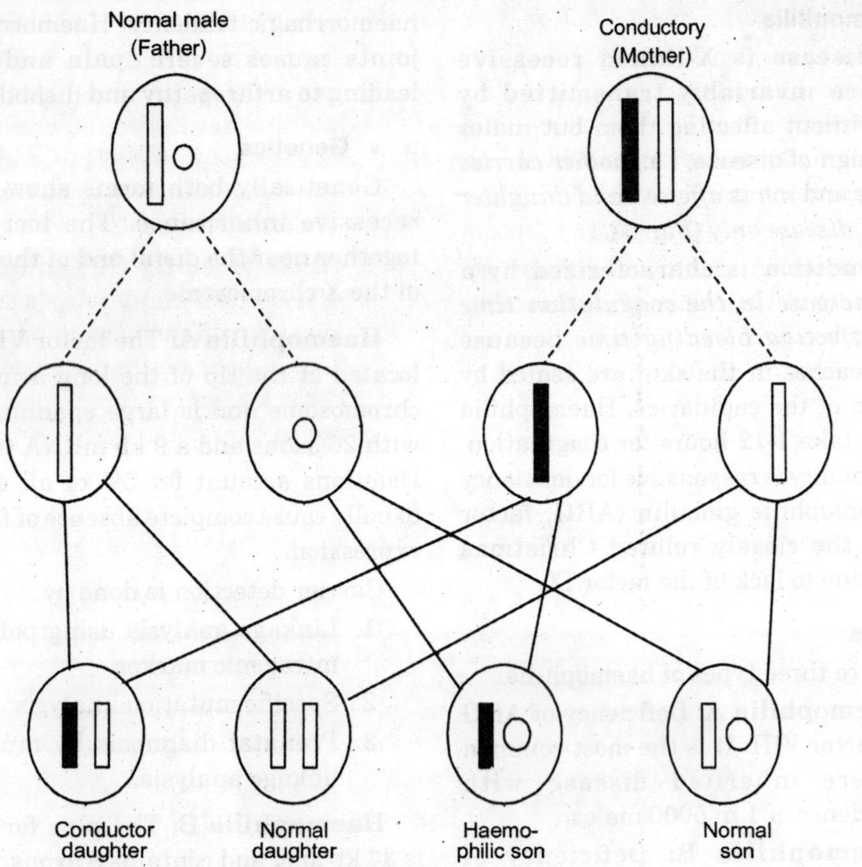


**Fig. 33.1:** Transmission of Haemophilia.

- **Disadvantages of Factor VIII or IX Therapy**

1. Factor VIII has a half-life of 8 hours so that repeated infusions are necessary for elective surgery or major trauma.
2. Purification process does not prevent transmission of viral infection, e.g. hepatitis B and HIV. A recombinant factor VIII is available but expensive for wide use.
3. About 10 per cent patients of both forms of haemophilia develop inhibitory antibodies to the relevant factor which their immune systems recognize as a foreign agent.

This problem may be overcome by using porcine factor VIII or by immunosuppression.

- **Other Congenital Haemorrhagic Diseases**

1. **Congenital parahemophilia and Hageman trait:** These are bleeding diseases, which may be caused by failure of thrombin formation due to a congenital deficiency of proaccelerin and Hageman factor respectively.
2. **Von Willebrand's disease:** Haemophilia A is a factor VIII protein defect that interrupts its clotting but not the oligosaccharide component or platelet adherent function. But this disease is due to an inherited defect of a specific oligosaccharide moiety or platelet adherent factor (von Willebrand's factor) plus deficiency of AHG.
3. **Afibrinogenemia:** It is an acute congenital shortage of fibrinogen due

to a non-X-linked recessive genetic defect, it may cause non-coagulability, severe hemophilia like bleeding syndromes may occur and may often occur and may often end in death unless treated.

4. **Hypoprothrombinemia:** Low blood prothrombin may develop due to a hereditary deficiency of prothrombin (or more frequently due to a dietary deficiency of vit. K or liver disease).

## Mutational Analysis of Single Gene Disorders

Analysis of mutation for single gene disorders can be divided into:

I. Direct Mutation Deletion Methods
II. Mutation Tracking Methods

### I. Direct Mutation Deletion Methods

1. *Deletion Detection*
2. *Mutation Specific RFLPs*
3. *Allele Specifi: Oligonucleotides*

1. **Deletion Detection:** This method was first used in haemoglobinopathies. Immature RBC, reticulocyte comtain only globin mRNA, from which a radioactive cDNA probe for globin genes can be prepared. This probe can be used to identify and localize DNA fragments which contain the globin genes in an autoradiograph of a Southern blot. In the case of δβ thalassaemia, the DNA probe complementary to the β-globin gene, does not hybridize with all the same DNA fragments in two or more different restriction digests as in normal person consistent with a deletion being the cause of this disorder.
2. **Mutation Specific RFLPs (restriction fragment length polymorphism):** Mutation responsible in some single gene disorders, eliminates a restriction enzyme recognition site. This direct method has been used in sickle-cell disease using the restriction enzyme Mst-II. This method has limitation clinically as it is rare for mutation to occur within the recognition sequence of restriction endonucleases.
3. **Allele Specific Oligonucleotides:** If it is a point mutation of a single gene disorder, then there is possibility to make short synthetic oligonucleotide DNA sequences corresponding to the normal and mutant sequences. By using appropriate reaction conditions, the normal and mutant DNA can be distinguished by demonstrating differential hybridization of the normal and mutant oligonucleotide sequences to the normal and mutant DNA sequences.

### II. Mutation Tracking Method

If a gene has been mapped but not get cloned so that mutation screening or direct mutation detection is not available, it is still possible to use inherited DNA sequence variation to track or follow a mutation within a family. There are a number of different types of *DNA sequence or polymorphisms* which can be used.

**DNA Sequence Polymorphisms:** In the hukan genome, there is an enormous amount of DNA sequence variation clinically the variation can be grouped into main types:

1. *Restriction Fragment Length Poly-morphisms (RFLPs).*
2. *Hypervariable Tandem Repeat DNA Length Polymorphism.*

1. **Restriction Fragment Length Polymorphisms:** Nucleotide sequence variation of the human genome commonly occurs once every 200 bp. These single base pair differences in DNA nucleotide sequences are inherited in a Mendelian codominant manner and have no phenotypic effects as they

usually occur in intergenic non-coding DNA. If a difference in DNA sequence occurs within the recognition sequence of a restriction enzyme, the fragments produced by that restriction enzyme will be of different lengths in different people. This can be recognized by the altered mobility of the restriction fragments on gel electrophoresis, so-called restriction fragment length polymorphisms, RFLPs.

2. **Hypervariable Tandem Repeat DNA Length Polymorphism:** The different classes of tandemly repeated DNA sequences in the human genome have proven to be useful clinically in the indirect methods of mutation detection or mutation tracking.

- **Clinical Applications of Mutation Tracking**

**Locus-linked DNA sequence tracking:** If a locus for a genetic disorder is shown to be linked to a DNA sequence variant or polymorphism, this can be used:

(a) For prenatal diagnosis;
(b) To detect the presence of a gene in a person before the onset of symptoms, i.e. presymptomatic diagnosis;
(c) To identify female carriers of X-linked disorders.

- **Allele specific DNA sequence variants**

Certain DNA sequence variants have been shown to occur in association with specific disease mutations more commonly than would be predicted by their individual frequencies in a population. This type of RFLP can be termed an allele specific DNA sequence variant and is said to be in linkage disequilibrium with the mutation.

- **DNA haplotypes**

When the family does not possess the necessary variation at closely linked polymorphic sites to determine on which chromosome the disease allele is located. It is often possible, however, by looking at a number of different closely linked DNA sequence polymorphisms, for the family to be made informative by establishing DNA haplotype, i.e. the pattern of DNA sequence polymorphisms flanking the disease and normal alleles in that particular family. This can be used clinically in carrier detection, presymptomatic and prenatal diagnosis.

## Congenital Abnormalities or Anomalies and Genetics (Dysmorphology)

Congenital abnormalities or birth defects are being used in general sense which encompasses all types of structural abnormalities which occurs in an embryo, foetus or newborn infant. But these terms cannot provide the insight into the underlying mechanism. More specific definition has been selected called *dysmorphology*, i.e. field of study of malformation which has been classified for added advantage of providing a combined clinical and aetiological classification.

- **Dysmorphology**

*Dysmorphology* is the part of genetics which is concerned with the diagnosis and management of *congenital anatomic abnormalities.* Congenital abnormalities are classified in the following way:

I. *Malformation*
II. *Disruption*
III. *Deformation*
IV. *Dysplasia*

### I. Malformation

A malformation is a morphological defect of an organ, or a part of an organ, resulting from an *intrinsically* abnormal developmental process, e.g. cleft lip, cleft palate, congenital heart defects such as ventricular or septal defects, neural tube defects such as anencephaly or lumbo-sacral myelomeningocele. Most malformations involving only a single organ show multifactorial inheritance, implying an interaction of many genes with environmental factors.

### II. **Disruption**

Disruption means an abnormal structure of an organ or tissue as a result external factors disturbing the normal developmental process. This is also known as secondary or extrinsic malformation. The extrinsic factors causing disruption of normal development are trauma, infection, ischaemia and teratogens. The examples are radial aplasia or failure of normal development of the radius and associated structures, can occur as a malformation, occurring in association with cardiac abnormalities in the Holt-Oram syndrome.

This is an autosomal dominant disorder resulting from a mutation in the TBX5 gene encoding a transcription factor. Radial aplasia may also occur, however, as a disruption, as part of the embryopathy caused by the drug thalidomide.

### III. **Deformation**

Deformation is an abnormal form, shape or position of a part of the body caused by mechanical forces. The forces may be:

(a) *Extrinsic to the foetus:* Alterations in the amount of amniotic fluid causing compression of the foetus, e.g. equinovarus foot, or club foot during oligohydramnios (less amniotic fluid).

(b) *Intrinsic to the foetus:* Diminished foetal movement caused by neurologic or muscular disorders.

### IV. **Dysplasia**

Dysplasia is an abnormal organization of cells into tissues and its morphologic consequences. Because the defect can involve all anatomic sites in which the affected tissue is found, dysplasias may show widespread involvement, e.g. achondroplasia, abnormality in chondrocytes; skeletal dysplasia such as thanatophoric dysplasia, caused by mutations in FGFR3; ectodermal dysplasia. Most dysplasias are caused by single gene defects and are associated with high recurrence risks for siblings and/or offspring.

- **Sequence**

Multiple anomalies in a subject may be causally or pathogenically related. The pattern of multiple anomalies is the concept of *sequence.* A *sequence* is a pattern of multiple anomalies derived from a single prior anomaly or mechanical factor. This can often be a single organ malformation. The 'Potter sequence' is caused by oligohydramnios, which can result from either diminished fetal urine output, secondary to failure of normal kidney development or to urethral obstruction, or chronic leakage of amniotic fluid. In either case, the oligohydramnios causes foetal compression, resulting in the characteristic flattened face, abnormal position and development of hands and feet, and pulmonary hypoplasia with ensuing respiratory insufficiency and death.

- **Syndrome**

A *syndrome* is a pattern of multiple anomalies thought to be pathogenically related, but not known to represent a simple single sequence, e.g. Marfan's syndrome.

- **Association**

*Association* means recognition of the fact that certain malformations tend to occur together yet this non-random occurrence of abnormalities cannot be explained on the basis of a sequence or a syndrome. The differences from a syndrome are the lack of consistency of abnormalities from one affected individual to another and the absence of a satisfactory underlying explanations. The name of associations are often expressed by the first letters of the organs or systems most commonly involved, e.g. VATER association features Vertebral, Anal, Tracheo-esophageal and Renal abnormalities. Associations are not usually genetic in origin though the underlying cause is generally unknown.

**Table 33.8:** Congentital abnormalities

| Major congenital abnormalities | Minor congenital structural abnormalities |
|---|---|
| **I. Cardiovascular system** | 1. Preauricular pit or tag |
| 1. Ventricular septal defect | 2. Epicanthus folds |
| 2. Atrial septal defect | 3. Lacrimal duct stenosis |
| 3. Patent ductus arteriosus | 4. Brushfield spots in the iris |
| 4. Fallot's tetralogy | 5. Lip pits |
| **II. Central nervous system** | 6. Single palmar crease |
| 1. Anencephaly | 7. Fifth finger clinodactyly |
| 2. Hydrocephaly | 8. Syndactyly between second and third toe |
| 3. Microcephaly | 9. Supernumerary nipple |
| 4. Lumbo-sacral spina bifida | 10. Umbilical hernia |
| **III. Gastro-intestinal** | 11. Hydrocele |
| 1. Cleft lip | 12. Sacral pit or dimple |
| 2. Cleft palate | |
| 3. Diaphragmatic hernia | |
| 4. Oesophageal atresia | |
| 5. Imperforated anus | |
| **IV. Limb** | |
| Transverse amputation | |
| **V. Urogenital** | |
| 1. Bilateral renal agenesis | |
| 2. Polycystic kidneys | |
| 3. Urinary bladder existrophy | |

## Causes of Malformation

I. **Genetic**
   A. Chromosomal
   B. Single gene
   C. Multifactorial

II. **Environmental**
   A. Drugs and chemicals
   B. Infections (maternal)
   C. Maternal illness
   D. Physical agents

III. **Unknown**

### I. Genetic

A. *Chromosomal defects*: **About 6% cases of chromosomal abnormalities are found. These are usually due to**

**Table 33.9:** Congenital abnormalities may be due to single gene defects

| Isolated | Inheritance |
|---|---|
| **I. Central nervous system** | |
| 1. Hydrocephalus | XR |
| 2. Megalencephaly | AD |
| 3. Microcephaly | AD/AR |
| **II. Ocular** | |
| 1. Anindia | AD |
| 2. Cataracts | AD/AR |
| 3. Microphthalmia | AD/AR |
| **III. Limb** | |
| 1. Brachydactylyl | AD |
| 2. Ectrodactylyl | AD |
| 3. Polydactyl | AD |
| **IV. Other** | |
| Infantile polycystic kidneys | AR |

| Syndromes | Inheritance | Abnormalities |
|---|---|---|
| 1. Apert syndrome | AD | Craniosynostosis, syndactyly |
| 2. EEC syndrome | AD | Ectodermal dysplasia, ectrodactyly cleft lip/palate. |
| 3. Meckel syndrome | AR | Encephalocele, Polydactyly, Polycystic kidneys |
| 4. Roberts syndrome | AR | Celft lip/palate phocomelia |
| 5. van der Woude syndrome | AD | Cleft lip/palate, lip pits. |
| AD—autosomal dominance<br>XR—sex-linked recessive | | AR—autosomal recessive |

autosomal imbalance such as duplication, deletion, trisomy or monosomy. These will result in severe structural and developmental abnormality.

B. *Single gene defects:* About 7.5% cases are involved. Some of these are isolated, i.e. they involve only one organ or system. Other single gene defects result in multiple congenital abnormality syndromes involving many organs or systems which do not have any definite underlying embryological relationship. There are sufficient evidences from dysmorphology that several different mutations, allelic or non-allelic, can cause similar if not identical malformations.

C. *Multifactorial:* The multifactorial factor accounts for the majority of congenital abnormalities in which genetic factors can be clearly implicated such as isolated malformations involving the heart, central nervous system and kidneys.

## II. Enviromental (Teratogens)

An agent which may be chemical or physical, produces or raises the incidence of congenital malformations by interfering with normal embryonic or foetal development is known as *teratogens.*

**Table 33.10:** Isolated malformations which may show multifactorial inheritance

A. **Cardiac**
1. Atrial septal defect
2. Fallot's tetrology
3. Patent ductus arteriosus
4. Ventricular septal defect

B. **Central nervous system**
1. Anencephaly
2. Encephalocoele
3. Spina bifida

C. **Genitourinary**
1. Hypospadias
2. Renal agenesis
3. Renal dysgenesis

D. **Other**
1. Cleft lip/palate
2. Congenital dislocation of hips
3. Talipes

### A. Drugs and Chemicals

**Table 33.11:** Drugs have been proved to be teratogenic

| Drugs | Effects |
|---|---|
| 1. Alcohol | Cardiac defects, microcephaly, characteristic facies |
| 2. Chloroquine Pyrimethamine | Chorioretinitis, deafness |
| 3. Diethylstilbesterol | Uterine malformations, vaginal adenocarcinoma |
| 4. Lithium | Cardiac defects (Ebstein's anomaly) |

*Contd.*

| Drugs | Effects |
|---|---|
| 5. Phenytoin | Cardiac defects, cleft palate, digital hypoplasia |
| 6. Retinoids | Ear and eye defects, hydrocephaly |
| 7. Streptomycin | Deafness |
| 8. Tetracycline | Dental enamel hypoplasia |
| (a) Isotretinoin | " |
| 9. Thalidomide | Phocomelia, cardiac and ear abnormalities |
| (b) Radiation | |
| 10. Valproic acid | Neural tube defects, characteristic facies |
| 11. Warfarin | Nasal hypoplasia, stippled epiphyses |
| 12. Androgens | |
| 13. Mercurials | |

## B. Infections (meternal)

Various infections interfere the embryogenesis and foetal development and causing damage particularly the brain, eyes and ears which are susceptible to infections.

### Intrauterine Infections

1. Rubella
2. Cytomegalovirus
3. Toxoplasmosis
4. Herpes simplex
5. Varicella zoster
6. Syphilis.

## C. Maternal illness

1. *Diabetes mellitus (insulin-dependent):* It causes—

**Table 33.12:** Infections as teratogenic effect

| Infections | Teratogenic effect |
|---|---|
| A. **Viruses** | |
| 1. Cytomegalovirus | Chorioretinitis, deafness, microcephaly |
| 2. Herpes simplex | Microcephaly, microphthalmia |
| 3. Rubella | Microcephaly, cataracts, retinitis, cardiac defect |
| 4. Varicella Zoster | Microcephaly, chorioretinitis, skin defects |
| B. **Bacteria** Syphilis | Hydrocephalus, Osteitis, retinitis |
| C. **Parasites** Toxoplasmosis | Hydrocephalus, microcephaly, cataracts, chorio-retinitis, deafness. |

1. Congenital heart disease
2. Nural tube defects
3. Sacral agenesis
4. Femoral hypoplasia
5. Sirenomelia (mermaidism)

Non-insulin-dependent diabetes and gestational diabetes do not convey an increased risk of congenital malformations in the offspring.

## D. Physical agents

1. The physical agents are—
   (a) Radiowaves
   (b) Ultrasound
   (c) Magnetic fields
   (d) Ionizing radiation
   (e) Hyperthermia.
2. **Phenylketonuria**
   It causes—
   (a) Mental retardation
   (b) Microcephaly
   (c) Congenital heart defects
3. **Epilepsy:** Epilepsy itself does not produce any congenital defects but the drugs used for it are teratogenic, e.g. (a) phenytoin which produces cleft lip/palate, (b) sodium valproate produces anencephaly, spina bifida.

**E. Unknown cause:** Clear cause is not known in some cases such as—

1. Isolated diaphragmatic hernia
2. Tracheo-oesophageal fistula
3. Anal atresia
4. Single limb reduction defects
5. Isolated limb defect, e.g. absence of a hand

## Polygenic or Multifactorial Disorders

The disorders result from the interaction of *multiple genes,* some of which may have a major effect but many of which may have a relatively minor effect.

## • Polygenic or Multifactorial Diseases

I. **Congenital malformation**

Cleft palate
Cleft lip
Congenital dislocation of the hip
Congenital heart diseases
Neural tube defects
Pyloric stenosis
Talipes

II. **Adult onset**

Diabetes mellitus
Epilepsy
Glaucoma
Hypertension
Ischaemic heart disease
Manic depression
Schizophrenia

## • Polygenic Inheritance and the Normal Distribution

Polygenic inheritance or quantitative inheritance which involves the inheritance and expression of a phenotype being determined by many genes at different loci with each gene exerting a small additive or interactive effect. Additive effect means the effects of the genes are cumulative, i.e. no one gene is dominant or recessive to another. Many human characteristics such as—

1. Skin
2. Height
3. Intelligence
4. Head circumference
5. Blood pressure
6. Dermatoglyphies (ridge count)

show a continuous distribution. If it is plotted then this takes a form of symmetrical bell-shaped curve distributed evenly about a mean (Fig. 33.2).

It can be shown that a phenotype with a normal distribution in the general population can be generated by polygenic inheritance involving the action of many genes at different loci, each of which exerts an equal additive effect. Thus, the clinical impact of multifactorial diseases is important in both the neonatal period and in adult life. Conceptually, this group of diseases poses the challenge of sorting out the ways in which the additive effects of several to many genes create the predisposition to disease, which in turn is manifested only in the presence of appropriate environmental triggers. It is expected that a combination of molecular genetics approaches, gene mapping, and genetic epidemiology will allow a clearer definition of these genetic determinants and of the genetic heterogeneity underlying disease susceptibility. Models for how such interactions can cause disease, and methods for identifying the nature and contribution of genetic factors in such diseases will be discussed in the following discussions.

## • Multifactorial Inheritance—the Liability-Threshold Model

It has been tried to extend the polygenic theory of inheritance of *continuous* traits to

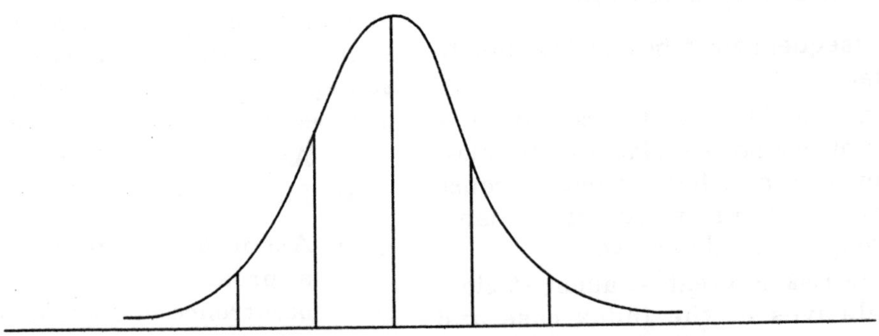

**Fig. 33.2:** The normal distribution.

try to account for *discontinuous* multifactorial disorders. According to the *liability-threshold* model, all of the factors which influence the development of a multifactorial disorder, whether genetic or environmental, can be considered as a single entity known as *liability.* The liabilities of all individuals in a population form a continuous variable, which has a normal distribution in both the general and in relative of affected individuals. However, the curves for these relatives will shifted to the right, with the extent to which they are shifted being directly related to the closeness of their relationship to the effected index case.

For a discontinuous phenotype, i.e. affected or not affected with an underlying continuous distribution, it is suggested that a threshold exists above which the abnormal phenotype is expressed. In the general population the proportion beyond the threshold is the *population incidence* and among relatives the proportion beyond the threshold is the *familial incidence.*

Liability includes all factors which contribute to the cause of the condition. Deleterious liability can be viewed as consisting of a combination of several bad genes and adverse environmental factors. Liability cannot be measured but the mean liability of a group can be determined from the incidence of the disease in that group using statistics of the normal distribution. The units of measurement are standard deviations and these can be used to estimate the correlation between relatives.

- **Consequences of the Liability/Threshold Model**

1. The incidence of the condition is greatest amongst relatives of the most severely affected patients because they are the most extreme deviants along the liability curve.
2. The risk is greatest amongst close relatives of the index case and decreases rapidly in more distant relatives.
3. If there is more than one affected close relative, then the risks for other relatives are increased.
4. If the condition is more common in individuals of one particular sex, then relatives of an affected individual of the more frequently affected sex.
5. The risk of recurrence for first degree relatives, i.e. siblings and offspring, approximates to the square root of the general population incidence, i.e. if the incidence is 1 in 1000 the sibling and offspring risk will equal approximately 1 in 32 or 3 per cent.

- **Heritability**

Heritability means the proportion of the total phenotypic variance of a condition which is caused by additive genetic variance.

It is difficult to assess an individual's liability for a particular disorder, but there is possibility to estimate what proportion of the aetiology can be correlated to genetic factors as opposed to environmental factors which is referred to as heritability.

Estimates of the heritability of a condition or trait provide an indication of the relative importance of genetic factors in its causation, so the greater the value for the heritability the greater the role of genetic factors.

## Identification of Genes which are Responsible for Multifactorial Disorders

With the characterization of highly-polymorphic markers scattered throughout the human genome, it has become possible to search for the genes responsible for susceptibility to common multifactorial diseases. The two major approaches are:

I. Associations of Disease
II. Linkage Analysis

I. **Associations of Disease:** This approach seeks the correlated occurrence of specific alleles at a genetic locus and the disease in population. Usually this involves the use of candidate genes, genes that encode proteins known or thought to be involved in the disease process. Finding that a specific allele is present in significantly higher frequency in individuals with the disease than in unaffected individuals suggests that the product of that allele, plays a role in the pathogenesis of the disease e.g. the epsilon 4 (ε4) allele of the apolipoprotein E gene may be an important risk factor for late-onset Alzheimer disease.

II. **Linkage Analysis:** Linkage studies seek for the cotransmission, in families, of a marker locus, regardless of the specific allele at the locus, with the disease. This approach provides direct evidence for the role of a genetic locus, or one very close to it, in the disease process.

This has proved valuable in mapping single gene disorders by studying the cosegregation of polymorphisms with the disease until a linked marker has been identified.

Animal models of human multifactorial diseases have also been used to identify genetic factors, whose human homologues can then be studied.

- **Disadvantages**

This type of method is very difficult in multifactorial disorders for the following reasons:

1. If a multifactorial disorder has a polygenic underlying genetic susceptibility, then it is unlikely that alleles at a single locus will make a major contribution. It is very difficult to develop strategies for detecting linkages of additive polygenes each of which makes only a small contribution to the phenotype.
2. There is a variable age of onset of diseases, so the genetic status of unaffected family members cannot be assessed with certainty.
3. One or two living affected members are present in most families with multifactorial diseases hence the number of informative meises available for study is small.
4. Some multifactorial diseases like coronary artery disease and schizophrenia, may be aetiologically heterogeneous with different genetic and environmental mechanisms involved in different subtype which can not be easily distinguished at the phenotypic level which makes the analysis very difficult.

# 34 TREATMENT OF HUMAN GENETICAL DISEASES AND GENE THERAPY

1. Conventional therapy, 2. Various types of treatment of genetic diseases.

## Introduction

Genetic disorders show progressive disability or chronic ill-health and there is no effective treatment. But due to recent development of biotechnology has inspired to treat the genetic disorders by gene therapy and also production of some proteins biosynthetically to treat the other genetic and non-genetic disorders. To keep a perspective on the limitations of the gene therapy, an approach to the conventional treatment in genetic diseases is discussed.

- **Conventional Therapy**

Conventional therapy cannot cure or ameliorated because the underlying gene or gene product were not identified for basic metabolic or molecular disorders. If it is observed, then the following may be applied such as Table 34.1.

(a) Dietary restriction in phenylke-tonuria;
(b) Hormone replacement in congenital adrenal hyperplasia;
(c) Vit $B_6$ in homocystinuria (Fig. 34.1).

Table 34.1 shows the various types of conventional therapy for genetic diseases.

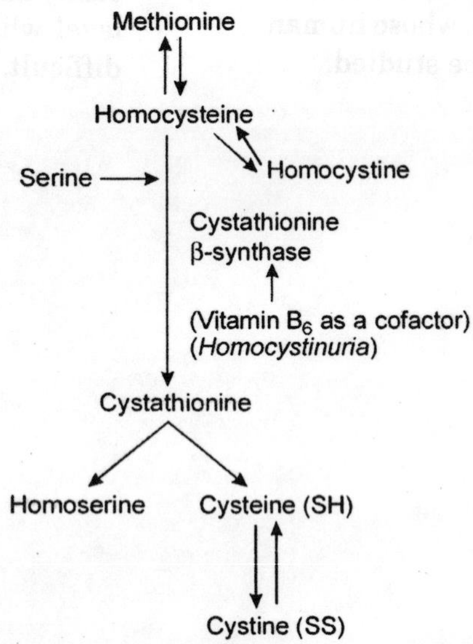


**Fig. 34.1:** Flow chart of homocysteine metabolism—most cases of homocystinuria are due to deficiency of cystathionine β-synthase which contains vitamin $B_6$ (Pyridoxine) as a cofactor.

**Table 34.1:** Various types of treatment for genetic diseases

| Types of treatment | Diseases |
|---|---|
| 1. **Enzyme induction by drugs** | |
| Phenobarbitone | Congenital non-haemolytic jaundice |
| 2. **Replacement of deficient enzyme or protein** | |
| (a) Blood transfusion | SC1D due to adenosine deaminase deficiency |
| (b) Bone marrow transplantation | Mucopolysaccharidoses |
| (c) Enzyme/Protein preparations | |
| i. Trypsin | Trypsinogen deficiency |
| ii. $\alpha_1$-Antitrypsin | $\alpha_1$-Antitrypsin deficiency |
| iii. Cryoprecipitate/Factor VIII | Haemophilia A |
| iv. $\alpha$-Glucosidase | Gaucher's disease |
| 3. **Replacement of deficient vitamins or coenzymes** | |
| i. $B_6$ | Homocystinuria |
| ii. $B_{12}$ | Methylmalonicacidaemia |
| iii. Biotin | Proprionicacidaemia |
| iv. Vit. D | Vitamin D-resistant rickets |
| 4. **Replacement of deficient product** | |
| i. Cortisone | Congenital adrenal hyperplasia |
| ii. Thyroxine | Congenital hypothyroidism |
| 5. **Substrate restriction in diet** | |
| (a) Amino acids | |
| i. Phenylalanine | Phenylketonuria |
| ii. Leucine, Isoleucine, Valine | Maple syrup urine disease |
| (b) Carbohydrate | |
| i. Galactose | Galactosaemia |
| (c) Lipid | |
| i. Cholesterol | Familial hypercholesterolaemia |
| (d) Protein | Urea cycle disorders |
| 6. **Drug therapy** | |
| i. Aminocaproic acid | Angioneurotic edema |
| ii. Dantrolene | Malignant hyperthermia |
| iii. Cholestyramine | Familial hypercholesterolemia |
| iv. Pancreatic enzymes | Cystic fibrosis |
| v. Penicillamine | Wilson's disease cystinuria |
| 7. **Drug/dietary avoidance** | |
| (a) Suphonamides | G6PD deficiency |
| (b) Barbiturates | Porphyria |
| 8. **Replacement of diseased tissue** | |
| (a) Kidney transplantation | Adult polycystic kidney disease |
| (b) Bone marrow transplantation | X-linked SCID, Wiskott-Aldrich syndrome |
| 9. **Removal of diseased tissue** | |
| (a) Colectomy | Polyposis coli |
| (b) Splenectomy | Hereditary spherocytosis |

## I. Protein and Enzyme Replacement

(a) When a genetic disorder is observed due to deficiency or abnormality in a specific enzyme of protein, then treatment can be done by replacement of the deficient or defective enzyme or protein;. e.g. Factor VIII concentrate in the treatment of haemophilia A.

(b) In case of inborn error of metabolism where enzyme deficiency occur, simple enzyme replacement is not adequate until and unless the quantity is sufficient.

(c) Though recombinant DNA technology is being used for biosynthesis of deficient or defective enzyme or protein but infection of these products cannot achieve the predicted result because metabolic processes take place inside the cell and these gene products is not normally transported into the cells. To overcome the situation the following methods are employed:

1. Artificial delivery system through liposomes allow proteins to cross the cell membrane but with limited success in the genetic disorders, e.g. mucopolysaccharidoses.
2. Again modification of enzyme or protein biochemically allows the normal cellular transport mechanism to target the enzyme to its normal location e.g. modifi-cation of β-glucosidase in the treatment of Gaucher's disease, enters the lysosome for effective treatment.

## II. Drug Treatment

In certain genetic disorders, drug therapy is used such as:

1. Certain drugs help to lower blood cholesterol level in familial hypercholesterolaemia.
2. Avoidance of certain drugs or foods can prevent the manifestation of the disorder, e.g. sulphonamides and fava beans in glucose-6-phosphate (G6PD) dehydrogenase.

## III. Removal of Tissue or Transplant

Replacement of defective tissue is employed with the development of tissue typing, e.g. transplantation of kidney in adult polycystic kidney disease. Removal of defective tissue is also used to prevent complications as a prophylactic or therapeutic measure in many familial cancer —predisposing syndromes, e.g. breast/ovary families, breast families, familial adenomatous polyposis, Turcot's disease etc.

# 35

# GENETICS OF BLOOD GROUP AND HAEMOGLOBINOPATHIES

1. Introduction, concept of blood group, 2. Division and subdivision, Genotype and Phenotype, 3. Agglutination, 4. Secretors, 5. Genetics ABO blood group, 6. Inheritance and medicolegal value, 7. Rh-blood group system, 8. Inheritance of Rh-blood group.

## Genetics of Blood Group

- **Introduction**

If there is blood loss in a moderate haemorrhage amounting within 1 litre, there is replacement of RBC within 7 days, provided the diet contains protein, vitamins and iron. But there is severe haemorrhage and blood haemoglobin falls below 40 per cent, then blood transfusion is necessary.

## Concept of blood group

Blood groups were discovered by Landsteiner. The principal blood group is based on the presence or absence of two major *antigens* which are *mucopolysaccharides* (glycoprotein) *in the red cell membrane* known as *agglutinogen.* These agglutinogens are known as A and B. With two active antigens, four possible combinations exist: the cell membrane may have one or the other, both, or neither of them. This means the red cells of an individual may contain:

1. A agglutinogen (antigen) only called type A or group A blood;
2. B agglutinogen (antigen) only called type B or group B blood;
3. A and B agglutinogens called type AB or group AB blood;
4. Neither agglutinogen A nor agglutinogen B called type O or group O blood.

The blood groups are *inherited* from one's parents according to *Mendelian laws.*

A lipoprotein complex called *elenin* present in the stroma which seems to contain at least the A, B and Rh antigens.

## Subdivisions of A, B and AB groups

The red cells of A blood group have been subdivided into two subgroups known as $A_1$ and $A_2$, and $A_1$ is less common. $A_1$ corpuscles are not agglutinated by $A_2$ serum and vice versa. But both $A_1$ and $A_2$ are agglutinated by B serum and O serum. Apart from $A_1$ and $A_2$, two more subgroups of A have been identified which are $A_3$ and $A_4$ but both groups are rare than $A_2$. Each of the A subgroups is determined by a *separate gene* and the genes for all the four subgroups are *alleles.*

The group B serum contains at least two kinds of antibodies, one agglutinates the corpuscles of both $A_1$ and $A_2$ groups and other agglutinates only $A_1$. AB blood group has also been divided into $A_1B$, $A_2B$, $A_3B$ and $A_4B$.

The gene 'I' is a multiple allele which determines the antigen production and can produce 15 genotypes and 10 phenotypes of blood groups (Table 35.1).

**Table 35.1:** Genotypes and phenotypes

| Phenotype | Genotype |
|---|---|
| O | $I^O/I^O$ |
| $A_1$ | $I^{A_1}/I^{A_1}$, $I^{A_1}/I^{A_2}$, $I^{A_1}/I^{A_3}$, $I^{A_1}/I^{A_4}$, $I^{A_1}/I^O$ |
| $A_2$ | $I^{A_2}/I^{A_2}$, $I^{A_2}/I^{A_3}$, $I^{A_2}/I^{A_4}$, $I^{A_2}/I^O$ |
| $A_3$ | $I^{A_3}/I^{A_3}$, $I^{A_3}/I^O$ |
| $A_4$ | $I^{A_4}/A_4$, $I^{A_4}/I^O$ |
| $A_1B$ | $I^{A_1}/I^B$ |
| $A_2B$ | $I^{A_2}/I^B$ |
| $A_3B$ | $I^{A_3}/I^B$ |
| $A_4B$ | $I^{A_4}/I^B$ |
| B | $I^B/I^B$, $I^B/I^O$ |

## Agglutination and Agglutinin

Agglutination occurs when red cells in the whole blood of donor is transfused to recipient which is due to presence of *agglutin* in plasma of recipient. The *nature of reaction* is similar to that between any *antigen and the antibody*. As for example, when donor's cells containing agglutinogen A (i.e. A group person) are transfused into a recipient whose

**Table 35.2:** Blood group having agglutinogen (antigen) and agglutinin (antibody) and percentage distribution

| Blood group of person | Antigen, i.e. agglutinogen present in R.B.C. | Antibody, i.e. agglutinin in serum |
|---|---|---|
| A (42%) | A | β or Anti-B agglutinin |
| B (9%) | B | α or Anti-A agglutinin |
| AB (3%) | A and B | — |
| O (46%) | — | a β or Anti-A and Anti-B |

plasma contains antibody called agglutinin α (alpha) or Anti-A, agglutination occurs though the person of group A contains β or Anti-B agglutinin (or antibody) in his plasma. Agglutination differs from clumping or rouleaux formation in case of erythrocyte sedimentation rate (ESR). Red cells of donor are responsible for agglutination.

Thus, α (alpha) agglutinin is present in the plasma of person Group O and Group B. The antibody agglutinin β (beta) or Anti-B is found in the plasma of Group O and Group A. These agglutinins are gammaglobulins. These *agglutinins appear shortly after birth*. They may be present at birth, but are usually produced at between 2-8 months of age in response to A and / or B antigens which are taken in with foods or that person are exposed to. They decrease with advancing age. The percentage distribution of the four groups and their agglutinins among the population are given in the Table 35.2.

- **Molecular Basis**

Persons with blood groups A, B and AB possess *enzymes with glycosyltransferase* activity which convert the basic blood group, is known as *H-antigen* (which possess little antigenic activity which has been identified as capital letter O) into A or B antigens by bonding specific sugars to the H-antigen. The alleles for blood groups A and B differ in seven single base substitutions which result in different A and B transferase activities, the A allele being associated with the addition of *N-acetylgalactosaminyl groups* and B allele with the addition of *D-galactosyl groups*. The O allele results from a critical single base pair deletion which results in an inactive protein incapable of modifying the H-antigen (Fig. 35.1).

- **H-antigen**

Another antigen known as H-antigen has also been identified on the erythrocytes which can be demonstrated by agglutination by Anti-H serum. This antigen seems to be an intermediate between antigen A and B.

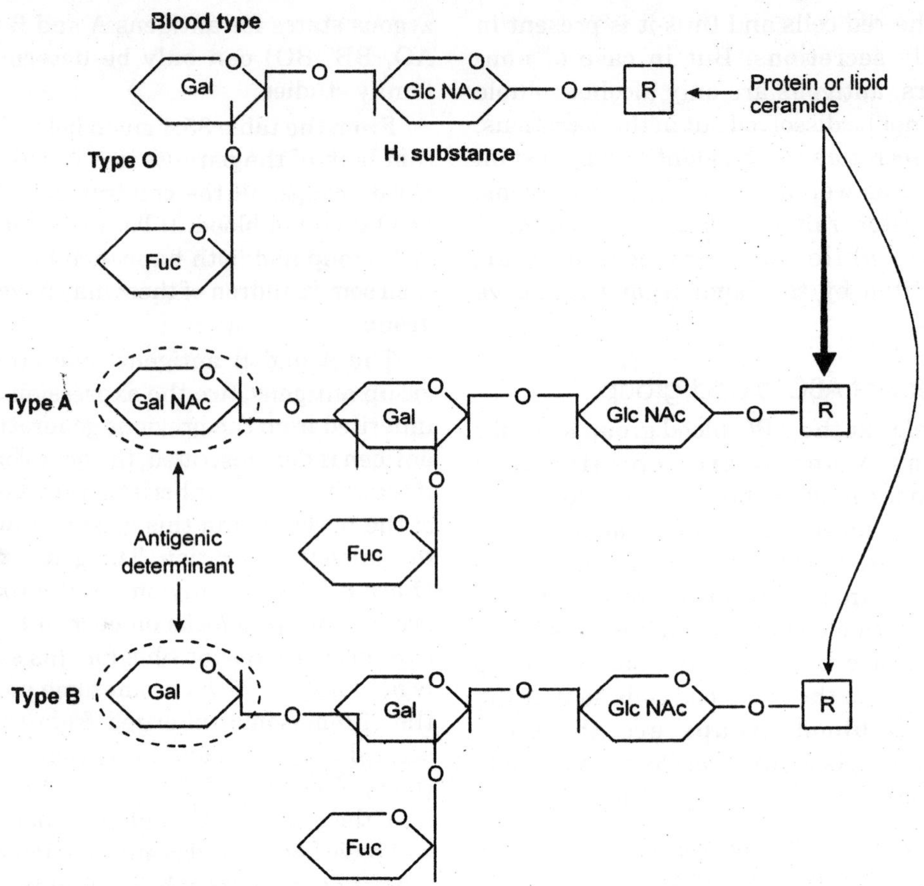


**Fig. 35.1:** Structure of blood group substances. These structures are same except type H having N-acetylgalactosamine (Gal NAc) at the nonreducing end, type B has galactose and type O has neither. Fuc = Fucose; Glc NAc = N-acetylglucosamine; Gal = Galactose.

The dominant gene H is responsible for the production of H-antigen and the genotypes are given below (Table 35.3).

**Table 35.3:** Genotype and Phenotype of H-antigen

| Genotype | Phenotype |
|---|---|
| HH | H-antigen |
| Hh | H-antigen |
| hh | No H-antigen |
| A group + H/H or H/h | A + H-antigen |
| B group + H/H or H/h | B + H-antigen |
| O group + H/H or H/h | None + H-antigen |
| O group + h/H | None |
| AB + H/H or H/h | A + B + H-antigen |

If any person whose blood does not given any reaction with Anti-A or Anti-B or Anti-H, it is considered as rare group and known as "Bombay phenotype" because it was found in small group of people in Bombay.

## Secretors

A and B antigens also occurs in the body secretions, e.g. saline, pancreatic juice, semen, amniotic fluid, eye secretion, nose secretion and mammary gland known as *secretors*. H, A and B antigens are also present in blood platelets, endothelial cells, plant products, the cell walls of bacteria, and in house dust the persons who are secretors have water-soluble antigen which can pass out of the red cells and thus it is present in the body secretions. But in case of non-secretors, antigens are only alcohol-soluble and cannot be dissolved out in the secretions. So, the secretors can be identified by test on the blood as well as on the body secretions. This secretor trait is inherited as a dominant gene 'S' while the non-secretor trait is inherited by the homozygous recessive allele 's'.

## Genetics of ABO blood group

The alleles at the ABO blood group locus for antigens A and B are inherited in a codominant manner but are both dominant to the gene for the antigen O codominance is the term used for two traits which are both expressed in the heterozygous state. In persons with blood group AB it is possible to demonstrate both A and B blood group substances on the red blood cells so that the A and B blood groups are therefore codominant. There are, therefore, six possible genotypes. The homozygous and heterozygous states for antigens A and B (i.e. AA, AO, BB, BO) can only be determined by family studies.

**Table 35.4:** Mode of inheritance of the ABO blood groups to the children

| | Mother X Father Blood group | Child Blood group |
|---|---|---|
| 1. | OXO | O |
| 2. | OXA | A |
| 3. | OXA | O, A |
| 4. | OXB | B |
| 5. | OXB | O, B |
| 6. | OXAB | A, B |
| 7. | AXA | A |
| 8. | AXA | O, A |
| 9. | AXA | A |
| 10. | AXB | AB |
| 11. | AXB | A, AB |
| 12. | AXB | B, AB |
| 13. | AXB | A, B, AB, O |
| 14. | ABXAB | A, B, AB |

From the table 35.4 given below it is seen that both of the parents in a family are of O blood groups, all the children of them must be O group of blood. If both the parents are of A group and both happened to be hybrid, then some children of them may have O blood group.

The A and B antigens, like other blood group antigens, are the expression of genes inherited from the previous generation. If an antigen is demonstrated, the gene controlling it must have been inherited from one or both of the parents, and this gene can be passed to the next generation. The genes A, B and O are *alleles*, i.e. any one of the three may occupy the ABO *locus* on each of the pair of *chromosomes* responsible for this system. It is not known which chromosome pair it is. If the chromosome inherited from the father carried the gene A and the chromosome from the mother carried the gene B, the child will be of the *genotype* AB and his or her red cells will have both A and B antigen. Persons who inherit O genes from both parents belong to group O. The O gene is an *amorph,* i.e. it does not produce a detectable antigen. Group O cells are recognised by the absence of A or B antigen. When the O gene is inherited with A, only the A gene expresses itself; red cells from an individual of the genotype AO will react in the same way as cells from someone of the genotype AA in tests will anti-A. Similarly, it cannot be differentiated BO from BB in tests with anti-B. The symbols A and B denote *phenotypes,* whereas AA, BO, etc. are *genotypes.*

When the same gene has been inherited from each parent, for example OO, the individual is said to be *homozygous* for that gene. When different genes are inherited, as is evident in the case of AB, the individual is *heterozygous.* The genotype of someone of the phenotype A (or B) may be deduced from a study of his family; the person who is homozygous AA will pass an A gene to each of his children while the heterozygous AO individual may pass either an A or an O gene to his offspring. As for all chromosomes, chance alone determines which of the pair ends up in the *zygote* from which the new individual result. Since each of the pair of chromosomes of an AB person carries either an A or B gene, an AB parents cannot normally have a group O child. By the same reasoning, someone who is group O cannot have an AB child because the child must inherit one O gene of the parent, and the gene from the other parent could not transmit both A and B (Table 35.5).

**Table 35.5:** Inheritance of classical blood group

| If peson receives gene | Blood group |
|---|---|
| 1. A + A or A + O | A |
| 2. B + B or B + O | B |
| 3. A + B | AB |
| 4. O + O | O |

## Inheritance and medicolegal value

It must be borne in mind that the child's blood may not be set in its true ABO type until as late as one year after birth. It follows that:

**Table 35.6:** Inheritance of blood group and medicolegal value

| If the baby having this group | Parents must have given | If mother was | Father could not have been |
|---|---|---|---|
| 1. O | O + O | Does not matter | AB |
| 2. AB | A + B | " | O |
| 3. A | A + O or A + A | B or O | B or O |
| 4. B | B + O or B + B | A or O | A or O |

The *medicolegal value* of blood grouping tests is greatly made *more valuable if the MN and the various Rhesus factors* are also studied. The M and N factors depend on two minor blood genes. It is very rare indeed for these to elicit an output of agglutinins when injected into man and they may therefore be ignored in carrying out transfusions. They are however antigenic to rabbits and agglutinating sera can be prepared by injecting human M or N cells into rabbits.

Each person carries two of the genes of the M and N group, i.e. M + M ( = M), N + N ( = N), or M + N (= MN). As a result.

**Table 35.7:** Inheritance of MN blood groups and medicolegal value

| If a baby's supplementary group is | Parents must have given it | If mother's supplementary group was | Fathers could not have been |
|---|---|---|---|
| 1. M | M + M | Does not matter | N |
| 2. N | N + N | " | M |
| 3. MN | M + N | N | N |
| 4. MN | | | |

It is fact that blood-grouping tests can never prove that any suspected person is the actual father; they can only show that he could not possibly have been the father.

## Rh blood group system

Rh blood group system was discovered in Rhesus monkey. There are three pairs of Rh agglutinogen Cc, Dd and Ee. C, D, E are Mendelian dominant and c, d, e are recessive. D is very strongly antigenic and persons are for practical purposes, either Rh positive possessing D antigen or Rh negative lacking D antigen. Three genes coding for these antigens are always present on a chromosome, but a gene and its alternative, i.e. allelomorph are never present simultaneously. This means CDE, or cde, or Cde or cDE etc. may be present but not CCD, CDD, cdd, etc. genes.

Regarding genotypes, that there are eight possible combinations, any one may be carried by both the parents. Based on calculation there are 64 possible combinations, i.e.

genotypes. Out of these 28 being identical, 36 subgroups are biologically available. Out of these, only 5 are commonly found, i.e. CDc/CDe, CDe/cDe, CDe/cDe. cDe/cde and cde/cde, but others are rare.

The $Rh^+$ groups containing the dominant agglutinogen, i.e. C, D, E will be $Rh^+$. But C and E cannot remain without D, so all $Rh^+$ contain D, i.e. belong to group D. The $Rh^-$ group cases contain the recessive agglutinogens c, d, e and belong to group d. Each person carries some Rh agglutinogen and maximum cases have D and are $Rh^+$. The remaining persons carry d and are $Rh^-$. Incompatible reactions of Rh is due to group D of donors with d group of recipient.

## Inheritance

Gene coding for Rh antigens are carried by both ova and spermatozoa. An antibody may be produced to any one of these antigens. Rh genes are inherited from both parents. If the genes carried by the sperm and ovum are identical, the gene composition of the offspring is homozygous in respect of the Rh factor(s)—if not, the offspring is heterozygous in respect of the Rh factor(s) (Table 35.8).

**Table 35.8:** Inheritance pattern of the Rh-group to the children

| Father | Mother | Children |
|---|---|---|
| $Rh^+ / Rh^+$ | $Rh^+ / Rh^+$ | $Rh^+$ (100%) |
| $Rh^+ / Rh^+$ | $Rh^+ / Rh^-$ | $Rh^+ / Rh^+$ (50%)<br>$Rh^+ / Rh^-$ (50%) |
| $Rh^+ / Rh^-$ | $Rh^+ / Rh^-$ | $Rh^+ / Rh^-$ (50%)<br>$Rh^+ / Rh^+$ (25%)<br>$Rh^- / Rh^-$ (25%) |
| $Rh^+ / Rh^-$ | $Rh^- / Rh^-$ | $Rh^+ / Rh$ (50%)<br>(incompatibility of blood might occur)<br>$Rh^- / Rh^-$ (50%) |
| $Rh^+ / Rh^+$ | $Rh^- / Rh^-$ | $Rh^- / Rh^-$ (100%)<br>(incompatibility of blood might occur) |
| $Rh^- / Rh^-$ | $Rh^- / Rh^-$ | $Rh^- / Rh^-$ (100%) |

Ovum and sperm carry the D gene, the offspring will be DD, and hence homozygous $D^+$. If one sex cell carries D and the other one d, the offspring will be heterozygous $D^+$ ($Rh^+$). If the ovum as well as the sperm carry a d gene, the offspring will be homozygous $D^-$($Rh^-$).

- **Rh antibody**

Blood for *normal individuals does not contain anti-Rh antibodies.* However, antibody production is evoked by:

(a) Transfusion of an Rh negative individual with Rh positive blood. Transfusion of 0.5 ml blood is sufficient to evoke.

(b) If father is $Rh^+$ and mother is $Rh^-$ and foetus will become $Rh^+$ inherited from father. Some foetal blood enters the maternal circulation through the placenta. The $Rh^+$ antigen, usually $D^+$ antigen evokes production of Anti-Rh antibody in the mother. Some of the antibodies from mother enter the foetal circulation haemolyses the red cells of foetus.

The $Rh^-$ antibody in maternal blood usually does not reach high enough levels with the first pregnancy to cause significant complications. However, in second or later pregnancy $Rh^+$ pregnancies, serious complications may occur because anti-Rh antibodies that have passed from maternal blood to the foetal blood haemolyse the foetal red cells. The clinical condition that develop in the foetus is known as *haemolytic disease of the newborn* or *erythroblastosis foetalis* (Fig. 35.2).

The syndromes are:

1. *Hydrops foetalis*—The foetus is oedematous and either dies in utero or within hours after birth.
2. *Icterus gravis neonatorum*—The infant is born at term and is jaundiced (haemolytic jaundice) or

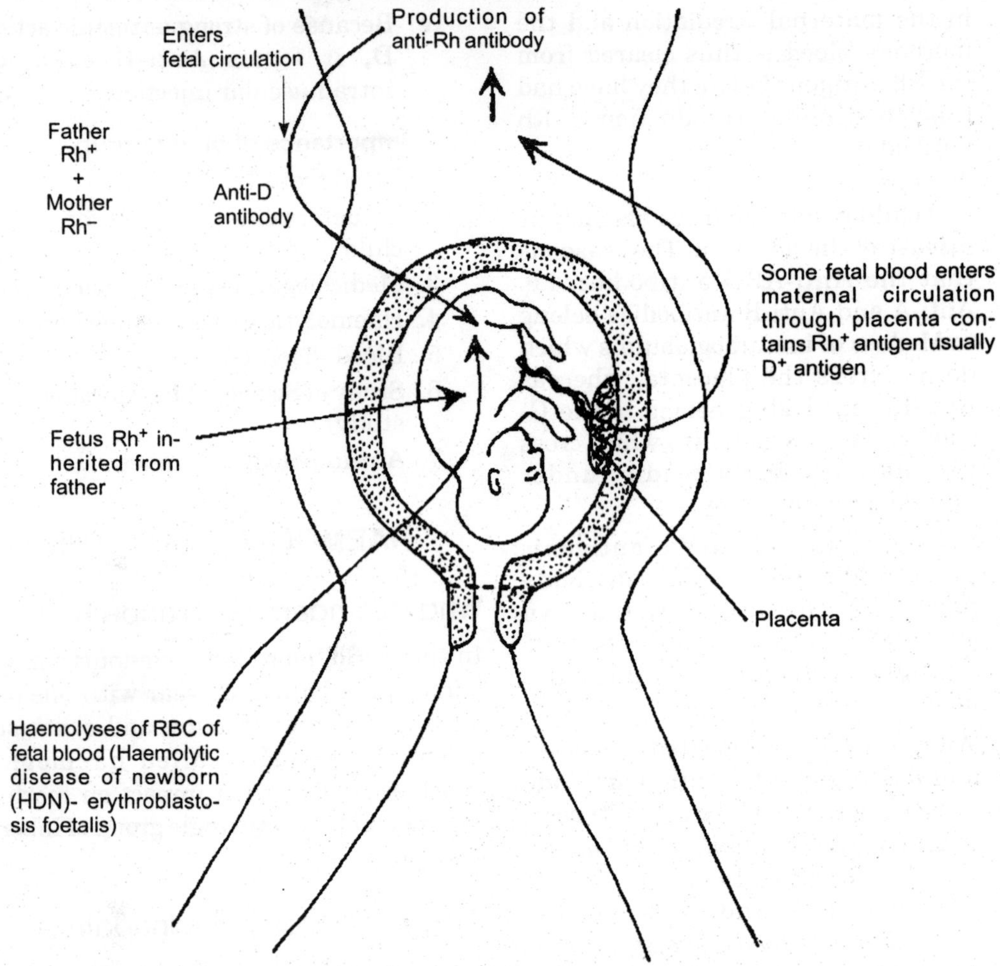


**Fig. 35.2:** Schematic representation of mechanism of haemolytic disease of newborn.

becomes so within 24 hours. Many nucleated red cells (erythroblasts) are present in blood—hence the name erythroblastosis faetalis.

3. *Kernicterus*—Bile pigments may pass into the brain of the foetus and newborn hypoxia makes the basal nuclei more accessible to the bile pigments and is deposited staining it bright yellow.

**Treatment**

Successive withdrawal of small quantities of the infant's blood and replacement of this blood by an equal volume of compatible Rh negative. This type of treatment does not change the inherited blood group of the individual, but removes the doomed red cells from the circulation as well as most of the antibodies—newly produced $Rh^+$ cells are not destroyed.

The foetal blood that sensitises the mother enters the mother's circulation during the final stage of labour, probably through injury created by separation of placenta. For desensitisation, the mother to remove these antigens, massive doses of anti-D gammaglobulin, i.e. anti-Rh antibody may be given within 72 hours after each childbirth. The antibody reacts with the foetal Rh antigens present in the maternal circulation and the mother's blood is thus cleared from the Rh antigens before they have had time to stimulate production of Rh antibodies.

ABO-incompatibility between mother and child rarely produces haemolytic disease of the newborn. The reason is that the Anti-ABO antibodies, i.e. Anti-A and Anti-B antibodies belong to M-class of immunoglobulins which do not cross the placenta, whereas Anti-Rh antibodies belong to the G-class of immunoglobulins which cross the placenta fairly readily under normal circumstances.

(c) Six agglutinogens have antigenic property stimulating antibody production. The antibodies are known as Anti-C, Anti-D, Anti-E, etc. D is strongly antigenic and others are mild.

(d) When D cells are repeatedly injected into a $Rh^-$ subject, Anti-D antibody is developed which are of two types:

1. *Early type or complete antibody or saline agglutinin*: Early Anti-D forms first and is known as *complete antibody*. It agglutinates D cells in vitro when they are suspended either in saline or albumin solution which is also known as *saline agglutinin.*
2. *Late or incomplete or blocking antibody*: The late Anti-D forms later and known as *incomplete antibody*. It agglutinates D cells in vitro, when they are suspended in albumin solution only but not in saline solution and known as *albumin agglutinin.* Though D cells are not agglutinated, but they are modified. Because these cells, once treated in this way, will not be agglutinated by early Anti-D serum even when they are suspended in albumin solution for which late Anti-D is also known as the *blocking antibody.*

(e) Because of strong antigenic action of D, it forms Anti-D even after intramuscular injection.

- **Importance of blood group**

1. Blood transfusion
2. To determine paternity for disputed child
3. Medicolegal test in forensic medicine
4. Haemolytic disease of newborn
5. Blood diseases
6. Study of Science of Races (Ethnological study)
7. Anthropological studies

## HAEMOGLOBINOPATHIES

### What are haemoglobinopathies?

In the world, more than one-fourth milliton persons are born each year with one of the disorders of the *structure* and *synthesis* of haemoglobin (Hb) called haemoglobin opathies. It has great impact on morbidity and mortality of any single group of inherited disorders.

### Groups of Haemoglobinopathies

The disorders of haemoglobin, i.e. haemoglobinopathies are divided into two groups:

I. *Disorders of Haemoglobin Structure* i.e. structural variants, e.g. sickle-cell Hb or HbS.

II. *Disorders of Haemoglobin Synthesis*, e.g. thalassaemia.

The *Group I* can subdivided by the way in which they interfere with (1) the normal function of Hb and or (2) the red blood cell, e.g. abnormal oxygen affinity or haemolytic anemia.

The *Group II* can be subdivided according to which globin chain is synthesized abnormally, i.e. α, β or δβ thalassaemia.

Family studies of the various disorders of human Hb and analysis of the mutations responsible for these at the protein and DNA level have led to an understanding of the normal structure, function and synthesis of Hb. This has allowed demonstration of the molecular pathology of these disorders and prenatal diagnosis of a number of the inherited disorders of human Hb is possible.

I. **Disorder of Haemoglobin Structure**
   A. *Structural globin chain* variants e.g. sickle-cell disease.
   B. *Synthesis* of the globin chain, e.g. Thalassaemia.

### A. Structural Variants

About 200 Hb electrophoretic variants have been observed. The types of mutations responsible for these Hb variants is as diverse as those observed in microbial genetics. Some 200 of the electrophoretic variants are *single amino acid* substitutions due to *point mutation.* The majority of these are rare but a few are polymorphic in certain populations, i.e. occurring in more than 1% of persons.

- **Types of Mutation**

1. **Point mutation:** It results in substitution of one amino acid for another can lead to an altered Hb such as HbS, C or E which are missense mutations.
2. **Deletion:** There are a number of Hb variants in which one or more amino acids of one of the globin chains is missing or deleted, e.g. Hb Freiburg.
3. **Insertion:** There are variants in which the globin chains are longer than normal due to insertion, e.g. Hb Grady.
4. **Frameshift mutation:** It involves disruption of the normal triplet reading frame, i.e. addition or removal of a number of bases which are not a multiple of three. In this instance translation of the mRNA continues until a termination codon is read in frame. The variants can result in either an elongated or shortened globin chain, i.e. functionally they are nonsense mutation.
5. **Chain termination:** A mutation in the termination codon itself can lead to an elongated globin chain, e.g. Hb constant spring.
6. **Fusion polypeptides:** These types of structural variants are the fusion polypeptides. HbS Lepore and Kenya which result from unequal cross-over events in meiosis.

- **Clinical Manifestation**

1. Many have diseases but are harmless.
2. These do not interfere with normal function.
3. They are identified during population survey of Hb-electrophoretic studies.
4. A number of variants interfere in a variety of ways, the normal functions the functional abnormalities of structural variants of haemoglobin such as:

**Table 35.9:** Functional abnormalities of structure variants of haemoglobin

| Clinical features | Functional abnormality | Example |
|---|---|---|
| 1. Haemolytic anemia | Sickling disorders | HbS<br>HbC<br>HbE |
| | Unstable haemoglobin | Hb Köln<br>Hh Gun Hill<br>Hb Bristab |
| 2. Cyanosis | Haemoglobin M (methaemo-globinaemia) | HbM (Boston)<br>HbM (Hyde Park) |
| 3. Polycythemia | Low $O_2$ affinity<br>High $O_2$ affinity | Hb Kansas<br>Hb Chesapeake<br>Hb Heathrow |

1. When the mutation is on the inside of the globin subunits, in close proximity to the haem or the interchain contact areas this can produce an unstable Hb molecule which precipitates in the

RBC, damaging the membrane and resulting in haemolysis of the RBC.

2. Mutation may interfere with the normal $O_2$ transport of Hb resulting either an enhanced or reduced $O_2$ affinity or a stable Hb in reduced form called methaemoglobin.

The structural variants of Hb is identified by electrophoresis. And the electrophoretic technique may reveal only the less number of variants than the actual number exist. The probable explanation may be than one-third of the possible Hb mutations which could occur, produce an altered change in the Hb molecule and thereby be detectable by electrophoresis.

## Sickle-Cell Disease

RBC of a person having sickle-cell disease appears birefringent under polarized microscope and also distorted shape under deoxygenated conditions, hence known as *sickling*. It has been observed that Hb of sickle-cell disease has different mobility than the normal Hb, i.e. Hb A electrophoretically, hence called HbS.

- **Clinical Features**

1. Cerebral symptoms, kidney failure, pneumonia, heart failure, weakness and lassitude. These are the manifestations of the mutant allele HbS.
2. HbS is less soluble than Hb A and tends to crystallize out, causing the sickle-shaped deformation of the RBC.
3. Sickle cells are less stable, having a shorter survival time in the circulation, leading to a more rapid RBC turnover with consequent anemia.
4. The sickle cells have a reduced deformability tending to obstruct small arteries, resulting in an inadequate $O_2$ supply to the tissues (Fig. 35.3).

### Mutational basis of sickle-cell disease

In sickle-cell disease, the genetic code indicates that the substitution of valine at the sixth position β-globin chain was due to an alteration in the second base of the triplet coding for glutamic acid, i.e. CAG to GTC. Using the restriction enzyme, Mst II, whose nucleotide recognition sequence is abolished by the point mutation in HbS, it can be demonstrated the difference between the persons with sickle-cell disease and normal persons in the restriction fragments when hybridized with a radioactively labelled β-globin probe.

Use of a mutation-specific RFLP is possible with a few other Hb structural variants, e.g. HbE using the restriction enzymes Hph-I or Mn-II. Family studies of linked polymorphic DNA sequence variants are used for carrier testing and prenatal diagnosis of the other haemoglobinopathies.

Gene for sickle-cell disease

↓

Mutation of gene

↓

Abnormal β-globin chain, i.e. substitution of valine for glutamic acid in HbS

↓

Less soluble than HbA, crystallize out

↓

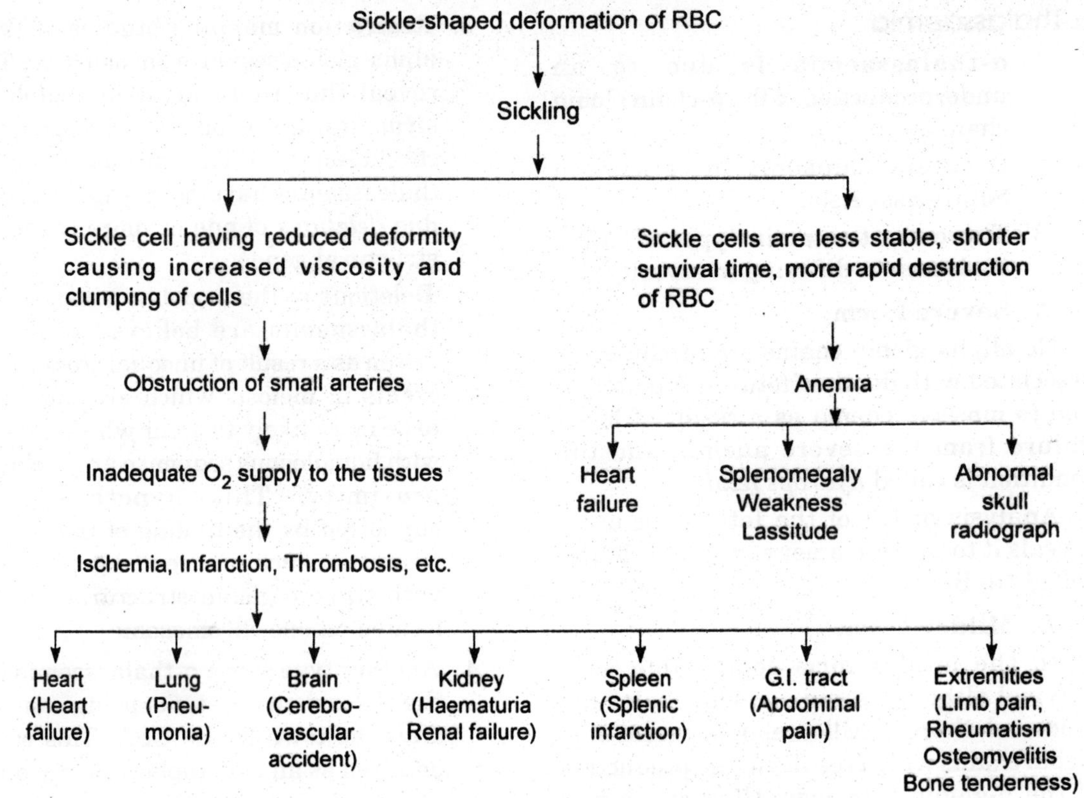


**Fig. 35.3:** The sequence of events of effect of gene for sickle-cell disease.

## II Disorders of Haemoglobin Synthesis

The *thalassaemias* are the commonest single group of inherited disorders in humans. *Occurence* persons in the following regions:

Mediterranean area, the Middle-East, the Indian Subcontinent and South-East Asia.

- **Pathophysiology**

They are heterogeneous group of disorders and are classified according to the particular globin chain or chains synthesized in reduced amounts (Fig. 35.4).

The pathophysiology is similar in all forms of thalassaemia. An imbalance of globin chains in the RBC precursors, which being insoluble precipitate, resulting in haemolysis of the RBC, i.e. a haemolytic anemia, with consequent compensatory hyperplasia of the bone marrow.

## Classification

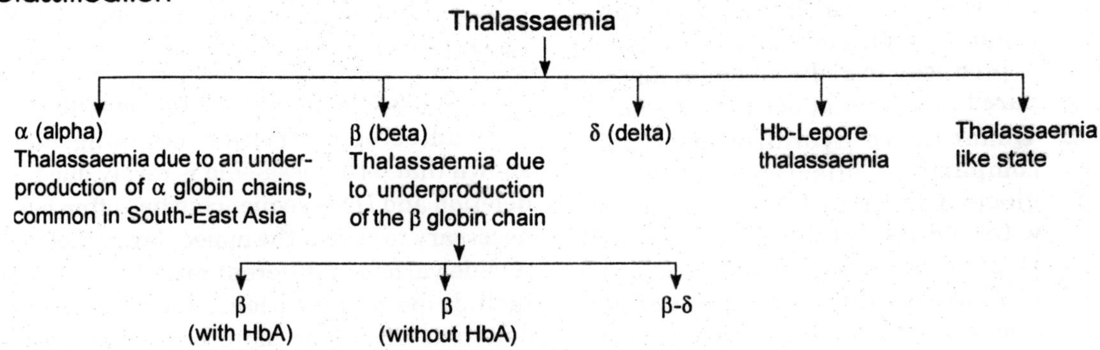

## α-Thalassaemia

1. α-thalassaemia is due to an underproduction of the α-chain globin chain.
2. It occurs commonly in persons in South-East Asia.
3. There are two main types of alpha thalassemia differing in severity.

A. **Severe Form**

No allpha globin chains are produced, is associated with death of fetus in the uterus, due to massive edema as a result of heart failure from the severe anemia and the condition is called *hydrops fetalis.*

Analysis of Hb of the fetuses in utero reveals it to be a tetramer of γ-globin chains, called Hb Barts.

B. **Milder Form**

In the milder form compatible with survival although some α-globin chains are produced, there is still a relative excess of β-globin chains which results in the production of the β-globin tetramer, HbH or what is known as HbH disease.

Both Hb Barts and Hb H globin tetramers have an $O_2$ affinity similar to that of myoglobin and do not release $O_2$ normally in the peripheral tissues in addition HbH is unstable and precipitates resulting in haemoglobin of the RBC.

- **Mutational basis of α-thalassaemia**

1. Using mRNA from reticulocyte from fetuses of hydrops fetalis, *in vitro* translation studies show no synthesis of alpha-globin whereas mRNA studies with the milder form, alpha-thalassaemia, HbH disease, α-globin is produced in reduced amounts compared to mRNA of normal person.
2. Quantitative hybridization studies comparing radioactive labelled α-globin CDNA to DNR from fetuses with hydrops fetalis, person with Hb H disease and normal person are consistent with α-thalassaemia being due to *deletion of the α-globin genes.*
3. Restriction mapping studies of the alpha-globin region of chromosome 16 reveal that there are two α-globin structural genes on the short arm of chromosome 16. The various forms α-thalassaemia have been shown to be due *deletions* of one or more of these structural genes.

   Deletions of the α-globin genes in α-thalassaemia are believed to have arisen as a result of unequal cross-over events in meiosis which are thought to be more likely to occur where genes with homologous sequence are in close proximity. This hypothesis is supported by the finding of the other product of such an event, i.e. persons with three α-globin structural genes located on one chromosome.
4. Another type of the α-thalassaemia is the Hb variant constant spring named after the town in the USA. This was detected as an electrophoretic variant in a person with HbH disease, i.e. α-thalassaemia Hb constant spring is due to an abnormally long α-chain which is the result of a mutation in the normal termination codon at position 142 in the α-globin gene.

## β-Thalassaemia

It is due to underproduction of the β-globin chain of haemoglobin. In vitro translation studies using mRNA from persons with β-thalassaemia due to these different types of mutations show either reduced or absent production of β-globin chains, $\beta^+$ and $\beta^0$ respectively.

- **Mutational basis of β-thalasaemia**

Studies of restriction mapping have shown that β-thalassaemia is rarely due to a deletion and DNA sequencing has often been necessary to reveal the molecular pathology. A wide variety of different mutations, which include point mutations, insertions, and deletions of one or more bases have been shown to be responsible. These occur at a number of places, both within the coding and noncoding portions of the β-globin genes as well as in the 5′ flanking promoter region, 5′ capping sequences and the 5′ polyadenylation sequences.

The various types of mutations causing β-thalassaemia are often unique to certain population groups and can be considered to fall into fine main functional types such as:

(a) Transcription mutations
(b) mRNA splicing mutations
(c) RNA modification mutations
(d) Chain termination mutations
(e) Missense mutations

- **Clinical Features**

1. Children affected with thalassaemia major or Cooley's anemia as it was originally known usually present in the first year of life with a severe transfusion-dependent anemias. Unless the child is adequately transfused, compensatory expansion of the bone marrow results in an unusually shaped face and skull.
2. Although persons with thalassaemia major used to die in their late teens or early twenties as a result of complications due to iron overload from repeated transfusions, the regular daily use of iron-chelating drugs, such as desferrioxamine has improved their long-term survival.
3. Individuals heterozygous for β-thalassaemia, *thalassaemia trait* or *thalassaemia minor,* usually have no symptoms or signs. They do have, however a mild hypochromic microcytic anemia.

## δβ-Thalassaemia

1. There is underproduction of both the δ- and β-globin chains.
2. Persons homozygous for δβ-thalassaemia produce no δ- or β-globin chains. Although one would expect such persons to have a fairly profound illness, they are only mildly anemic due to increased production of γ-globin chains with HbF levels being much higher than the mild compensatory increase seen in homozygotes for β-thalassaemia.

## Mutational Basis

It is to be due to extensive deletions in the β-globin region involving the δ- and β-globin structural genes. Some deletions extend to include the $^{A}\gamma$-globin gene so that only the $^{G}\gamma$-globin chain is synthesized.

*Hereditary Persistent of Fetal Haemoglobin (HPFH)*:

In HPFH, there is persistence of the production of fetal Hb into childhood and adult life is included in the thalassaemias. Most forms of HPFH are in fact, a form of δβ thalassaemia in which continued γ-chain synthesis compensates for the lack of production of δ- and β-globin chains. Persons with HPFH continue to produce significant amounts of fetal Hb after birth accounting for 20-30 per cent of total Hb in heterozygotes and 100% in homozygotes. This is not associated with any symptoms and was originally considered more of a scientific curiosity than a medical problem.

- **Mutational Basis**

Some forms of HPFH have been shown to be due to deletions of the δ- and β-globin genes. Analysis of the nondeletion forms of HPFH has shown point mutations in the 5′ flanking promoter region of either the $^{G}\gamma$ or $^{A}\gamma$-globin genes near the CAT box sequences involved in the control of expression of the haemoglobin genes.

- **Clinical Variation of the Haemoglobinopatheis**

1. The marked mutational heterogeneity of β-thalassaemia means that affected individuals are often compound heterozygotes, i.e. they have different mutations in their β-globin genes

leading to a wide range of severity of the disorder. One form of β-thalassaemia of intermediate severity, known as thalassaemia, requires less frequent transfusions.

2. In certain populations many of the haemoglobinopathies are relatively common and not unexpectedly, persons are reported who have two different disorders of Hb. Understandably, in the past, recognition of such individuals was often quite difficult.
3. Due to the high prevalence in certain populations of some of the structural variants of Hb, such as HbS, individuals can present with severe anemia who are found to be heterozygous for both HbS and β-thalassaemia i.e. be double heterozygotes.
4. Certain combinations of haemoglobinopathies can also result in a previously unexplained mild form of what is normally thought to be an invariably severe haemoglobinopathy. For example, deletion of one or two of the α-globin genes in a person homozygous for β-thalassaemia results in a milder illness because there is less of an imbalance in globin chain production. Similarly, the presence of one of the forms of HPFH in a person homozygous for β thalassaemia can contribute to amelioration of the disease as the increased production of γ-globin chain compensates for the deficient β-globin chain production.

## QUESTION

1. State briefly the pattern of inheritance of blood group A, B, O.
2. State briefly the pattern of inheritance of Rh group.
3. Describe briefly about Rh group.
4. Describe briefly with diagram the reactions occur in case of father $Rh^+$ and mother $Rh^-$.
5. Discuss medicolegal importance of blood groups.

# 36 GENE THERAPY

1. Genetic and non-genetic diseases treated by gene therapy or Diseases amenable to gene therapy.

## What is gene therapy?

Gene therapy is the replacement of a deficient gene product or correction of an abnormal gene.

## How gene therapy is done?

Gene therapy can be done either:

1. In *vitro* by treatment of cells or tissue from an affected individual in culture, with reintroduction to the affected individual.
2. In *vivo* if cells cannot be cultured or be replaced in the affected individual.

- **Approval of Ethical Committees**

Considering the technical, therapeutics and safety aspects, Recombinant DNA Advisory Committee on the Ethics of Gene Therapies has recommended that all gene therapy programmes must be approved by hospital reseach committees.

There are two lines of therapy:

1. **Germ line gene therapy:** In which genetic changes could be distributed to both somatic and germ cell, and therefore be transmitted to future generations, and morally and ethically not acceptable.
2. **Somatic cell gene therapy:** It is universally agreed and all programmes are focussed only on this line of therapy in which the alteration in genetic information is targeted to specific cells, tissues or organs in which the disorder is manifested.

- **Technical Consideration**

There are some technical aspects which must be considered before gene therapy. Such as:

1. **Gene characterization:** Gene involved should be *cloned*, which include not only the structural gene but also the DNA sequences involved in the control and regulation of expression of that gene.
2. **Target cells or Tissues or Organs:** The specific cells or tissues or organs with genetic disorder which is to be treated should be identified and accessible such as treatment of inherited disease of haemoglobin like β-thalassaemia, it involves removal of red bone marrow from the affected person and treating it in *vitro* and then returning it to the patient by transfusion (Fig. 36.1). Here the

targeted cells, i.e. bone marrow stem cells are small in number from which immature red blood cells or reticulocytes are developed.

3. **Vector system**
   (a) Foreign gene which is to be introduced must be efficient and safe.
   (b) Before gene therapy, there should be trials in animals to prove that the inserted gene functions adequately with appropriate regulatory promoter and enhancer sequences.
   (c) In the trial it should be shown (1) that the treated tissue or cell population has a reasonable life span, (2) the gene product continues to be expressed and (3) that the body does not react adversely to the gene product, e.g. by producing antibodies to the protein product and (4) *insertional mutagenesis* should be averted, i.e. introduction of the foreign gene or DNA sequence does not have harmful effects which inadvertently leading to a malignancy or a mutagenic effect on either the somatic or the germ cell lines. It arises due to mistake as a result of the insertion of the gene or DNA sequence into the host DNA. (5) There is possibility to affect the germ cells though the therapy is targeted to somatic cells or tissues which should be evaluated.

## Methods of Gene Therapy

There are two methods of gene therapy:

I. Viral Agents;
II. Physical Agents.

### I. Viral Agents

**Retroviruses:** These are *retroviruses* derived from a family of viruses which includes the human immuno-deficiency virus and viruses which causes cancer to some species. These are most useful and developed viral vector systems. The retroviruses integrate into the host DNA by making a copy of their RNA molecule using the enzyme reverse transcriptase. The provirus so formed is the template for the production of the mRNA for the various viral gene products and the new genomic RNA of the virus. If the provirus is stably integrated into dividing stem cells, all subsequent progeny cells will inherit a copy of the viral genome.

Retroviruses are rendered incapable of independent replication to avert any adverse effects associated with infectivity. For this reason, retroviruses for gene therapy need two elements:

(a) Packaging cell line;
(b) Vector or helper virus.

(a) **Packaging cell line:** The packaging cell line is a cell line which has been infected with a retrovirus in which the provirus is genetically engineered to lack the region of the proviral DNA called the packaging sequence, i.e. the cells will produce all of the normal viral proteins but provirus cannot be packaged into infectious virus particles.

(b) **Vector or helper virus:** Another retroviral provirus is engineered with more than 90% of the viral genetic material removed leaving only the minimal DNA sequences necessary to produce copies of the viral RNA sequences along with the sequences necessary for packaging of the viral genomic RNA. This is the vector backbone into which the foreign DNA or gene can be inserted. If this vector, or helper virus is introduced into the packaging cell line which contains the provirus missing the packaging sequences, the RNA produced by the vector provirus can be packaged into virus. These viruses can be used to

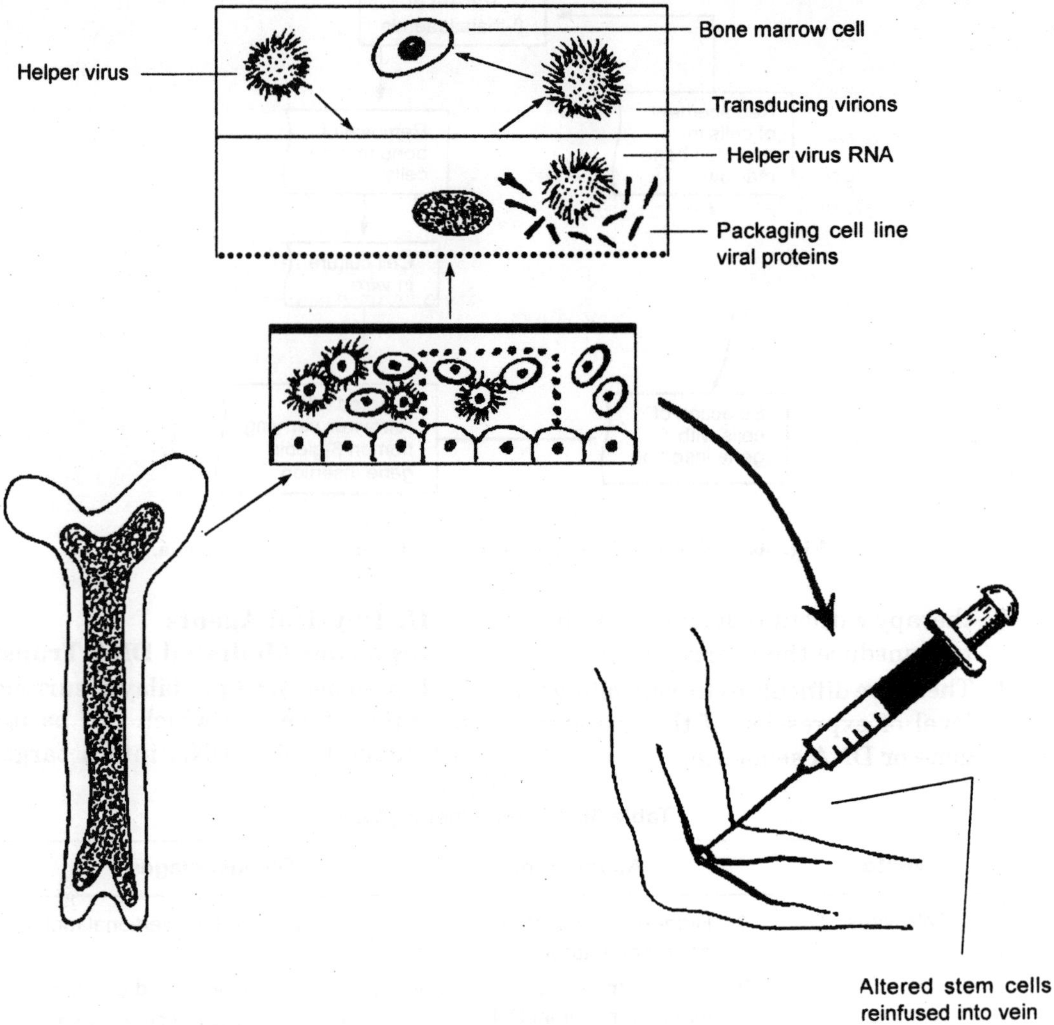


**Fig. 36.1:** Retrovirus gene therapy using bone marrow cells.

infect, or what is commonly known as *transfect,* the particular target cells or tissue. The virons do not replicate to produce further infective virions since the retrovirus vector cannot replicate because it lacks the sequences necessary for this to occur (Fig. 36.1).

- **Disadvantages of Retrovirus**

1. Only a small DNA sequence can be introduced into the target cells, usually less than 7 kb, limiting their use in gene therapy. Attempts are being made by inserting a modified dystrophin gene in which a large amount of the gene has been deleted but which still has relatively normal function, which is known as a *minidystrophin gene.*
2. These retroviruses can only infect dividing cells which means that genetic disorders which affect the system, e.g. CNS cannot be treated by retroviruses.
3. The retroviruses are unstable and cannot be purified for use in gene

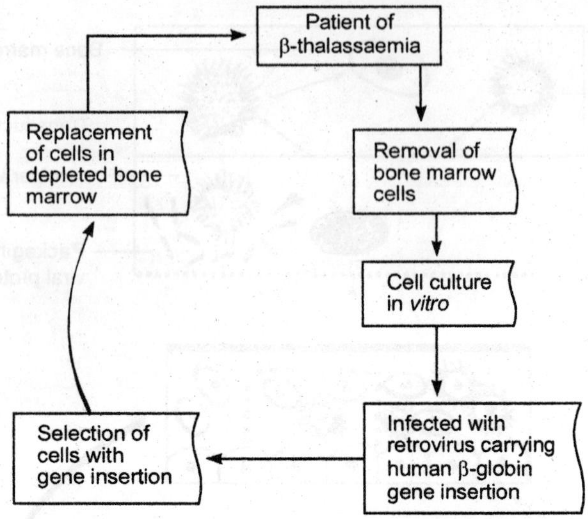


**Fig. 36.2:** Showing the sequence of gene therapy in β-thalassaemia.

therapy without reducing their ability to transduce their target cells.

4. There are difficulties in controlling the level of expression of the introduced gene or DNA sequence.

## II. Physical Agents

### Liposome Mediated DNA Transfer

Liposomes are lipid bilayer surrounding an aqueous vesicle which can be used to introduce foreign DNA into a target cell

**Table 36.1:** Gene therapy vectors

| Vector | Advantages | Disadvantages |
|---|---|---|
| I. Retrovirus | 1 High–efficiency transduction of appropriate target cells.<br>2. Long–term expression-integration into chromosomal DNA. | 1. Potential for insertional mutagenesis.<br>2. Requirement for dividing cells.<br>3. Limited size of DNA insert. |
| II. Adenovirus | 1. High transduction efficiency.<br>2. Broad range of target cells<br>3. Does not require cell division<br>4. Low risk of insertional mutagenesis. | 1. Transient expression.<br>2. Immunogenicity<br>3. Direct cytopathic effect of virus. |
| III. Adeno-associated virus | Does not require cell division, site-specific integration. | 1. Potential for insertional mutagenesis if integration not site-specific.<br>2. Limited size of DNA insert low efficiency limited target cell range, transient expression. |
| IV. Non-viral vectors | No infection risk, completely synthetic, No limitation on insert size. | Low efficiency, Limited target cell range.<br>Transient expression. |

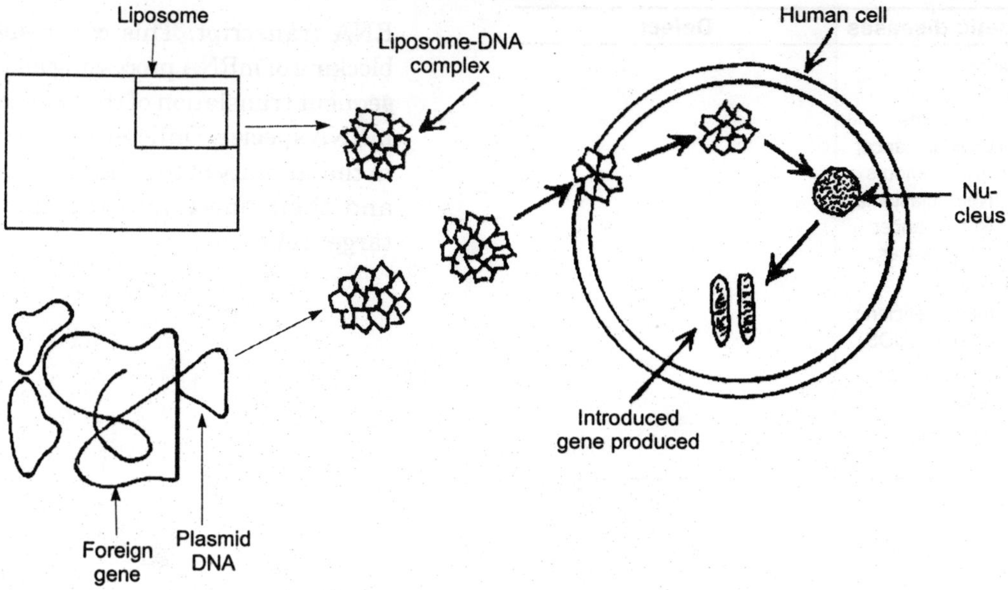


**Fig. 36.3:** Schematic representation of Liposome mediated gene therapy.

(Fig. 36.3). The disadvantage is that the liposomes are not very efficient and the expression of the foreign gene is transient, so that the treatment has to be repeated. The advantage of the method is that a much larger DNA sequence can be introduced into the target cells or tissues than with viral vector systems (Fig. 36.3).

(b) **Receptor-mediated Endocytosis:** A variation of liposome-mediated gene transfer is to target the DNA to specific receptors on the surfaces of cells. A complex is made between plasmid DNA containing the foreign gene or DNA sequence and specific polypeptide ligands for which the cell has a receptor on its surface.

(c) **Oligonucleotides:** There is possibility to synthesize antisense oligonucleotides, which bind to specific mRNA sequences, or specifically engineered RNA molecules known as ribozymes, which have intrinsic RNA catalytic properties resulting in cleavage of the

**Table 36.2:** Genetic and non-genetic diseases treated by gene therapy or diseases amenable to gene therapy.

| Genetic diseases | Defect |
|---|---|
| 1. Immune deficiency | Adenosine deaminase deficiency purine nucleoside phosphorylase deficiency chronic granulomatous disease. |
| 2. Hypercholesterolaemia | Low density lipoprotein receptor abnormalities. |
| 3. Haemophilia | Factor VIII deficiency (A)<br>Factor IX deficiency (B) |
| 4. Mucopolysaccharidosis VII | β-glucuronidase deficiency |
| 5. Emphysema | $\alpha_1$-Antitrypsin deficiency |
| 6. Cystic fibrosa | CFTR mutations |
| 7. Phenylketonuria | Phenylalanine hydroxylase deficiency |
| 8. Urea cycle abnormalities | |
| (a) Hyperammonaemia | Arnithine transcarbamylase deficiency |
| (b) Citrullinaemia | Argininosuccinate synthetase deficiency |
| 9. Muscular dystrophy | Dystrophin mulation |
| 10. Thalassaemia/ Sickle-cell disease | α- and β- globin mutation |

*Contd.*

| Genetic diseases | Defect |
|---|---|
| 11. Cancer<br>(a) Malignant melanoma<br>(b) Ovarian cancer<br>(c) Brain tumours<br>(d) Neuroblastoma<br>(e) Renal cancer<br>(f) Lung cancer | |
| 12. Acquired immune deficiency syndrome (AIDS) | |
| 13. Cardiovascular diseases | |
| 14. Rheumatoid arthritis. | |

RNA transcript. This can result in blocking of mRNA processing and subsequent translation of that particular mRNA species. Difficulties arise both in the stability of the oligonucleotides and their entry into the relevant target cells.

# 37

# NATURE VS. NURTURE CONTROVERSY IN MAN—IDENTICAL TWINS STUDIES

1. Types of twins—identical or monozygous, fraternal or dizygous, 2. Cause and occurrence, 3. Distinction between identical and fraternal twins, 4. Concordance and discordance, 5. Twins reared apart, 6. Inheritance of criminality.

## Nature

Nature problem has been studied in plants by dividing the same individual into parts and subjecting these parts with identical genotypes to different environments. Plants that grow from the parts of the same individual are members of a clone; barring mutation, they have the same genotype. The differences between the plants in the vertical rows are, accordingly genotypic. The horizontal rows show how the same plant behaves in the different environments; those differences are environmental.

- **Twins**

Twins may be either:

1. **Identical or Monozygous:** Arise from a single fertilized ovum or zygote formed by fertilization of a single egg by a single sperm, usually occurs after a period of cell division, separates into two distinct embryos. Twins arising in this way are members of a clone and should have the same genotype.

   Thus, the two embryos will have in their cells identical sets of chromosomes, since they are derived by mitosis from a single zygote. The twins often share a placenta, although they may be enclosed in separate amnions. Such twins usually resemble each other very closely and are invariably of the same sex, though variations in their position and blood supply while in the uterus may produce differences at birth.

Since the one-egg twins carry identical sets of genetic instructions, it can be argued that any differences between them are due, not to their genes, but to the effects of their environment. Identical twins, therefore, are a valuable source of evidence for assessing the relative importance of heredity and environment. It was observed by Freeman *et al.* that the average difference in height of 50 pairs of identical twins reared together was only 1.7 cm while the average difference for the same number of non-identical twins was 4.4 cm. Freeman *et al.* observed the scores of identical twins in intelligence tests showed a greater correlation than those of non-identical twins even when the former have been brought up in different environments, but the results showed that educational background can make a considerable difference.

2. **Fraternal or Dizygous:** Fraternal twins come from two different eggs, each fertilized by a different sperm. Since different gametes from the same parent will in general have different sets of genes, fraternal twins have different genotypes. The genotypes of such twins need not be any more or any less different than the genotypes of siblings—brothers or sisters who are not twins. Fraternal twins are simply siblings born simultaneously. Though they develop simultaneously in the uterus, will not necessarily be any more alike than if they were brothers or sisters born at different times, e.g. they can differ in sex.

### • Causes and Occurrence

The causes of multiple births are not clear. *Fraternal twins* arise from two eggs released simultaneously from the ovaries of the mother fertilized by two different spermatozoa. In many animals like dogs, cats, mice, etc. multiple ovulation is a rule. In some human races like Japanese it is less frequent than the Whites and Negroes. Thus it is partially conditioned by the genotype.

*Identical twins,* appear sporadically in man and some other animals but the cause is not clearly known.

In the armadillo, at least, a single egg regularly produces four embryos resulting in identical quadruplets, the process is thus under genotypic control.

### How to distinguish fraternal and identical twins?

Fraternal twins of different sexes are often seen like of the same sex. When twins are of the same sex, the problem is to distinguish the identicals from the fraternals. Fraternals usually are enclosed in separate birth membranes and have separate placentae, whereas identicals are generally enclosed in one membrane and are attached to one placenta but there exceptions to both rules. For a definite diagnosis, the comparison of the pairs in traits known to be determined by genes, e.g. blood group genes which express themselves wherever they are present. If members of twin pairs differ in such a known genotypic trait such as:

(a) Having different blood groups, or

(b) Having different sexes.

Then they are fraternal. If alike in all such traits, they are probably identicals.

### • Concordance and Discordance

When members of twin pair both show or fail to show the trait in question, they are called *concordant* and symbols are (+ +) or (– –); when only one of the twins shows the trait, they are called *discordant* and the symbols are (+ –). Identical twins are always concordant pertaining to fully penetrant hereditary characteristics, whereas fraternal twins are sometimes discordant with respect to such characteristics. This fact is used to distinguish the identical from the fraternal twins. If the variation in a trait is due to the environment, the frequency of concordance and discordance should be alike among identical and fraternal twins, within the limits of statistical errors of observation.

### • Diseases in Twins

In case of measles, scarlet fever tuberculosis, tumorous (all types) tumorous (specific kinds), diabetes mellitus, feeble-mindedness etc, the identical twins have shown higher percentage of concordance with respect to certain traits in pairs than the discordance. In fraternal twins, higher percentage concordance was seen only in measles infection but higher percentage of discordance was seen in case of other diseases.

### • Twins Reared Apart

For study, twins are usually taken from the same family and they usually receive similar treatment from their parents and other people. But objection was raised against the use of twins for nature-nurture studies because of the similarity of environment may be greater for identical than for fraternal twins since the former are so nearly identical in appearance, habits and experience. This is not so serious objection because if the identical really have more nearly similar environments because if the identicals actually have more nearly similar environments than the fraternals, this greater similarity of environments is itself an indirect result of their greater genotypic similarity. It is also essential to observe how great a phenotypic difference can arise in persons of the same genotype.

Newman *et al.* studied the mean differences between identical and fraternal twins in various traits and found that identical twins resemble one another more in traits like stature, head length, and head width than fraternal twins, regardless of whether the identicals were raised together or apart. It indicates that these traits and incidentally, other physical traits by which people recognize each other, such as facial appearance and expression—are more strongly influenced by hereditary variations than by the environmental variations that commonly differentiate people from one another.

### • Inheritance of criminality in twins

Criminality is usually referred as "psychic" traits and it has been doubted and denied by some workers about inheritance of criminality. Several investigators in different countries have studied in prison persons who were members of twin pairs. Then it was determined whether the co-twin also had a criminal record at anytime. If so, the twins were scored as concordant and if not, they were entered as discordant. Identical and fraternal twins were distinguished, using fraternal twins only of like sex. The results have shown that the proportions of concordant twin pairs is in every case higher among the identicals than among the fraternals.

## QUESTION

1. What are the different types of twins? Describe the identical or monozygous twins.
2. Describe the fraternal twins. How to distinguish fraternal and identical twins?
3. What do you mean by concordance and discordance?
4. What results do you expect if the twins are reared apart? What is your opinion about inheritance of criminality in twins?

# 38

# EUGENICS OR FUTURE OF HUMAN GENETICS

1. Introduction, 2. Types, 3. Measurement, 4. Glossary.

## Introduction

Eugenics is the branch of biological science which deals with the improvement in the hereditary qualities of future generations of mankind, and it is the science of improving the qualities of human race. The term eugenics was coined by *Francis Galton* (1822-1911) as "the study of eugenics under social control that may improve or impair the racial qualities of future generations either physically or mentally".

*Galton* first concluded that the evolution of a trait through natural selection could be substituted by the evolution of a trait through social selection and education. *Galton* proposed many observations of human genetics mathematically. In eugenics, he analysed hereditary characters of man and animals mathematically by selective marriage and breeding method respectively. His aim was to improve the human species by selective breeding but this measure remains incomplete today.

## Types of Eugenics

I. **Negative Eugenics:** To diminish the incidents of undesirable traits by discouraging the procreation of carriers of certain harmful genes.

II. **Positive Eugenics:** Encourage the carriers of desirable genotypes. Thus selection might act upon harmful mutant genes in human and other populations.

## Measurement in Eugenics

**I. Eugenic Measurement**

A. **Negative Eugenics or Restrictive Approach**

(a) Control of immigration by laws.

(b) Prohibition of consanguinous marraige by laws.

(c) Social and lawful sexual separation of the defective men or women to inhibit the inheritance of their genome which can be done by sterilization, termination of pregnancy, etc.

(d) Genetic counselling—couple should take genetic counselling and proper law should be formulated.

### B. Positive Eugenics or Constructive Approach

(a) Selective marriage and formulation of rules and laws.

(b) Teaching of fundamental knowledge of heredity and genetics from school level.

(c) Removal of social hindrance by laws.

(d) Muller and others proposed the utilization of spermatozoa bank contains the preserved, frozen sperm of outstanding creative individual and the method is known as *germinal choice.*

(e) Genetic Engineering: Many harmful genes are removed by genetic engineering to control the inheritance of inborn error.

## II. Euthenic Measurements

Environment has great effect on the development of human traits apart from genetics. If the identical twins are nurtured in different environmental conditions, will develop differential traits though the twins have same genome. Therefore, the euthenic measurement should be considered in the following ways:

(a) Children should be provided good environment to develop of their traits at the optimum level.

(b) Knowledge of hygiene and health should be discussed all the people.

## III. Euphenic Measurements

These measurements include:

1. Symptomatic treatments of genetic diseases;
2. Diagnosis of different genetic diseases.

For symptomatic treatment the following points should be considered:

(a) Prenatal and postnatal diagnosis of genetic diseases should be done by geneticist either by:

1. Aminocentesis, or
2. Karyotype judgement.

(b) In case of thalassaemics and genetic anemia need careful treatment and efforts should be made to suppress the synthesis of abnormal haemoglobin chain.

(c) In case of inborn error of metabolism, the subject should missing enzymes and immunological reactions should be investigated before the treatment.

## Glossary

**Euthenics:** The science which deals with improvement of the traits of the individuals after birth by the influence of environment.

**Euphenics:** The science which deals with the symptomatic treatment of genetic diseases of man by different methods.

# 39 GENETIC COUNSELLING

## Why Genetic Counselling is necessary?

Couple who have had a child with serious abnormalities must upset them with question in their mind why this has happened and in future whether their child (or children) would be similarly affected.

Individuals who have a family history of such disorders are seriously concerned about developing the disorders or transmitting it to future generations.

Considering the needs and importance of providing them with accurate and appropriate information has led to introduce genetic counselling.

## What is Genetic Counselling?

Genetic Counselling is a communication process which deals with the risk of developing or transmitting a genetic disorder. The person seeking genetic counselling is known as *consultant*. During counselling process, the counselling should try to ensure that the person is provided with relevant and valuable information which enables the person to understand the following:

(a) The medical diagnosis and its implication in terms of prognosis and possible treatment.

(b) The mode of inheritance of the disorder and the risk of developing or transmitting it.

(c) The choices or options available for dealing with the risks.

- **Steps in Genetic Counselling**

The important steps in genetic counselling are:

I. **Establishment of a Diagnosis:** Based on history, examination and investigations.

II. **Estimation of a Recurrence Risk:** The provision of a recurrence risk does not simply involve conveying stark risk figure in isolation. It is very important that information which is provided is understood and that parents are given as much background information as possible to help them reach their own decision. As a working rule of thumb, recurrence risks should be:

(a) quantification

(b) qualification

(c) placed in context.

(a) **Quantification**

1. Most prospective parents will be familiar to some degree with the

concept of risks but experience indicates that a risk of 1 : 6 can easily be misinterpreted or remembered as 6 : 1 or 16 per cent.

2. Another alarming but entirely understandable misconception is that this risk applies only to every sixth child, so that having had an affected child there should be no problems with the next five children.
3. It is therefore vital to emphasize that the risk applies to each child, and that chance does not have a memory.

(b) **Qualification:** Several studies have indicated that the factor which most influence parents when deciding whether or not to have another child is the nature of the long term burden associated with a risk rather than its precise numerical value.

Other factors such as whether a condition can be successfully treated, whether it is associated with pain and suffering and whether prenatal diagnosis is available, will all be relevant to the decision-making process.

(c) **Placing risks in context:** Prospective parents seen at a genetic counselling clinic, should be provided with information which enables them to put their risks in context so as to be able to decide for themselves whether a risk is high or low.

III. **Discussing the options:** Having established the diagnosis and discussed the risk of occurrence or recurrence, the counsellor is obliged to ensure that the consultant are provided with all the information necessary for them to arrive of their own information decision.

IV. **Communication and Support:** Communication is a two-way process. Not only does the counsellor provides information, but he or she also has to be receptive to the fears and aspirations, expressed or unexpressed of the consultant.

V. **Genetic Counselling—Directive or Non-directive:** Genetic counselling should be non-directive and the genetic counselling should be non-judgmental. The goal of genetic counselling is to provide accurate information which enables those who have received counselling to make their own fully informed decision.

Marriage between blood relatives conveys an increased risk of an autosomal recessive disorder in future offspring. The probability that first cousins will have a child with an autosomal recessive disorder is approximately 3 per cent, although this risk can be greater if there is a family history of a specific genetic disorder.

The most sensitive technique for paternity testing is genetic fingerprinting. Other polymorphic systems can disprove paternity but cannot confirm with a high statistical probability that a particular male is the biological father.

# GLOSSARY

## A

**Amber mutation:** A mutation in which a polypeptide chain is terminated prematurely; results from an alteration in a codon such that the codon becomes UAG, which signals chain termination.

**Anticodon:** The triplet of nucleotides in a tRNA molecule that associate by complementary base pairing with the codon in the mRNA during translation.

**Antiparallel:** Molecules that are parallel but point in opposite direction, e.g. the strands of DNA are antiparallel.

**Antisense:** Strand of DNA that has the same sequence as mRNA.

**AUG:** Codes for the first amino acid in protein sequences, which is formyl-methionine (f Met). f Met is often removed post-translationally.

**Blunt end:** DNA end with no unpaired bases.

**Cap:** The structure found at the 5′ end of many eukaryotic mRNAs. It consists 7′-methylguanosine-pppX, where X is the first nucleotide encoded in the DNA. It is not present in prokaryotic mRNAs; it is added post-transcriptionally near the TATA (Hogness) box.

**CAP:** CAP is catabolite gene activator protein which participates in the initiation of transcription in prokaryotes.

**Capsid:** The protein coat of a virion or virus particle.

**cDNA** (Complementary DNA): A single-stranded DNA molecule which is complementary to an mRNA molecule and is synthesized from it by the action of reverse transcriptase.

**CH:** Constant portion of an immunoglobulin heavy chain.

**Chimeric molecule:** A molecule, e.g. DNA or RNA containing sequences derived from two different species.

**Cistron:** A DNA fragment or portion that specifies or codes for a particular polypeptide.

**Class switch:** A switch in the expression of a B-lymphocyte from one antibody class to another.

**Clone:** A large number of cells or molecules which are identical with a single parental cell or molecule.

**Codon:** A group of 3 nucleotides that codes for an amino acid.

**Cohesive ends or Cohesive termini:** DNA molecules with single-stranded ends that show complementarity, making it possible, e.g. to join end to end with introduced fragments.

**Consensus sequence:** An average sequence, each nucleotide of which is the most frequent at that position in a set of examples; used for RNA splice sites and other sites.

**Cosmid:** A plasmid into which the DNA sequences from bacteriophage lambda that are necessary for the packaging of DNA have been inserted.

**Cross hybridization:** Hybridization of a probe to imperfectly matching molecules.

**Crossing-over:** Exchange of genetic material between chromosomes that pair during meiosis.

**Cut:** A double-strand scission in the duplex polynucleotide—in distinction to the single-strand "nick".

**Episome:** A circular gene fragment.

**Endonuclease:** An enzyme which cleaves internal bonds in DNA or RNA.

**Exon:** Portion of DNA that codes for the final mRNA.

**Exonuclease:** An enzyme which cleaves nucleotides from either the 3′ or 5′ ends of DNA or RNA.

**Exon shuffle:** The alternative RNA processing patterns that lead to expression of different combinations of exons from the same gene.

**Fe:** An antibody fragment from $C_H$ that is crystallizable.

**Fe receptor:** The receptor for Fe fragment.

**Gap:** A double-stranded DNA is said to be gapped when one strand is missing over a short region.

**Gene library:** Random collection of cloned fragments in a vector that ideally includes all the genetic information of the species which is also known as shotgun collection.

**Gene cleavage:** Taking out some portion of gene by restriction enzyme.

**Gene splicing:** Gene splicing is the process of combination of one molecule of DNA with another.

**Genome:** All the genes of an organism or individual.

**Genomic blotting or Southern blotting:** Method of transferring DNA fragments that have been separated by gel electrophoresis (agarose) to a nitrocellulose filter such that the relative positions of the DNA fragments are maintained. The DNA is usually visualized by hybridization with a $P^{32}$-labelled DNA or RNA.

**Grünstein-Hogness assay:** Colony hybridization procedure for identification of plasmid clones. Colonies are transferred to a filter and hybridized with a probe.

**H Chain:** Heavy chain of immunoglobulin molecule (e.g. $V_H$ and $C_H$).

**Heteroduplex:** A DNA molecule with two strands, which come from two different individuals, so that there may be some base pairs of blocks of base pairs that do not match.

**Hinge:** Short flexible amino acid sequence of an immunoglobulin protein, permitting one portion to move relative to the other. When present, the hinge separates the antigen-combining site from the Fc portion of the molecule.

**Hogness box (TATA box):** The hypothesized eukaryotic RNA polymerase II promoter, analogous to the Pribnow box.

**Hybridization:** The specific reassociation of complementary strands of DNA with DNA, DNA with RNA or RNA with RNA.

**IF-1:** Initiation factor 1 (also IF-2, IF-3) for protein synthesis.

**Immunoglobulin class:** If we consider the case of mouse, there are 8 classes of immunoglobulins (Ig) viz. IgM (μ chain); IgD (δ chain); four IgGs (γ chain); IgA (α chain) and IgE (ε chain).

The class is determined by the constant region of the heavy chain. Class is associated with generic properties of the antibody, cellular and tissue localization, complement binding, and other factors. Class is independent of the variable region and of the light chain.

**Initiation codon:** Codes for the first amino acid in protein sequences, which is formylmethionine. f-Met is often removed post-translationally.

**Insert:** An additional length of base pairs in DNA.

**Integration and excision:** Integration means a recombination in which a genetic element is inserted; excision is the reverse of integration.

**Intervening sequence:** A portion of a gene that is transcribed but does not appear in the final mRNA transcript.

**Intron:** Intervening sequence in DNA.

**Inversion:** The alteration of a DNA molecule made by removing a fragment, reversing its orientation, and pulling it back into place.

**Inverted repeat in DNA:** A self-complementary nucleic acid sequence, i.e. a sequence identical to its complementary strand (both read in the same 5′ and 3′ direction). Perfect palindromes (e.g. GAATTC) frequently occur as sites of recognition for restriction enzymes. Less perfect palindromes, e.g. TACCTC-TGGCGTGATA freqently occur in binding sites for other proteins, such as repressor. Interrupted palindromes, e.g. an inverted repeat such as GGTTXXX-AACC afford the possibility in single-stranded nucleic acids for the loop stem (hairpin) structure as in tRNA.

**Lac operon:** An operon in *E. coli* that codes for 3 genes involved in the metabolism of lactose.

**Library:** A collection of cloned fragments that represents the entire genome. Libraries may be either genomic DNA or cDNA. Sometimes called shotgun collection.

**Ligation:** The enzyme-catalyzed joining in phosphodiester linkage of two stretches of DNA or RNA into one. The respective enzymes are DNA ligase and RNA ligase.

**Ligase, DNA ligase:** Catalyzes the formation of phosphodiester bond at the site of a single-strand break in duplex DNA. RNA also acts as a substrate to some extent.

**mRNA:** Messenger RNA.

**Nick:** A single–strand scission of the DNA which can be done with deoxyribonuclease and ethidine bromide.

**Nonsense mutation:** A mutation that results in the termination of a polypeptide chain, e.g. ochre and amber.

**Nucleotide replacement site:** Position in a codon where a point mutation has occurred.

**Ochrea mutation:** A mutation in which a polypeptide chain is terminated prematurely; results from an alteration in the codon such that the codon becomes UGA, which signals chain termination.

**Operator:** A region of DNA that interacts with a repressor protein to control the expression of an adjacent gene or group of genes.

**Operon:** A gene unit consisting of one or more genes that specify a polypeptide and an "operator" that regulates the transcription of the structural gene.

**Palindrome:** See inverted repeat in DNA.

**Plasmid:** Extrachromosal, autonomously replicating, circular DNA segment.

**Polyadenylation:** Nontranscriptive addition of poly (A) (polyadenylate) to the 3′ end of eukaryotic RNA.

**Polymerase:** Enzyme that catalyzes the assembly of nucleotides into RNA and of deoxynucleotides into DNA.

**Pribnow box:** TATA ATG; consensus sequence near the RNA start point of prokaryotic promoters.

**Probe:** A molecule used to detect the presence of a specific fragments of DNA or RNA in common probes are cDNA molecules.

**Promoter:** A DNA sequence at which RNA polymerase binds and then initiate transcription.

**Pseudogene:** A sequence that looks like a gene but does not function as one, it appears to have no phenotype and could be the vestigial remains of a gene.

**Reading:** One-way linear process by which nucleotide sequences are decoded, e.g. by protein-synthesizing systems.

**Readthrough:** The transcription of a region beyond a normal termination sequence, due to occassional failure of RNA polymerase to recognize the termination signal.

**Regulatory gene:** A gene whose product is involved in the regulation of another gene, such as a repressor gene.

**Regulatory sequence:** A DNA sequence involved in regulating the expression of the gene, e.g. promoters, operators.

**Recombinant DNA:** The altered DNA that results from the insertion of a sequence of deoxynucleotides not previously present into an existing molecule of DNA (i.e. foreign DNA molecule) by enzymatic or by chemical means.

**Restriction enzyme:** An endodeoxy-nuclease that causes cleavage of both strands of DNA at highly specific sites dictated by the base sequence, usually isolated from bacteria. There are many, e.g. Eco R1; Bam I, Hind III.

**R loop:** Three-stranded structure in which an RNA-DNA hybrid displaces the other strand of DNA, leaving a DNA loop with characteristic appearance in the electron microscope.

**RNA splicing:** See splicing.

**Reverse transcription:** RNA-directed synthesis of DNA catalyzed by reverse transcriptase.

**SD sequence:** Shine-Dalgarno or ribosome recognition sequence, begins 3-11 nucleotides upstream from the AUG in mRNA. It is complementary to the 3′ end of 16S ribosomal RNA.

**Sequence ladder:** Bands in gel corresponding to DNA sequence.

**Shot gunning:** See library.

**Silent site mutation, silent mutation:** Mutation in a codon or sequence that does not cause an amino acid change.

**Southern blot technique:** Method of transferring DNA fragments that have been separated by gel electrophoresis (agarose) to a nitrocellulose filter such that the relative positions of the DNA fragments are maintained. The DNA is usually visualized by hybridization with a $P^{32}$-labelled DNA or RNA probe.

**Signal:** The end product observed when a specific sequence of DNA or RNA is detected by autoradiography or some other method.

**Sheroplast:** A bacterium whose wall is partially, or nearly completely, removed, so that the cell assumes a spherical shape.

**Splicing:** (a) Method of attachment of one DNA molecule to another. (b) RNA splicing—Removal of introns from mRNA precursors.

**Split gene:** One that is not continuous but has been interrupted; interrupted gene.

**Start codon:** See initiation codon.

**Start point (RNA technology):** First nucleotide of a transcript.

**Sticky ends (Cohesive ends):** See cohesive terminal.

**Structural gene:** A gene that determines the primary structure (i.e. the amino acid sequence) of a polypeptide (see operon; regulatory gene).

**Suppressor gene:** A gene that can reverse the effect of a specific type of mutation in other genes.

**Suppressor mutation:** A mutation that totally or partially restores a function lost by a primary mutation and is located at a genetic site different from the primary mutation.

**Termination codon:** A codon that specifies the termination of translation.

**Termination sequence:** A DNA sequence at the end of a transcriptional unit signals the end of transcription.

**Transcription:** Formation of the RNA from the DNA template.

**Transduction:** The transfer of genetic material from one cell to another by means of a viral vector (for bacteria, the vector is bacteriophage).

**Transfection:** Infection of a cell with isolated DNA or RNA from a virus or viral vector.

**Transformation:** The introduction of an exogenous DNA preparation (transforming agent) into a cell.

**Tandem:** It is described as the multiple copies of the same sequence (e.g. DNA) that lie adjacent to one another.

**Transgenic:** It describes the introduction of new DNA into germ cells by its injection into the nucleus of the ovum.

**Translation:** The process in which the genetic code contained in the nucleotide sequence of mRNA directs the order of amino acids in the formation of peptide.

**Transposable element:** A segment or fragment of DNA that can move from one position in the genome to another.

**Transversion:** A mutation caused by the substitution of a pyrimidine for a purine or vice versa.

**tRNA:** Transfer RNA.

**tRNA gene:** Region of DNA that is transcribed to produce tRNA.

**tRNA suppressor:** A mutation in a tRNA gene that alters its anticodon to a sequence that is complementary to a termination codon. This allows the suppression of amino acid chain termination (nonsense mutation).

**Vector:** An agent consisting of a DNA molecule (plasmid or bacteriophage) into which a foreign DNA can be introduced for the purpose of cloning.

**$V_H$:** Variable portion of immunoglobulin heavy chain.

## B

**Genetics:** The Science of Genetics seeks to account for the resemblance or difference between the young and its ancestors. It is concerned with the heritable or genotypic variations of an individual and with laws governing the process of inheritance.

**Heredity:** Heredity may be defined as the occurrence in the living organisms of qualities that are derived from their ancestors. Heredity includes not only physical but also physiological qualities. These qualities may be either *"expressed"* or they may be *"latent"* in the offspring.

**Inheritance:** Transmission of characteristics, e.g. physical, mental or physiological from parents to the offsprings.

**Genetic variation:** Differences among members of any particular species are spoken of as *"variations"*. These may be—

(a) *Depending on environmental factors* —(1) fluctuating or pheno-typic variation.

(b) *Without depending on environmental factor*—(1) Germinal or genotypic variation.

The phenotypic variations are not heritable whereas the genotypic variations are handed down to the progeny from one generation to another.

**Haploid:** In meiosis, the number of chromosomes is halved, the daughter cells receiving only one of each type of chromosome instead of two. The daughter cells are said to be haploid state, haploid coming from the Greek word *"haploos"* meaning single. The mature sperm and ova contains half the normal number called *haploid number* of chromosomes.

**Diploid:** In mitosis, the chromosomes duplicate themselves and then divide in such a way that each daughter cell receives a full complement i.e. *diploid number* of chromosomes. When a sperm and ovum unite the resultant cell, i.e. zygote has a full complement of chromosomes, i.e. diploid number, one-half from the female parent and one-half from the male.

**Chromosomes:** The nucleus contains thread-like structures called chromosomes. It has the following features and functions:

1. Its number in somatic cell is 46 and in germ cell 23. Germ cell contains additional sex chromosome either X or Y.
2. In mitosis original number of chromosome is present in daughter cell whereas in meiosis half number is present in gamete.
3. It contains bands called genes having DNA molecules which transmits hereditary characters and sends message to the cytoplasm for protein synthesis according to genetic code.

**Giant chromosome:** Discovery of giant chromosome had an enormous impact on the study of gene. It was found that the chromosomes in the salivary glands of Drosophila larvae are relatively enormous, about a hundred times larger than those found elsewhere in the body. Moreover, far from being uniform in appearance, these *giant chromosomes* are traversed by numerous darkly-staining bands spaced at irregular intervals and of characteristic shape. Some are narrow and rather difficult to detect, others are thicker and clearer. Each type of chromosome has its own characteristic pattern of banding and this is constant in all specimens.

**Chromatids:** In prophase of mitosis certain clearly visible events can be seen of which is the condensing of the chromatin threads to form distinct chromosomes. Long and thin at first, they gradually become shorter and fatter. As the chromosomes shorten, it becomes increasingly clear that each consists of a pair of bodies lying close to each other. These are called chromatids. They tend to lie parallel along most of their length but are particularly closely associated in the vicinity of a specialized region called *centromere*. The chromatids of one chromosome are usually referred to as sister chromatids.

**Chromatin granules or chromatin:** It is the precursors of the chromosomes which make their appearance when the cell undergoes division. Each chromosome is made up of a giant molecule of DNA that in multicellular organism is covered with protein. The molecule looks like a string of beads, the beads are balls of histone proteins with DNA wrapped around them, and the string is more DNA with other proteins attached to it. The complex of DNA and proteins is called *chromatin.*

**Haploid state:** In meiosis the number of chromosomes is halved, the daughter cells receiving only one of each type of chromosome instead to two. The daughter cells are said to be in the haploid state, haploid coming from Greek word "haploos" meaning single.

**Diploid state:** A cell normally has a fixed number of chromosomes and these occur in pairs and is known as diploid state.

**Homozygous:** When an organism contains identical genes (i.e. gene for tallness or genes for dwarfness), it is said to be homozygous.

**Heterozygous:** When an organism contains two dissimilar genes, it is said to be heterozygous (in contrast to the homozygous condition where the two genes are identical).

**Monohybrid inheritance:** Inheri-tance of just one pair of contrasting characteristics, what is known to geneticist as monohybrid inheritance.

**Dihybrid inheritance:** Inheritance of two pairs of characteristic is known as dihybrid inheritance.

**Particulate inheritance:** When inherited two characteristics are combined in one generation but separated in the next generation. For this reason inheritance may be described as *"particulate"*.

**Dominant and recessive:** When one gene for tallness combines with gene for dwarfness and if the characteristic of tallness appears in the generation, the gene for tallness is called *"dominant"* and the dwarfness is known as *"recessive"*.

**Genotype and Phenotype:** The genetical constitution of an organism is known as its *"genotype"*. The outward appearance of the organism, i.e. the way the genes express themselves in the structure of the organism, is known as the *"phenotype"*.

**Alleles:** Gene controlling size exists in two varieties. One functions normally and is responsible for producing tallness. The other, however, influences development in such a way that, if two are present together, a dwarf is produced. Such genes are known as *"alleles"*. Allele comes from the Greek word *"allelon"* meaning of one another.

**Breeding true:** A homozygous organism, when self-fertilized, produces offspring all of which are identical with the parent. Exactly the same result is given if a homozygous organism is crossed with another homozygous. This is known as *"breeding true"*.

**"Pure-line" generation:** If the off-springs are self-fertilized or crossed amongst themselves, another gene-ration of identical organism is produced. If this is continued, a *"pure line"* is established.

**Synapsis and Bivalent:** In mitosis homologous chromosomes do not associate with each other in any way. But in meiosis they come to lie side by side, a process known as *"synapsis"*. In this condition each pair of homologous chromosomes constitutes a bivalent.

**Syngamy:** In most organisms sexual reproduction involves the union of gametes, a process known as *"syngamy"*. Syngamy takes place by a process of fertilization.

**Heterogametes:** Generally the gametes differ from each other in structure, size and behaviour, for which reason they are known as hetero-gametes.

**Isogametes and Anisogametes:** In primitive organisms, such as the green algae and fungi, often produce gametes that are identical with each other (isogametes), or only very slightly different (anisogametes).

**Punnett square:** A convenient graphic method is to employ a chequer board or *"Punnett square"*, so called because it was first used by the geneticist R.C.

| | ♂ gametes | | |
|---|---|---|---|
| | | $\frac{1}{2}$T | $\frac{1}{2}$t |
| ♀ gametes | $\frac{1}{2}$T | $\frac{1}{4}$TT | $\frac{1}{4}$Tt |
| | $\frac{1}{2}$t | $\frac{1}{4}$tT | $\frac{1}{4}$tt |

fusions

Punnett. In this device the gametes of one parent are written along the top of a series of boxes, and the gametes of the other parent are written down the side. The products of the various fusion are written in the appropriate boxes and their relative numbers can be estimated

**Congenital diseases:** Diseases inherited by the offspring from their parents called congenital diseases, e.g.

(a) Albinism
(b) Achondroplasia (chondrodystrophic dwarfism)
(c) Haemophilia
(d) Colour blindness
(e) Sicklecell anemia Sickle cell
(f) Alkaptonuria.

**Heterosomes and Autosomes:** The sex chromosomes are obviously an exception to the general rule that homologous chromosomes are identical in appearance. Being different, they are known as *"heterosomes"* (which literally means "different bodies"), in contrast to all the other pairs of chromosomes which are identical and are called *"autosomes"*.

**Homogametic and Heterogametic sex:** The sex chromosomes are trans-mitted in a normal Mendelian manner. Clearly, a female will produce only one kind of gamete as far as the sex chromosomes are concerned (XX): all her egg will contain X chromosome. For this reason the female is termed homogametic. The male, however, will produce two kinds of gamete: half the sperm will contain the X chromosomes, the other half Y (XY). The male is therefore the heterogametic sex.

The male is not invariably the heterogametic sex in different species, e.g. in birds the male is XX and the female is XY. In some insects the female is XX and the male is XO, the Y chromosome being absent.

**Congenital Anomalies or defects:** Body defects during birth, e.g. (1) cleft palate, (2) hare lip.

**Inborn error of metabolism:** Metabolic defect inherited due to defective gene, e.g.

| | |
|---|---|
| 1. Albinism: | due to defect in tyrosine metabolism |
| 2. Alcaptonuria: | due to defect in tryptophan metabolim |
| 3. Phenylketonuria (Folling's disease): | failure to convert phenylalanine to tyrosine: |

**Prokaryotic cell and Eukaryotic cell**

| Prokaryotic cell | Eukaryotic cell |
|---|---|
| 1. **Example:** Eubacteria, blue-green algae, spirochetes, rickettsiae, mycoplasma, pleuropneumonia-like organism (PPLO). | 1. **Example:** Cell of higher organisms in both plant and animal. |
| 2. **Size:** Very small. | 2. **Size:** Much larger. |
| 3. **Type:** Simple. | 3. **Type:** More complex. |
| 4. **Membrane:** Single membrane usually surrounded by a rigid cell wall. No other membrane. | 4. **Membrane:** Surrounded by a membrane and nucleus and other organalles are also surrounded by membrane. |
| 5. **Nucleus:** Absent | 5. **Nucleus:** Present. |
| 6. **Organalles:** Absent | 6. **Organalles:** Present. |
| 7. **Chromosomes, gene, DNA:** Only one chromosome consists of a single molecule of double helica DNA densely coiled to form the nuclear zone. | 7. **Chromosome, gene, DNA:** The genetic material is divided into several or many chromosomes which undergo mitosis during cell division. |
| 8. **Reproduction:** Asexual division. | 8. **Reproduction:** Sexual. |
| 9. **Metabolic reaction:** Metabolic reactions are segregated within structural compartments. | 9. **Metabolic reaction:** No. |
| 10. **Biological evolution:** First cells to arise in biological evolution. | 10. **Biological evolution:** More recent in evolutionary origin presumably derived from prokaryotes. |
| 11. **Genetic information:** Lesser. | 11. **Genetic information:** Contains more genetic information, e.g. thousand times DNA than *E.coli* and hundred thousand DNA than λ phage. |
| 12. **Gradation of DNA:** DNA of lower forms. | 12. **Gradation of DNA:** DNA of higher forms is associated with basic proteins called *histones.* These basic proteins are a device for making DNA more compact so that a contour length of many centimeters can be accommodated within a few micrometers. |
| 13. **Degree of structural organization:** Lower. | 13. **Degree of structural organization:** Much higher. |
| 14. **Shape of chromosomes:** No change. | 14. **Shape of chromosomes:** Changes markedly during the cell cycle. |
| 15. **Nuclear membrane:** Absent. | 15. **Nuclear membrane:** Bounded by a nuclear membrane. |
| 16. **Transcription and translation:** Transcription and translation are closely coupled. | 16. **Transcription and translation:** Transcription and translation are separated in space and time. |
| 17. **Gene expression:** Poor and not complex. | 17. **Gene expression:** Gene expression is far richer and more complex. |

**Genome:** Genome is a complete set of hereditary factors as contained in the haploid assortment of chromosomes. A genome therefore contains genetic codes from both parents of the individual. There are about 100,000 odd genes in a genome and 4550 out of them have been identified so far. Out of these about 1500 have been properly located on the respective chromosomes.

**Plasmid:** Some bacteria harbour circular DNA molecules that replicate autonomously in the bacterial protoplasm. Such an autonomous DNA molecule is called a plasmid. Plasmid can carry genetic information to confer resistance antibiotics on the host bacterium. Using recombinant DNA technology, a specific segment of DNA can be in vitro into a plasmid.

**Karyotype:** The individual chromo-somes are arranged in a arbitrary pattern called Karyotype. The individual autosome pairs are identified by the numbers 1-22 on the basis of their morphologic characteristics.

**Barr body:** The inactive X chromosome condenses and can be seen in various types of cells, usually near the nuclear membrane called Barr body. Thus, there is a Barr body for each X chromosome in excess of one in the cell. The inactive X chromosome is also visible as a small "drumstick" of chromatin projecting from the nuclei of 1-15% of the polymorphonuclear leucocytes in females but not in males. The Barr body is also called *sex chromatin*.

**F body:** Another form of sex chromatin is the *F body,* a portion of the Y chromosomes that can be stained with modern fluorescence techniques. Thus, the number of Y chromosomes in a cell can be estimated by the number of F bodies it contains.

**Restriction enzymes or Restriction endonucleases:** A series of enzymes capable of (a) *recognizing specific symmetric sequences* of a nucleotides in DNA and (b) subsequently *catalyzing the cleavage* of *phosphodiester backbone* at *symmetric sites within the specific sequences*. These enzymes are called restriction endonucleases. They are isolated from bacteria.

**Importance:**

1. Analysis of chromosome structure.
2. Mapping of DNA molecules.
3. Recombining DNA molecules from different sources to create new genetic sequences which is referred as *restriction endonucleases.*
4. Isolation of genes.
5. Sequencing very long DNA.
6. Mapping of chromosomes by series of restriction enzymes.
7. The pattern of restriction fragments can serve as a fingerprint of DNA.

**Characteristics:**

1. The cell's own DNA is not degraded because the site recognized by its restriction enzyme is methylated.
2. The CG postdiester bond is each strand distal to the symmetry axis is cleaved by the endonuclease.

**Types:** More than 80 restriction enzymes have been purified and characterized. Their names consist of 3-letter abbreviation for the host organism (e.g. Eco for *E.coli*), followed by a strain designation (e.g. R) and a Roman numeral (e.g. I) i.e. Eco RI.

# BIBLIOGRAPHY


1. Aberts, B., Sternglanz, R., Recent excitement in the DNA replication problem, Nature, 259, 655-661, 1977.
2. Alkinson, D.E., and FOX, C.F. (eds), Modulation of Protein Function, New York, Academic Press, 1979.
3. Avery, O.T., Macleod, C.M., and McCarty, M., J. Exp. Med., 79, 144, 1944.
4. Barr, M.L., *et al.*—The XX XY sex chromosomes abnormality, Can. Med. Ass. J., 87, 891-901, 1962.
5. Barrell, B.G. and Clark, B.F.C., Handbook of Nucleic Acid Sequences, Johnson-Bruvvery, Oxford, 1974.

5A. Bauer, W.R., Structure and reaction of closed duplex DNA, Annu. Rev. Biophys. Bioeng., 7, 287, 1978.

6. Baum, H. and Gergely, J. (eds), Molecular Aspects of Medicine, New York, Pregamon Press, 1978.
7. Bauer, W.R., Crick, F.H.C., and White, J.H., Supercoiled DNA, Sci. Amer., 243(1), 118-133, 1980.
8. Beerman, W., and Clever, U., Chromosomal puffs, Sci. Amer., 210(4), 50-58, 1974.
9. Bertrand, K., Korn, L., *et al.*, New features of the regulation of the tryptophan operon, Science, 189, 22-26, 1975.

9A. Bergsma, D(ed), Birth Defects, 8(4) : 1-116.

10. Bird, R.E., *et al.*, Origin and sequence of chromosome replication in *E.coli.*, J. Med. Biol., 70, 549-566, 1972.

10A. Bishop, J.M., Sci. Amer., 246(3), 90-92, 1982.

11. Brenner, S., Jacob, F., and Meselson, M., An unstable intermediate carrying information from genes to ribosomes for protein synthesis, Nature, 190, 576-581, 1961.
12. Britten, R.T., and Kohne, D.E., "Repeated sequences DNA", Science, 161, 529-540, 1968.
13. Britten, R., and Kohn, D., Repeated segments of DNA, Sci. Amer., 222(4), 24-31, 1968.
14. Brimacombe, R., Staffler, G., and Wittmann, H.G., Ribosome structure, Ann. Rev. Biochem., 47, 217-249, 1978.
15. Brown, D.D., "The Isolation of Genes", Sci. Am., 229, 20-29, 1973.

15A. Carter, C.O., Progr. Med. Genetics, IV, 59-84, 1965.

16. Catens, J., Stent, G.S. and Watson, J.B. (eds), Phage and the Origins of Molecular Biology, Cold Spring Laboratory of Quantitative Biology, 1966.

17. Cavalli-Sforza, I.I., The genetics of human population, Sci. Am., 231(3), 80, 1974.
18. Campbell, P.N., and Sargent, I.R., Techniques in Protein Biosynthesis, 3 vols., Academic Press, New York, 1967.
19. Chamberlin, M.J., The selectivity of transcription, Ann. Rev. Biochem., 43, 721-776, 1974.
20. Clark, B.F.C., *et al.* (eds), Gene Expression : Protein Synthesis and Control, RNA Synthesis and Control, Chromatin Structure and Function, New York, Pregman Press, 1978.
21. Cohen, S.N., The manipulation of genes, Sci. Am., 233(1), 24, 1975.
21A. Cooper, G.M., Science, 217, 801-806, 1982.
21B. Court-Brown, W.M., Human Population Cytogenetics, North Holland Publication, Amsterdam, 1967.
22. Crick, F.H.C., The genetic code III, Sci. Amer., 215(4), 55-62, 1966.
23. Crick, F.H.C., *et al.*, General nature of the genetic code for protein, Nature, 192, 1227-1232, 1961.
24. Crick, F., Split genes and RNA splicing, Science, 204, 264-271, 1979.
25. Crick, F.H.C., "The origin of the genetic code", J. Mol. Biol., 38, 367-379, 1968.
26. Davidson, J.N., The Biochemistry of the Nucleic Acids, 7th ed., Academic Press, New York, 1972.
27. Davidson, J.N., and Cohn, W.E., (eds), Progress in Nucleic Acid Research and Molecular Biology, vol. I, Academic Press, New York, 1963.
28. Darnell, J.E., Jr., The Processing of mRNA, Sci. Am. (Oct.), 1983.
29. Davidson, J.N., The Biochemistry of the Nucleic Acids, Academic Press, 1977.
30. Davidson, E.H., and Britten, R.J., Regulation of gene expression : possible role of repetitive sequences, Science, 204, 1052-1059, 1979.
31. Davidson, E.H., and Britten, R.J., "Organization, Transcription and Regulation in the Animal Genome".
32. De Robertis, E.M., and Gurdon, J.B., Gene transplantation and the analysis of development, Sci. Amer., 241(6), 74-82, 1979.
33. DePamphilis, M.L., and Wassarman, P.M., Replication of eucaryotic chromosomes : a close-up of the replication fork, Ann. Rev. Biochem, 49, 627-666, 1980.
34. Dickson, R.C., *et al.*, Genetic Regulation : the lac control region, Science, 187, 32, 1975.
34A. De Grouchy, J. and Jerleau, C., Clinical Atlas of Human Chromosome, John Wiley and Sons, New York, 1977.
34B. Dorfman, A. (eds), Antenatal Diagnosis, Chicago University Press, Chicago, 1972.
35. Drlica, K., Understanding DNA and Gene Cloning, a guide for the curious, Wiley, 1984.
36. Drake, J.W., *et al.*, The Biochemistry of Mutagenesis, Ann. Rev. Biochem., 45, 11-38, 1976.
37. Drake, J.W., The Molecular Basis of Mutation, Holden-Day, 1970.
38. Dupraw, E. J., DNA and chromosomes, Holt, New York, 1970.
38A. Emery, A.E.H. (eds), Modern trends in Human Genetics, Butterworth, London, 1970.
39. Fiddes, J.C., The nucleotide sequence of a viral DNA, Sci. Am., 237(6), 54, 1977.
39A. Ford, C.E. and Clegg, H.M., Britain Med. Bull., 25, 110-114, 1969.
39B. Fraser, F.C., Amer. J. Human Genetics, 22, 336-352, 1970.

40. Ford, C.E., *et al.*, The chromosomes in a patient showing both mongolism and the Klinefelter syndrome, Lancet, 1, 709-710, 1959.

41. Freeman, Newman and Halzinger, Twins. A study of heredity and environment, Univ. Chicago Press, 1937.

42. Garen, A., Sense and nonsense in the genetic code, Science, 160, 149-159, 1968.

42A. Gardner, E.J., Human Heredity, John Wiley and Sons, New York, 1983.

43. Gordon, J.W. and Ruddle, F. H., Mammalian Gonadal determination and gametogenesis, Science, 211, 1265-1271, 1981.

44. Grumbach, M.M., and Barr, M.L., Cytologic tests of chromosomal sex in relation to sexual anomalies in man, Recent Prog. Horm. Res., 14, 255-334, 1958.

45. Haynes, R.H., and Hanawalt, P.C. (editors), The Molecular Basis of Life, 1968.

46. Harris Henry, Nucleus and Cytoplasm, Oxford University Press, 3rd ed., 1974.

46A. Harris, H. and Hirshhorn, K. Advances in Human Genetics, I-V, Plenum Publishing Company, New York and London, 1971-1975.

47. Hanawalt, P.C., et al., DNA repair in bacteria and mammalian cell, Am. Rev. Biochem., 48, 783-836, 1979.

48. Hamkala, B., and Miller, I. (Jr.), "Electron Microscopy of Genetic Activity", Ann. Rev. Biochem., 42, 379-396, 1973.

49. Hanawalt, P.C., "Repair of Genetic Material in Living Cells", Endevour, 31, 83-87, 1972.

50. Haselkorn, R., and Rothman-Denes, "Protein Synthesis", Ann. Rev. Biochem., 42, 397-438, 1973.

51. Hess, E.L., Origins of Molecular Biology, Science, 168, 664-669, 1970.

52. Hirsh, J., and Schleif, R., Electron Microscopy of Gene Regulation : the L-arabinose operon, Proc. Nat. Acad. Sci., 73, 1518-1522, 1976.

52A. Igo-Kemenes, Tet. al., Ann. Rev. Biochem., 51, 89-121, 1982.

53. Jacob, F., and Monod, J., Genetic regulatory mechanism in the synthesis of proteins, J. Mol. Biol., 3, 318-336, 1961.

54. Kaziro, Y., The role of GTP in polypeptide chain elongation, Biochem. Biophysics. Acta., 505, 95-127, 1978.

54A. Kissmeyer-Nielsen, F. (eds), Histocompatibility Testing, Proc. Int. Histocompatibility Workshop and Conference, Copenhagen, 1975.

55. Lacey, J.C., and Pruitt, K.M., "Origin of the Genetic Code", Nature, 223, 799-804, 1969.

55A. Lawler, S.D. *et al*—Medical Bull., 15, 145-149, 1957.

56. Lewis, K.R., and John, B., The Matter of Mendelian Heredity, 1964.

57. Lewin, B.M., The Molecular Basis of Gene Expression, Wiley -Interscience, New York, 1970.

58. Lewis, B.M., Gene Expression, vol. 2, Eucaryotic chromosomes, Wiley, New York, 1974.

59. Lehman, I.R., DNA ligase : structure, mechanism, and function, Science, 186, 790-797, 1974.

59A. Lewin, B., Genes, John Wiley and Turpin, Sons, New York, 1983.

59B. Lewin, B., Gene Expression-2, Eukaryotic Chromosomes, John Wiley and Sons, New York, 1980.

60. Loewen, P.C., Sekiya, T., and Khorana, H.G., "The Nucleotide Sequence Adjoining the CCA End of an Escherichia coli Tyrosine Transfer Ribonucleic Acid Gene", J. Biol. Chem., 249, 217-226, 1974.

61. Long, E.C., and Daurid, I.B., Repeated genes in eucaryotes, Ann, Rev. Biochem., 49, 727-766, 1980.

61A. Luning, K.G. and Searla, A.G., Mutation Res., 12, 291-304, 1971.

62. Lyon, M.F., X-chromosome inactivation and development patterns in mammals, Biol. Rev., 47, 1, 1972.

63. Maniatis, T., and Ptashne, M., A DNA Operator-Repressor system, Sci. Am., 232(1), 64, 1976.

64. McKusick, V.A., and Ruddle, F.H., The status of the gene map of the human chromosomes, Science, 196, 390-405, 1977.

65. McKusick, V.A., Mendelian Inheritance in Man, John Hopkins Press, Baltimore, 1975.

66. McGhee, J.D., and Felsenfeld, G., Nucleosome structure, Ann. Rev. Biochem., 49, 1115-1155, 1980.

67. Meselson, M., and Stahl, F.W., The replication of DNA in *Escherichia coli*, Proc. Nat. Acad. Sci., 44, 671-682, 1958.

68. Mittwoeh, U., Genetics of Sex Differentiation, New York, Academic Press, p. 1-253, 1973.

69. Mirsky, A.E., "The Discovery of DNA", Sci. Am. 218, 78-88, 1963.

70. Miller, O.L., The visualization of genes in action, Sci. Amer., 228(3), 34-42, 1973.

71. Mittwoeh, U., Genetics of Sex Differentiation, p. 1-253, Academic Press, New York.

72. Nirenberg, M.W., The Genetic Code II, Sci. Amer., 208(3), 80-94, 1963.

73. Nomura, M., The control of ribosome synthesis, Sci. Am. (Jan.), 250, 102, 1984.

74. Nyhan, W.L. (editor), Heritable Disorders of Amino Acid Metabolism, Patterns of Clinical Expression and Genetic Variation, Wiley, 1974.

75. Ohno, S., Sex chromosomes and sex-linked genes, New York, Springer-Verlag, 1967.

76. Ohno, S., A hormone-like action of H-Y antigen and gonadal development of XY-XX mosaic males and hermaphrodites, Hum. Genet., 35, 21-25, 1976.

77. Palade, G., Intracellular aspects of the process of protein synthesis, Science, 189, 347, 1975.

77A. Pawlowitzki, I.H., Humangenetik, 16 : 131, 1972.

78. Pedder, I.J. and Wyne, E.G., Genetics : A Basic Guide, Hutchinson Educational, 1972.

78A. Penrose, L.S., Outline of Human Genetics, Heinemann, London, 1973.

79. Prescott, D.M., and Kuempel, P.L., Bidircctional replication of the chromosome in *E.coli,* Proc. Nat. Acad. Sci., 69, 2842-2845, 1972.

80. Atashe, M., and Gilbert, W., "Genetic Repressor", Scientific American, June, 1970.

80A. Race, R.R. and Sanger, R., Blood Groups in Man, Oxford University Press, Oxford, 1975.

Ratnoff, O.D. and Bennett, B., Science, 179, 1291-1298, 1973.

81. Revel, M., and Groner, Y., Post-transcriptional and translational control of gene expression in eukaryotes, Ann. Rev., Biochem., 47, 1079-1126, 1978.

82. Rich, A., and Kim, S.H., Three-dimensional structure of transfer RNA, Sci. Amer., 238(1), 52-62, 1978.

82A. Roberts, J.A.F. and Pembrey, M.E., An Introduction to Medical Genetics, Oxford Univ. Press, Oxford, 1976.

83. Russell, T.R., *et al.* (eds), From Gene to Protein : Information Transfer in Normal and Abnormal Cells, Academic Press, New York, 1979.

83A. Sandberg, A.A. *et al.*, the Lancet, II, 488-489, 1961.

84. Schatz, G., and Mason, T.I., "The Biosynthesis of Mitochondrial Proteins, Ann. Rev. Biochem., 43, 51-87, 1974.

85. Shih, V.E., Laboratory Techniques for the Detection of Hereditary Metabolic Disorders, CRC Press, 1973.

86. Simpson, R.T., "Structure and Function of Chromatin", Adv. Enzymol., 38, 41-108, 1973.

87. Sperling, R., and Wachtel, E.J., The histones, Advan. Protein Chem., 29, 85-133, 1981.

88. Srb, A.M., and Owen, R.D., General Genetics, 2nd ed., W.H. Freeman, 1965.

89. Srb, A.M., Owen, R.D., and Edger, R.S., (editors), Facets of Genetics, W.H. Freeman, 1970.

90. Stent, G.S., Molecular Genetics, Freeman, San Francisco, 1971.

91. Stanbury, J.B., et al., The Metabolic Basis of Inherited Disease, 5th ed., Mc Graw-Hill, 1983.

92. Stanbury, J.B., *et al.* (eds), The Metabolic Basis of Inherited Diseases, 4th ed., McGraw-Hill, 1978.

93. Soll, D., *et al.* (eds), Transfer RNA, Biological Aspects, Cold spring Harbor, New York, Cold Spring Harbor Laboratory, 1979.

94. Stein, G.S., *et al.*, "Nonhistone Chromosomal Proteins and Gene Regulation", Science, 183, 817-823, 1974.

95. Stern, C., Principles of Human Genetics, San Francisco, 1949.

95A. Stern, G., Principles of Human Genetics, W.H. Freeman, San Francisco, 1973.

96. Sullivan, Navin., The Message of the Genes, Routledge and Kegal Paul, 1968.

96A. Tabin, C.J. *et al.*, Nature, 300, 142-149, 1982.

97. Taylor, J. H., (ed.), Selected Papers on Molecular Genetics, Academic Press, 1965.

97A. Thompson, J.S., and Thompson, M.W., Genetics in Medicine, 4th ed., Saunders, Philadelphia, 1986.

98. Tso, P.O.P., Basic Principles in Nucleic Acid Chemistry, vols. 1 and 2, Academic Press, 1974.

99. Watson, J.D., The Double Helix, Weidenfeld and Nicolson, 1965; Penguin, 1970.

100. Watson, J.D., Molecular Biology of the Gene, 3rd ed., Benjamin, Menlo Park, Calif., 1975.

101. Watson, J., Tooze, J., Kantz, D., Recombinant DNA, A Short Course, Freeman, 1984.

102. Watson, J.D., and Crick, F.H.C., Molecular structure of nucleic acid. A structure for deoxyribose nucleic acid, Nature, 171, 737-738, 1953a.

103. Watson, J.D., and Crick, F.H.C., Genetic implications of the structure of deoxyribonucleic acid, Nature, 171, 964-967, 1953b.

104. Wachtel, S.S., *et al.*, Possible role of H-Y antigen in the primary determination of sex, Nature, 257, 235-236, 1975.

105. Weatherall, D.J., The Thalassemia Syndrome, Blackwell Scientific publication, Oxford, 1965.

106. Weisman, S.M., Gene structure and function. In Bondy, R.K., and Rosenberg, I.F., (eds), Metabolic Control and Disease, 8th ed., Philadelphia, W.B. Saunders Co., p. 1, 1980.

107. Wetzel, R., Application of recombinant DNA technology, Am. Sci., 305, 77, 1980.

108. Weissbach, H., and Pestka, S., (eds.), Molecular Mechanism of Protein Biosynthesis, Academic Press, 1977.

109. Weatherhal, D.J., and Clegg, J.B., Recent developments in the

molecular genetics of human hemoglobin, Cell, 12, 467-479, 1979.

110A. Whitehouse, H.L.K., Towards the Understanding of Mechanism of Heredity, English Language Book Society, London, 1972.

110B. Wittinghill, M., Human Genetics and its Foundations, Reinhold Publications, New York, 1965.

111. Wool, I.G., Structure and function of eukaryotic ribosomes, Ann. Rev. Biochem., 48, 719-754, 1979.

112. Woese, C.R., The Genetic Code, Harper and Row, New York, 1967.

113. Yanofsky, C., Gene structure and protein structure, Sci., Amer., 216(5), 80-94, 1967.

114. Ycas, M., The Biological Code, North-Holland, New York, 1969.

# INDEX

*Some exciting gifts are up for grabs with this book!!*

# HEREDITY, GENETICS AND GENETICAL DISEASES

Dr R N Roy

**FREE Gifts Rs. 1000/-**

## Dear Readers

By now you must have gone through this book in the entirety. So we invite your assessment on this title with absolute impartiality and candid observations. After all, we publishers blindly depend on your benchmarking. Friends, please fill up the open spaces below with CAPITAL LETTERS. You may add some extra sheets of paper if the space provided is not sufficient for you.

i. What are the important topics of your syllabus which are not adequately covered in this book?

ii. In which chapters and on what topics the treatment of the subject-matter by the author is not satisfactory and why?

iii. Have you come across the typos, factual inaccuracies, inconsistencies, and other forms of mistakes in the book? If yes, then please specify the mistakes and page numbers.

iv. Name top three books on the same subject, which, in your consideration, are better than this book. Could you provide some reasons?

Your 1st Preference: ..........................................................................

Your 2nd Preference: ..........................................................................

Your 3rd Preference: ..........................................................................

v. Any suggestion to improve the quality of illustrations, diagrams, etc.?

vi. How did you like the cover design? Any suggestion on colours?

vii. Suggestions for further improvement of production-standard:

viii. Who recommended you this book:

☐ Teacher ☐ Friend ☐ Parents ☐ Bookseller ☐ Librarian

ix. Name of the person who has recommended:

Designation: ..........................................................................

Address: ..........................................................................

Phone No.: ..........................................................................

e-mail: ..........................................................................

x. Name and address of the bookseller from whom you have purchased this book:

...........................................................................................

...........................................................................................

...........................................................................................

xi. Name and address of the library where you would like to find the book:

...........................................................................................

...........................................................................................

xii. Name and address of your institution (Please mention the University or Board):

...........................................................................................

...........................................................................................

xiii. Your name, e-mail, and complete postal address:

...........................................................................................

...........................................................................................

...........................................................................................

xiv. A few other titles published by us that you would like to read:

...........................................................................................

...........................................................................................

The best assessment will be rewarded half-yearly. The award will be in the form of our publications, as decided by the Editorial and Marketing Operations Division, amounting to Rs. 1000/- (Rupees one thousand) only.

...........................................................................................

Please mail the filled up coupon at your earliest to :

*Editorial & Marketing Operations Division*


**NEW CENTRAL BOOK AGENCY (P) Ltd**

**8/1 Chintamoni Das Lane, Kolkata 700 009**

**West Bengal, INDIA**

***Phone + Fax:*** 033 + 2241 8411, 2241 8573, 2241 9094

***e-mail:*** ncbapvtltd@eth.net, ncba@cal2.vsnl.net.in

***Website:*** www.centralgroup.com